Springer

*Berlin
Heidelberg
New York
Barcelona
Budapest
Hong Kong
London
Milan
Paris
Singapore
Tokyo*

Structure of
Free Polyatomic Molecules

Basic Data

Editor: K. Kuchitsu

 Springer

ISBN 3-540-60766-8 Springer-Verlag Berlin Heidelberg New York

Library of Congress Cataloging in Publication Data

Die Deutsche Bibliothek - CIP-Einheitsaufnahme
Structure of Free Polyatomic Molecules - Basic Data / ed.: Kozo Kuchitsu
Berlin, Heidelberg, New York, Barcelona, Budapest, Hong Kong,
London, Milan, Paris, Santa Clara, Singapore, Tokyo: Springer, 1998
ISBN 3-540-60766-8

This work is subject to copyright. All rights are reserved, whether the whole or part of the material is concerned, specifically the rights of translation, reprinting, reuse of illustrations, recitation, broadcasting, reproduction on microfilm or in other ways, and storage in data banks. Duplication of this publication or parts thereof is permitted only under the provisions of the German Copyright Law of September 9, 1965, in its current version, and permission for use must always be obtained from Springer-Verlag. Violations are liable for prosecution act under German Copyright Law.

© Springer-Verlag Berlin Heidelberg 1998
Printed in Germany

The use of general descriptive names, registered names, trademarks, etc. in this publication does not imply, even in the absence of a specific statement, that such names are exempt from the relevant protective laws and regulations and therefore free for general use.

Product Liability: The data and other information in this handbook have been carefully extracted and evaluated by experts from the original literature. Furthermore, they have been checked for correctness by authors and the editorial staff before printing. Nevertheless, the publisher can give no guarantee for the correctness of the data and information provided. In any individual case of application, the respective user must check the correctness by consulting other relevant sources of information.

Typesetting: Redaktion Landolt-Börnstein, Darmstadt
Printing: Computer to plate, Mercedes-Druck, Berlin
Binding: Lüderitz & Bauer, Berlin

SPIN: 10058592   63/3020 - 5 4 3 2 1 0 – Printed on acid-free paper

Editor

K. Kuchitsu
Department of Chemistry
Josai University
Sakado, Japan

Contributors

G. Graner
Laboratoire de Physique Moléculaire
et Applications
C.N.R.S
Orsay, France

Infrared and Raman spectroscopic data
and references
(Introduction 1.2.1)

E. Hirota
The Graduate University
for Advanced Studies
Hayama, Kanagawa, Japan

Microwave spectroscopic data
Introduction

T. Iijima
Department of Chemistry
Gakushuin University
Toshima-ku, Tokyo, Japan

Electron diffraction data

K. Kuchitsu
Department of Chemistry
Josai University
Sakado, Japan

Electron diffraction data
Edition, Coordination, Introduction

D.A. Ramsay
Steacie Institute of Molecular Sciences
National Research Council of Canada
Ottawa, Canada

Molecules in excited electronic states
Electronic and photoelectron
spectroscopic data and references
(Introduction 1.5)

J. Vogt and **N. Vogt**
Sektion für Spektren- und
Strukturdokumentation
Universität Ulm, Germany

Compilation of microwave and
electron diffraction data and references
Assistance to trace back references
on all experimental methods
Expert assistance at the evaluation of
electron diffraction data
Drawing of ball-stick molecular models

Preface

The frequent use of well known critical data handbooks like Beilstein, Gmelin and Landolt-Börnstein is impeded by the fact that only larger libraries – often far away from scientist's working places – can afford such costly collections. To satisfy the urgent need of many scientists working in the field of molecular science to have at their working place a comprehensive, high quality, but inexpensive collection of at least the basic data of their field of interest, this volume summarizes the most important structures of free polyatomic molecules. All data have been selected from those listed in the data tables presented on more than 1600 pages in various volumes of the New Series of Landolt-Börnstein, II/7, II/15, II/21 and II/23.

We hope to meet with this volume the needs of researchers and students in the wider fields of molecular science and materials engineering, forming a bridge between the laboratory and the sources of additional information in the libraries.

My sincere thanks are due to the Editor in Chief, Professor W. Martienssen, for his thoughtful guidance and to all authors of the present volume for their friendly cooperation; to Dr. B. Mez-Starck and Sektion für Spektren- und Strukturdokumentation, Universität Ulm, for their valuable assicstance and support; and to Springer-Verlag, especially Dr. R. Poerschke, for continual support and care. I greatly appreciate the expert help of Professor K. Hata and M. Nakahara in improving the nomenclature and of Drs. J.H. Callomon and D.A. Ramsay with checking the English text.

I am also grateful for the former Editors in Chief of Landolt-Börnstein and the former Volume editors for II/7 and 15: Professor K.-H. Hellwege, Dr. A.M. Hellwege and Professor O. Madelung, for their iniative and permanent support, to the coauthors of the earlier volumes II/7, 15 and 21: Dr. J.H. Callomon, W.J. Lafferty, A.G. Maki and C.S. Pote, for their expert contributions, and to the Redaktion Landolt-Börnstein Darmstadt, especially Mrs. H. Weise and Mr. Dipl.Phys. T. Schneider, for their valuable technical assistance.

Sakado, May 1998   The Editor

Table of contents

1	Introduction	1
1.1	General remarks	1
1.2	Experimental methods	1
1.2.1	Spectroscopy	1
1.2.2	Electron diffraction	4
1.2.3	Combined use of diffraction and spectroscopy, theoretical calculations, or other methods	6
1.3	Significance of geometric parameters	6
1.3.1	Spectroscopy	7
1.3.2	Electron diffraction	9
1.3.3	Table of distance parameters and their definitions	10
1.4	Uncertainties	10
1.4.1	Microwave spectroscopy	10
1.4.2	Infrared and Raman spectroscopy	11
1.4.3	Electron diffraction	11
1.5	Geometric structures of molecules in excited electronic states	13
1.6	Arrangement of the tables	16
1.6.1	General remarks on the content	16
1.6.2	Presentation of the data and comments	16
1.6.3	Order of molecules	17
1.6.4	Nomenclature	17
1.6.5	Figures and structural formulae	18
1.7	References for 1.1···1.6	18
1.8	References to general literature	19
1.8.1	General	19
1.8.2	Microwave spectroscopy	20
1.8.3	Infrared, Raman and electronic spectroscopy	21
1.8.4	Electron diffraction	22
1.9	List of symbols and abbreviations	24
1.9.1	List of symbols	24
1.9.2	List of abbreviations	24
2	Inorganic molecules	25
3	Organic molecules	79

1 Introduction

1.1 General remarks

This volume of the Basic Data Series contains data on the geometric parameters (internuclear distances, bond angles, dihedral angles of internal rotation etc.) of free polyatomic molecules including free radicals and molecular ions. (For the diatomic structures measured by high-resolution spectroscopy, see [1], and for molecular structures in crystals, see [2].)

The geometric parameters are determined in the gas phase either by an analysis of the rotational constants (and sometimes the vibrational constants) obtained from microwave, infrared, Raman, electronic and photoelectron spectroscopy or by analysis of electron diffraction intensities. Most of the structures listed in the tables are for molecules in the electronic ground state, but structures for electronically excited states have also been included as far as they are available (see 1.5).

A number of basic structure data have been selected carefully from those listed in the data tables presented in the New Series of Landolt-Börnstein, II/7 [3], II/15 [4], II/21 [5], and II/23 [6]; these data were originally taken mostly from the literature published between 1960 and 1993. Other data listed in these volumes had to be omitted in this volume for the limit of space. The names of these molecules are cited here with their reference numbers for their original data tables. For the structure data published before 1960, see [7].

As a general principle, the structures reported in the original literature have been taken after critical reexamination of their uncertainties but without making any further numerical re-analysis. With few exceptions, those molecules for which geometric parameters were reported only qualitatively, e.g., when only equilibrium molecular symmetry was reported, have been omitted. The policy governing the selection of data and the assessment of their uncertainties differs slightly according to experimental method, as described in detail below. For molecules studied by more than one experimental technique, e.g., by microwave and infrared spectroscopy or by microwave spectroscopy and electron diffraction, a decision has been made in each case as to whether it is necessary or desirable to list more than one geometric structure. When spectroscopic and diffraction structures are listed together for the purpose of critical comparison, they are in general regarded as compatible with each other, to the best of our present knowledge, within their experimental uncertainties; otherwise, specific comments are given in each case. (Note that a slight difference exists in the physical significance of the structural parameters reported in spectroscopy and electron diffraction: see 1.3.)

1.2 Experimental methods

1.2.1 Spectroscopy

Molecular spectroscopy extends from the radiofrequency region down to the soft X-ray region. The rotational spectra of molecules normally appear in the microwave and far-infrared regions, the vibrational spectra in the infrared region and the electronic spectra in the visible and shorter-wavelength regions, although there exist a number of exceptions. In long-wavelength regions, microwave and far-infrared, the resolution is inherently high so that the rotational constants are readily determined with high precision. In the infrared, visible and ultraviolet regions large grating spectrographs were set up to resolve the rotational structures of molecules. However, recent developments of spectroscopic techniques, in particular the introduction of lasers and development of Fourier transform spectroscopy, have changed the situation in these regions considerably. In the infrared, some fixed-frequency lasers such as the carbon dioxide and nitrous oxide lasers have been combined with external electric or magnetic fields to bring molecular transitions into resonance with laser frequencies; the electric-field case is referred to as laser Stark spectroscopy and the magnetic-field case as laser magnetic resonance or LMR. LMR was originally developed in the far-infrared region. More recently, infrared spectroscopy with tunable laser sources has become more common. Diode lasers are available in nearly the entire infrared region and also, in recent years, from the near infrared to the visible region, the color center laser oscillates in the region $1.5\cdots3.5$ μm; and a difference frequency output can be generated in the $2\cdots4.2$ μm region by non-linear mixing of two visible lasers, one tunable and the other fixed in frequency. In the visible and shorter wavelength regions grating spectrographs have been traditionally employed; however, dye lasers are now widely used, and harmonic generation of the output or nonlinear mixing of two or more outputs allow us to cover the wavelength region down to the vacuum ultraviolet. Laser spectroscopy has improved not only the resolution, but also the sensitivity so that even a small number of molecules prepared in the form of a beam have been detected, eliminating the Doppler broadening from their spectra. Many complexes formed by weak intermolecular forces have also been detected in beams. The recent development of Fourier spectroscopy is remarkable; spectrometers of high performance are now commercially available and allow us to record the spectra of molecules from 10 cm^{-1} to 5×10^4 cm^{-1} (200 nm) with Doppler-limited resolution over most of this range. The method of Fourier transformation has been introduced in microwave spectroscopy in combination with free jet expansion of the sample and has resulted in much improvement in resolution and sensitivity.

During the last decade considerable improvements have occurred in infrared spectroscopic instrumentation. For high resolution studies, grating instruments have been replaced by Fourier transform spectrometers which are capable of resolving 0.002 cm^{-1} or better and have precision of measurement of 1×10^{-4} cm^{-1} (3 MHz) or even 5×10^{-5} cm^{-1}, the accuracy being dependent on calibration lines, typically 4×10^{-4} cm^{-1}. A joint effort by many laboratories is aimed at providing a coherent set of calibration lines so that accuracies near 1×10^{-4} cm^{-1} can be reached in most spectral regions. Recommendations to correct previous wavenumber standards have been issued by the International Union of Pure and Applied Chemistry (IUPAC) [8].

Diode laser and difference-frequency laser spectrometers have also been developed with instrumental resolution high enough that the Doppler width of the lines in the spectrum sets the practical limit to resolution. These spectrometers can achieve high precision in wavenumber measurements but the accuracy is again highly dependent on calibration lines. Since molecular jets are becoming more common nowadays, Doppler widths can be reduced, making laser techniques more promising. At present, Raman measurements for structural purposes are still made in the visible region of the spectrum using grating spectrometers and tend to be roughly 10 to 100 times less precise than infrared measurements.

In recent years, new schemes have been introduced in Raman studies, taking advantage of Fourier transform spectrometers and of lasers. For instance CARS techniques can now achieve nominal resolutions of 0.003 to 0.005 cm^{-1} with accuracies of 0.001 cm^{-1} or better. Unfortunately, these experiments are still limited to a small number of skilled laboratories. Nevertheless, joint works using both IR and Raman data are appearing, Raman being especially useful for infrared inactive bands.

Both infrared and Raman studies have an advantage over microwave spectroscopy in that a much larger number of lines can be conveniently measured, often at higher values of the rotational quantum numbers. Many papers now combine IR and microwave data, to derive a full set of rotational constants.

In modern papers ground-state constants are frequently reported with cited uncertainties of $\pm1\times10^{-7}$ cm^{-1} (3 kHz) from infrared work and $\pm1\times10^{-5}$ cm^{-1} (0.3 MHz) from Raman studies. In band spectra, two sets of rotational constants are obtained, those of the upper and lower states involved in the transition, and a statistical treatment allows the differences between the constants to be determined to precisions approaching or equal to microwave uncertainties (1 kHz or less). Thus equilibrium rotational constants of polar molecules can be quite precisely calculated by using microwave-determined B_0 constants and infrared-determined α constants. When the values of some of these α constants are missing, they can be substituted by reliable *ab initio* values. Despite the recent instrumental improvements, the resolution available from both infrared and Raman studies is still much lower than that from microwave spectroscopy, and therefore, studies are limited to fairly small and simple molecules. However, these techniques are not restricted to polar molecules as is the case for microwave spectroscopy, and thus infrared and Raman spectroscopy play an important role in the determination of the structures of small symmetric non-polar molecules.

Many early infrared and Raman papers have reported studies on polar molecules that subsequently have been reexamined in the microwave region. In most of these cases, the microwave work is clearly superior and the infrared results have not been included in these tables. In some cases, however, the addition of even relatively low-precision optical data, when combined with microwave data, will lead to improved structural estimates. For example, frequently the A_0 (or C_0) rotational constant of a symmetric top can be obtained either from perturbation-induced transitions in the infrared spectrum or from suitable combinations of transitions in a fundamental band, a combination band and a hot band, or else by the analysis of a perpendicular band in the Raman spectrum. It is not possible to obtain this rotational constant in the pure rotational spectrum of a symmetric top molecule, and therefore combining the optical and microwave data leads to much improvement in determining the positions of the off-axis atoms of such molecules.

These spectroscopic methods with high resolution provide us with *rotational constants*, from which we may extract information on the geometric structures of molecules. The rotational constants are inversely proportional to the principal moments of inertia. In a rigid molecule, the three principal moments of inertia (two for a linear molecule) are defined such that

$$I_a \leq I_b \leq I_c \tag{1}$$

where a, b, and c denote the three principal axes of inertia. The rotational constants are given by

$$A = h/8\pi^2 I_a, \quad B = h/8\pi^2 I_b, \quad C = h/8\pi^2 I_c \quad (A \geq B \geq C) \tag{2}$$

where h denotes Planck's constant. A *linear* molecule has two identical rotational constants, I_a being taken to be zero. A *spherical top* has three identical rotational constants. They are usually denoted by B. Therefore, only one piece of information on molecular geometry can be obtained for the two types of molecules per isotopic species. The definition of a *symmetric top* is that two of the three rotational constants are equal; $A > B = C$ for a prolate symmetric top, and $A = B > C$ for an oblate symmetric top. In these cases, at most two independent constants can be determined; usually only the B rotational constant is obtained from pure rotational spectra because of the selection rules. An *asymmetric top* molecule in general gives three independent rotational constants, but for a rigid, planar molecule the relation $I_c = I_a + I_b$ holds, and only two of the three are independent. The *inertial defect*, defined as $\Delta = I_c - I_a - I_b$, has a small and, in most ordinary cases, positive value for many planar molecules and is accounted for mainly by vibration-rotation interactions, as discussed below. For most molecules the number of independent geometric parameters exceeds that of the available independent rotational constants. Therefore, in order to determine the geometric parameters completely one has to work with isotopically substituted species as well.

The geometric significance of the rotational constants is somewhat obscured by the contributions of intramolecular motions. In most *quasi-rigid* molecules, because the frequencies of vibration are high compared with those of rotation, the rotational constants observed correspond to the inverse moments of inertia averaged over the vibrational amplitudes. Averaging is in general quite complicated. When all the intramolecular vibrations have small amplitudes and there are no accidental degeneracies or resonances, the rotational constants, for example the B constant, in the v-th vibrational state are given by expressions of the form

$$B_v = B_e - \sum_s \alpha_s^B (v_s + d_s/2) + \ldots \tag{3}$$

where v_s and d_s denote the vibrational quantum number and the degeneracy of the s-th normal mode respectively. Accordingly even for the ground vibrational state the rotational constant B_0 is not identical with the *equilibrium rotational constant* B_e, which can be interpreted purely geometrically by Eq.(2), namely,

$$B_e = h/8\pi^2 I_b^{(e)}. \tag{4}$$

The *vibration-rotation constants* α_s^B are complicated functions of the harmonic (quadratic) and anharmonic (mainly cubic) potential constants [9] and depend on the masses of the component atoms. Since a rotational constant is inversely proportional to a moment of inertia, α_s^B does not simply represent averaged vibrational contributions. It has, however, been proven [10] that the rotational constant corrected for the harmonic part of α_s^B gives the moment of inertia which corresponds to the real vibrational average.

The corrected rotational constant is often denoted as B_z, i. e.,

$$B_z = B_0 + \sum_s \alpha_s^B (harmonic) d_s / 2 = h / 8\pi^2 I_b^{(z)}. \tag{5}$$

When a *non-rigid* molecule has one or more intramolecular motions of large amplitude such as internal rotation, inversion, and ring puckering, it is much more difficult to interpret the rotational constants. However, by treating such large-amplitude motions separately from small-amplitude vibrations, one can often calculate from the observed spectra the effective rotational constant, which may then be analyzed as if there were only small-amplitude vibrations. In weakly bound complexes such as van der Waals complexes, component molecules execute labile motions with each other, so that it is fairly difficult to determine the geometric parameters. Even for such molecules we may determine the symmetries of the complexes at equilibrium, although the deviations from the equilibrium geometry can be quite large. In the following tables we add a comment "large-amplitude motion" and list "effective symmetry class" for molecules like complexes which execute large-amplitude motions.

The vibration-rotation spectra and/or the rotational spectra in excited vibrational states provide the α_s^B constants and, when all the α_s^B constants are determined, the equilibrium rotational constants can be obtained by extrapolation. This method has often been hampered by anharmonic or harmonic resonance interactions in excited vibrational states, such as Fermi resonances arising from cubic and higher anharmonic force constants in the vibrational potential, or by Coriolis resonances. Equilibrium rotational constants have so far been determined only for a limited number of simple molecules. To be even more precise, one has further to consider the contributions of electrons to the moments of inertia, and to correct for the small effects of centrifugal distortion which arise from transformation of the original Hamiltonian to eliminate indeterminacy terms [11]. Higher-order time-independent effects such as the breakdown of the Born-Oppenheimer separation between the electronic and nuclear motions have been discussed so far only for diatomic molecules [12].

Often the spectroscopic methods can show the presence of symmetry elements in a molecule without going into a detailed numerical analysis of the rotational constants. Because the spectrum of a symmetric top appears regular, it is easy to show that the molecule has a C_n axis with $n \geq 3$. The statistical weight due to degeneracy of nuclear spins can supply additional information. Even for an asymmetric top a C_2 axis causes intensity alternations in its spectrum. A plane of symmetry can be detected by isotopic substitution of one of two atoms located symmetrically with respect to the plane. For a molecule of the form H_2M where the two hydrogen atoms are symmetrically placed with respect to the symmetry plane, we will have only one singly-deuterated species, HDM. In cases where a C_{3v} symmetric internal rotor like the methyl group is attached to the framework such that one C–H bond is in a symmetry plane, we will have two kinds of singly-deuterated species: one species with the C–D bond in the symmetry plane and the other out of the plane, the latter having the statistical weight of two. If the internal rotation barrier is low, the energy levels of the latter species may be split into two by the tunneling. A small inertial defect suggests that a molecule is planar or very nearly planar at equilibrium.

1.2.2 Electron diffraction

The design of a gas electron diffraction apparatus is similar to that of an electron microscope, but the electron optical system is much simpler. The sample gas is introduced into the diffraction chamber through a nozzle, and the electrons scattered by the sample molecules near the nozzle tip are either collected on a photographic plate or measured directly by electron counting. Typical experimental conditions are as follows: electron accelerating voltage, ≈50 kV, corresponding to electron wavelength ≈0.05 Å; electron beam diameter, ≈0.1 mm; nozzle diameter, ≈0.3 mm; nozzle-to-detector distance, ≈20···60 cm; sample pressure at the nozzle tip, ≈20 Torr. For a sample having a much lower vapor pressure at room temperature, the sample is heated during the experiment.

Since the scattering intensity decreases very rapidly with the scattering angle, a rotating sector, which has an opening proportional to the cube of the radius, is used for photographic measurements to compensate for the rapidly falling intensity. The photographic density (usually ranging from about 0.2 to 0.8) is measured by a microphotometer. The photographic density is converted to electron intensity by use of a calibration function. Most of the electron diffraction data listed in the following tables have been obtained by this sector-microphotometer method. The intensity I_T is a function of the scattering variable,

$$s = (4\pi/\lambda) \sin(\theta/2) \tag{6}$$

where λ is the electron wavelength and θ is the scattering angle. When a smooth background,

$$I_B = I_{atomic} + I_{inelastic} + I_{extraneous}, \tag{7}$$

is subtracted from I_T, the molecular term,

$$I_M = I_T - I_B = \sum_{i \neq j}\sum |f_i||f_j|\cos(\eta_i - \eta_j)\exp(-\tfrac{1}{2}l_{ij}^2 s^2)\sin s(r_{aij} - \kappa_{ij}s^2)/sr_{aij} \tag{8}$$

is obtained. The molecular term is a sum of contributions from all the atom pairs $i - j$ in the molecule. The experimental process of gas electron diffraction is to measure and analyze I_M, in order to derive the internuclear distance r_a and the root-mean-square amplitudes l (sometimes denoted as u), often abbreviated as "the mean amplitude". The complex atomic scattering factor,

$$f_i(s) = |f_i(s)|\exp[i\eta_i(s)] \tag{9}$$

is usually calculated with sufficient accuracy by use of an atomic potential function. The mean amplitudes can also be calculated if the quadratic force field of the molecule is known. The asymmetry parameter κ represents a slight (often negligible) deviation of the argument from a linear function of s. The significance of the r_a distance is described in 1.3.2.

The molecular term I_M, or sometimes I_M/I_B, is analyzed by a least-squares method, and the bond distances, the bond angles (and other geometric parameters such as the dihedral angles of internal rotation and the angles related to ring geometry) or the nonbonded distances, and some of the mean amplitudes are determined. The rest of the mean amplitudes are usually constrained to either calculated or assumed values. The ratio, $(I_M/I_B)_{obs}/(I_M/I_B)_{calc}$, is often determined in the analysis as another empirical parameter. Whether this ratio, often called the index of resolution, is close to or appreciably less than unity is a good indicator of the quality of the experiment. However, the mean amplitudes and the index of resolution are not listed in the following tables, because they have no direct geometric consequence. The number of adjustable parameters depends on the accuracy of the experimental intensity and on the complexity of the molecule. The analysis often requires assumptions about molecular symmetry and/or the parametric values on which the molecular term depends only weakly. In studies of conformational and other equilibria, differences in the "frame" structures of component species are often assumed to be equal to those estimated by *ab initio* calculations, or simply ignored. The uncertainties originating from these assumptions, if significant, are included in the error estimates of the final parameter values (1.4.3).

In comparison with other methods for determining molecular geometry, such as spectroscopy (1.2.1) and X-ray crystallography, the merits of gas electron diffraction are the following:

1) The average values of internuclear distances (particularly bond distances) in free molecules can be determined directly and, in many cases, accurately.

2) In principle, the only restrictions on the sample molecule are a sufficient vapor pressure and chemical stability. Polar or nonpolar, light or heavy, small or large molecules can be studied.

On the other hand, electron diffraction has the following drawbacks:

1) Only the distance parameters averaged in thermal equilibrium, instead of those in individual quantum states, can be obtained. Therefore, the parameters derived from electron diffraction are subject to various vibrational effects. These effects can be considerable in non-rigid ionic molecules present in the vapors or salts at high temperatures.

2) An accurate measurement of the molecular term needs much caution. Any undetected accident in the experiment or analysis can cause serious, often undiscovered, systematic error in the geometric parameters derived.

3) The resolution of internuclear distances is limited by thermal vibration to only about 0.1 Å and cannot be improved experimentally except in special cases, such as PF_3, when nonbonded distances allow a resolution. Closely spaced, inequivalent distances are measured only as their weighted average values, their differences remaining very uncertain.

4) When the molecule under study contains elements of very different atomic numbers, the parameters relating to lighter atoms may be very uncertain. As a result of 3) and 4), in the worst cases even qualitative conclusions derived from electron diffraction on molecular geometry (such as the structural formula, equilibrium symmetry, and the presence of conformers) can be in error.

5) Besides the above errors, the distance parameters may be systematically uncertain by as much as a few parts per thousand. The source of this error is in the scale factor (the electron wavelength times the nozzle-to-detector distance).

1.2.3 Combined use of electron diffraction and spectroscopy, theoretical calculations, or other methods

The disadvantages of electron diffraction mentioned above can be appreciably reduced if data obtained from other methods of structure determination are used conjointly. Vibrational spectroscopy supplies information on the force field, which is used very frequently in analyses of ED data for the calculation of mean amplitudes, shrinkage corrections, etc. (1.3.2). The equilibrium molecular symmetry determined by spectroscopy may be assumed, and the rotational constants determined by high-resolution spectroscopy may be used to adjust the scale factor and to set constraints on the geometric parameters. In principle, the structure can thus be determined with less ambiguity and more accuracy than either by spectroscopy or by electron diffraction alone [13]. Many examples are listed in the following tables, e.g., electron diffraction plus microwave spectroscopy, abbreviated as ED, MW. For this purpose, the experimental data derived from different methods should be analyzed in terms of a consistent set of geometric parameters, usually in terms of the r_z parameters described in 1.3.1, so as not to introduce additional systematic error. See General References [E-9], [E-24] for details.

The geometric structure and the force field estimated by an *ab initio* molecular-orbital (MO) calculation [14] are often taken into the analysis as valuable sources of information [15, 16]. Molecular-mechanics calculations [17] are also used widely [16]. Many examples of the joint use of these theoretical structural data are found in the following tables. Sometimes an NMR spectrum of molecules dissolved in a liquid crystal provides information on the geometric structure [18], which can also be taken into the analysis of electron diffraction data [19]. Mass spectrometry is also used with electron diffraction for estimating the composition of the sample vapor [20].

1.3 Significance of geometric parameters

The internuclear distances and angles listed in the following tables are based on various different definitions. Some of them are defined on physical and geometrical principles, while others are defined operationally, i.e. by the method used for deriving the parameters from the experimental data. Numerically, the differences may not necessarily be important in comparison with experimental uncertainties, but it is always important to specify the definition of the parameters determined in order to make a precise and systematic comparison of experimental structures with one another or with the corresponding theoretical structures, such as those derived from *ab initio* calculations. A brief summary of the definitions is made in the present section. For a more detailed discussion of the significance of the structures and their relationship, see General References [E-10], [E-20].

1.3.1 Spectroscopy

In most spectroscopic studies, nuclear positions in the molecule-fixed coordinates are first derived from the rotational constants, and bond distances and angles are then calculated from the nuclear coordinates. This may be contrasted with the electron diffraction case, where internuclear distances are directly derived from experiment, and angles or nuclear coordinates are calculated from the internuclear distances (see 1.3.2). The equilibrium (r_e), and average (r_{av} or r_z) structures, which have well-defined physical significance, are reported occasionally, but most other cases quote the r_0 or r_s structures operationally derived from spectroscopic experiments.

(a) r_0 structure: A set of parameters obtained from the zero-point ("ground state"[1])) rotational constants is called the r_0 structure. Sometimes the rotational constants of a sufficient number of isotopic species are combined. In other cases, assumptions are made for some of the parameters. As pointed out in 1.2.1, the ground-state rotational constants do not correspond to the moments of inertia averaged over the ground vibrational state, but rather their inverses. It is therefore difficult to assess the exact physical meaning of the r_0 parameters. Furthermore, the r_0 parameters derived from different combinations of isotopic species are often widely different from one another. For example, r_0(C–H) distances can show deviations of as much as 0.1 Å. Note also that the relation $I_c = I_a + I_b$ for a planar molecule does not strictly hold for the ground-state rotational constants and this introduces supplementary ambiguities. This remark applies also to the r_s structure, below.

(b) r_s structure: As demonstrated by Costain [21], a set of geometric parameters that are much more consistent than the r_0 parameters can be obtained when only isotopic *differences* of rotational constants are used. Such a structure is called a substitution structure (r_s). Kraitchman's equations [22] are used most conveniently for calculating the r_s structure. For a linear molecule the coordinate of the i-th atom a_i is given by

$$a_i^2 = \Delta I_b / \mu \tag{10}$$

where ΔI_b is the change in the moment of inertia upon substitution of the i-th atom by its isotope. When the isotopic atom has a mass differing from the original atom by Δm_i, μ is given by $M\Delta m_i/(M+\Delta m_i)$, where M is the total mass of the parent molecule.

For a general asymmetric top, Kraitchman gave the following equation:

$$a_i^2 = \frac{\Delta P_a}{\mu} \left[1 - \Delta P_b / (P_a - P_b)\right]\left[1 - \Delta P_c / (P_a - P_c)\right]. \tag{11}$$

Equations for b_i^2 and c_i^2 are obtained by cyclic permutation of a, b and c. The moment P_a is defined by

$$P_a = (-I_a + I_b + I_c)/2, \tag{12}$$

P_b and P_c being defined in a similar way, and ΔP denotes the change of P on isotopic substitution. When a molecule has a plane or axis of symmetry, the above equations are simpler.

All the singly substituted isotopic species are needed to obtain a complete r_s structure. However, it is sometimes impossible, e.g. when the molecule contains atoms having only one stable nuclide such as ^{19}F, ^{31}P, ^{127}I, or difficult, e.g. for a chemical reason, to make complete isotopic substitutions. In these cases, one is forced to use a first-moment equation or a condition that the cross-products of inertia be zero. In

---

[1]) "ground state" can and often does also mean ground electronic state.

some cases even a part or all of the three moments of inertia of the parent species are used. The r_s structure thus obtained is in reality a hybrid of the r_s and r_0 structures. Nevertheless, even in these cases the structure is usually called r_s.

In the r_s scheme the coordinates of an atom located far from a principal inertial plane can be determined accurately, whereas those of an atom located close to an inertial plane are poorly defined, irrespective of the atomic mass. In the latter case the relative signs of the coordinates are difficult to determine, because Kraitchman's equations give only the absolute values. For small coordinates, doubly-substituted species may be of some use [23].

It has been shown for diatomic molecules that the r_s parameter is a simple average of the r_0 and r_e parameters. This relation holds only approximately, or may even fail, for polyatomic molecules. However, Watson [24] demonstrated that, if higher-order terms were neglected in the expansion of moments of inertia in terms of the isotope mass difference, the equilibrium moment of inertia was approximately given by $2I_s - I_0$, where I_s denoted the moment of inertia calculated using r_s coordinates.

(c) r_{av} (r_z) structure: When a set of the B_z constants defined in Eq. (5) are used in place of B_0, the average structure in the ground vibrational state is obtained. This structure is usually called r_{av} or r_z [10,25]. The r_z structure defines primarily the average nuclear positions during the vibrational motion, as remarked above. Hence, the r_z(X–Y) distance between the nuclei X and Y defines the distance between the average nuclear positions instead of the average of the instantaneous X–Y distance. A simple calculation shows that for a molecule with small vibrational amplitudes

$$r_z \cong r_e + \langle \Delta z \rangle_0 \tag{13}$$

where Δz denotes the instantaneous displacement, Δr, of r(X–Y) projected on the equilibrium X–Y axis (taken as a temporary z axis), and 0 denotes the average over the ground vibrational state. Because of the presence of nuclear vibrations perpendicular to this z axis, r_z does *not* agree with the real average X–Y distance $r_e + \langle \Delta r \rangle_0$. In general, it is not easy to estimate the difference between r_z and r_e, i.e. $\langle \Delta z \rangle_0$, even for a bond distance. On the other hand, the difference between $\langle \Delta r \rangle$ and $\langle \Delta z \rangle$ can be estimated with sufficient accuracy by the use of the quadratic force field. The average bond angle can be defined unambiguously in terms of the average nuclear positions.

The r_z nuclear position depends on the mass of the nucleus. Therefore, the isotope effect must be known precisely when the B_z constants of other isotopic species are needed for a complete determination of the r_z structure. The r_z structure from such isotopic substitution is highly sensitive to the estimated (or assumed) isotope effect on r_z positions. This presents a serious difficulty in the experimental derivation of the r_z structures for all but the simplest molecules [26,27].

The r_z structure can be generalized to excited vibrational states. The average nuclear positions for a vibrational state, characterized by a set of vibrational quantum numbers v may be defined in a similar way.

(d) r_e (r_m) structure: When all the α_s^A, α_s^B and α_s^C constants given in Eq.(3) have been determined, the equilibrium constants A_e, B_e and C_e are obtained. If a sufficient number of these constants are available, the equilibrium (r_e) structure, which defines the nuclear positions corresponding to the potential-energy minimum, can be determined. For a few simple molecules the α constants are given as functions of a limited number of third-order anharmonic potential constants which are common to all isotopic species. Therefore, if a number of rotational constants is determined for isotopic species in the ground as well as excited vibrational states, these anharmonic potential constants may be determined simultaneously with the r_e structural parameters [28]. The geometric significance of A_e, B_e and C_e (Eq.(4)) and r_e is unambiguous. However, because of the various experimental difficulties, accurate r_e structures are currently known for only a small number of simple molecules, as listed in the following tables.

In view of this situation, Watson [24] proposed what he called an r_m structure. As mentioned above, he pointed out that $2I_s - I_0$, which he referred to as I_m, was very close to the equilibrium moment of inertia I_e. Watson called the structure derived from a set of I_m the "mass-dependence" (r_m) structure. In a number of examples he has shown that the r_m structure is indeed very close to the r_e structure, except for some parameters involving hydrogen. A drawback of this method is that data for more isotopic species than are necessary for the r_s method are needed for structure determination. Nakata et al. [29–31] pointed out that there existed additivity relations for isotopic effects on the structure parameters and employed this fact to eliminate higher-order terms in the expansion of moments of inertia in terms of isotope mass differences which were neglected in the original treatment of Watson. In this way Nakata et al. have expanded the applicability of the r_m method. A similar approach was proposed by Harmony et al. [30–32]. They noted that the ratio $\rho = I_s/I_0$ is not significantly isotope-dependent species (but may be different for different inertial axes) and proposed to use $[I_m^\rho]_i = (2\rho - 1)[I_0]_i$ for the moment of inertia of the i-th isotopic species. The structure thus obtained is called the r_m^ρ structure.

1.3.2 Electron diffraction

An average internuclear distance can be regarded as the first moment of the probability distribution function of this distance, $P(r)$, which is approximately Gaussian unless the distance depends strongly on a large-amplitude vibration [35].

(a) r_a distances: When the asymmetry parameter κ in Eq. (8) is small, the distance parameter derived directly from an experiment of electron diffraction is r_a. This parameter is exactly equal to the center of gravity of the $P(r)/r$ distribution [36],

$$r_a = r_g(1) \equiv \left[\int P(r)dr\right] / \left[\int P(r)/r \cdot dr\right] . \tag{14}$$

(b) r_g distances: If the $P(r)$ function is approximately Gaussian, then r_a is related to the center of gravity of $P(r)$, r_g, as

$$r_g = r_g(0) \equiv \left[\int rP(r)dr\right] / \left[\int P(r)dr\right] \cong r_a + l^2/r_a \tag{15}$$

where l is the mean amplitude. In contrast with the r_z defined in Eq. (13), the r_g(X–Y) bond distance is a *real* (instead of *projected*) average, over thermal equilibrium, of the instantaneous X–Y distance, namely,

$$r_g = r_e + \langle \Delta r \rangle_T . \tag{16}$$

The r_e bond distance can be estimated if the bond-stretching anharmonicity is assumed [2]. For example, for a group of similar bonds (e.g., for the C–C bonds in hydrocarbons), the average displacements $\langle \Delta r \rangle_T$ are estimated to be nearly equal to one another, so that the observed differences in the r_g distances may well be approximated as those in the r_e distances. For this reason, many of the recent studies of electron diffraction report the r_g bond distances together with the r_a distances derived directly from experiment.

On the other hand, a set of the r_g bond distances and nonbonded distances cannot define a physically meaningful bond angle without corrections for linear or nonlinear "shrinkage effects" arising from the curvelinear path followed by atoms during a vibration [37]. The angle parameters determined by electron diffraction, either from the r_a distances or from the r_g distances, are quoted in the following tables as effective angles, when no corrections for the shrinkage effects have been made. The effective angles derived from the r_a distances are denoted as θ_a. The difference between an effective angle and the well-defined angles, such as the θ_e or θ_z angles derived from the r_e and r_z structures, respectively, depends on the amplitudes of bending, puckering, or torsional vibrations. This difference can often be estimated with sufficient approximation, since the shrinkage effect can be calculated if the quadratic force field of the molecule is known.

(c) r_α and r_α^0 structures: A number of recent studies of electron diffraction report the θ_α angles defined in terms of a set of the r_α (bonded and nonbonded) internuclear distances derived [38] from the corresponding r_g distances by

---

[2]) The r_e bond distance can also be estimated directly from experimental ED data on the basis of an assumed potential function model. See for example, [E–25], [E–42].

$$r_\alpha = r_g - \left(\langle \Delta x^2 \rangle_T + \langle \Delta y^2 \rangle_T\right) / 2r - \delta r \qquad (17)$$

where Δx and Δy denote the displacements perpendicular to the equilibrium nuclear axis (z) and δr denotes a small displacement due to centrifugal force. The shrinkage effects are eliminated in this structure.

The r_α distance corresponds to the distance between the thermal-average nuclear positions,

$$r_\alpha \cong r_e + \langle \Delta z \rangle_T \qquad (18)$$

and when it is extrapolated to zero Kelvin temperature,

$$r_\alpha^0 = \lim_{T \to 0} r_\alpha \cong r_e + \langle \Delta z \rangle_0 \qquad (19)$$

the r_α^0 structure is practically identical with the r_z structure. For a joint analysis of electron diffraction and spectroscopy discussed in 1.2.3, the r_a or r_g distances observed by electron diffraction should be converted to the r_α^0 distances in order to make a consistent analysis for the r_z nuclear positions. In practice, the extrapolation from r_α to r_α^0 can be made approximately on the basis of simple assumptions about anharmonicity. Except in a large-amplitude case, the uncertainty due to the r_α to r_α^0 conversion can be regarded as trivial.

1.3.3 Table of distance parameters and their definitions [a])

Symbol	Definition
r_e	Distance between equilibrium nuclear positions
r_{av}, r_z, r_α^0	Distance between average nuclear positions (ground vibrational state)
r_α	Distance between average nuclear positions (thermal equilibrium)
r_0	Distance between effective nuclear positions derived from rotational constants of zero-point vibrational levels
r_s	Distance between effective nuclear positions derived from isotopic differences in rotational constants
r_m	Distance between effective nuclear positions derived from the mass-dependence method of Watson, very close to r_e for molecules without hydrogen atoms. r_m^ρ is the distance obtained by a slightly modified method of Harmony et al.
r_g	Thermal average value of internuclear distance
r_a	Constant argument in the molecular term, Eq. (8), equal to the center of gravity of the $P(r)/r$ distribution function for specified temperature (no correction for the shrinkage effects normally being made)

[a]) Some authors distinguish between "mean value" and "average value". However, the two words are used interchangeably in the following tables.

1.4 Uncertainties

1.4.1 Microwave spectroscopy

In many cases, authors give uncertainties originating only from the experimental errors in the rotational constants. Since microwave spectroscopy gives rotational constants with six to eight significant digits, this source of error is very small in comparison with other systematic errors. In the r_s scheme, error estimates often rely on the consistency of the results when more than the necessary number of isotopic species are available. These estimates include some of the uncertainties due to vibration-rotation interactions, but probably not all.

In the following tables, an attempt has been made to list either the r_e or the r_z structures, because they are physically well-defined as mentioned in 1.3. In a few cases where the r_e structures are reported, bond distances have been determined to a few parts in 10^4. Higher-order effects are of this order of magnitude or less. For the r_z structure, the systematic error due to the B_0 to B_z conversion may contribute to the total uncertainties, which are roughly a few parts in 10^3 or less. On the other hand, when different isotopic species are combined, the above-mentioned isotope effect on the nuclear positions may cause errors in r_z distances of 0.01 Å or more.

Most recent microwave papers report r_s structures with judicious estimates of errors. Although the contributions of vibration-rotation interactions are taken into account only empirically in most of these cases, errors given in the original papers are reproduced in the tables. However, in cases where the number of isotopic species investigated is small or the vibration-rotation interactions are apparently not considered properly, errors are increased slightly and are so indicated.

Generally speaking, smaller errors are assigned to the r_s parameters than to the r_0 parameters because of the consistency of the data. For the r_0 parameters, errors given in the original papers have been increased in most cases. Reported errors have also been increased in cases where some of the parameters are assumed because of the shortage of experimental data. In these cases the assumed parameters are mentioned in the tables. Curl [39] has developed a least-squares method, called the diagnostic least squares, by which the uncertainties in the "assumed" parameters may be taken into account. By this method, more parameters than the number of input experimental data can be "determined". The results depend obviously on authors' estimates of the uncertainties in the "assumed" parameters. In these cases errors listed in the tables have been made somewhat larger than the original estimates.

In summary, the uncertainties represent the "reasonable limits of error" estimated by E. Hirota. In other words, the true parameter is expected to be in the range of the errors listed in the tables except under very unusual circumstances.

1.4.2 Infrared and Raman spectroscopy

Bond distances obtained in infrared and Raman studies are normally r_0 or r_e distances. Although data from the infrared are now, in general, precise enough to obtain r_s distances, the large amounts of pure rare isotopic species required to obtain these data preclude the substitution technique, and only a handful of partial r_s structures have been obtained by optical methods. In these tables, whenever both r_e and r_0 structures have been derived for a given molecule, only the r_e structure is listed. Uncertainties in r_e parameters are not easily estimated since for polyatomic molecules the study of several vibration-rotation bands is required and quite often the data used come from several different laboratories and have been obtained over a period of years. The derivation of r_e parameters also often necessitates certain assumptions regarding the effects of perturbations, especially Fermi resonances. The limitations of r_0 parameters have been detailed above in Section 1.4.1.

In the following tables the uncertainties given for the infrared and Raman-derived structures are those cited by the authors. In a few cases where the errors appear to be overly optimistic, a footnote is inserted to this effect. The structural parameters of a few molecules have been computed using rotational constants reported in the literature, and the errors cited are based on the experimental uncertainties and the effects of errors of model.

1.4.3 Electron diffraction

Purely random errors in the geometric parameters determined by electron diffraction are usually estimated in a least-squares analysis from differences between the observed and calculated intensities. However, the errors in the geometric parameters are by no means distributed randomly, and the following systematic errors often make much more important contributions.

(a) Experimental sources: Systematic errors in the measurement of I_M (1.2.2) are one of the most significant sources of error. Inaccurate sector calibration and nonuniform extraneous scattering are typical examples. An error in the scale factor increases or decreases all the distances in the same proportion, while the angles remain undisturbed. Systematic errors may be even larger under unfavorable experimental conditions, for example, when the experiment is done at very high temperature, when the sample is unstable and/or impure, or when only a small quantity of sample is available.

(b) Analytical sources: The estimation of the background (I_B in Eq. (7)) and various assumptions made in the analysis are other important sources of systematic error. In particular, the uncertainty in the difference between nearly equal, inequivalent internuclear distances depends so delicately on various sources of error that it is very difficult to estimate. These distances correlate strongly with mean amplitudes; when mean amplitudes are fixed at assumed values in a least-squares analysis, the systematic error in the distances caused by this correlation is sometimes overlooked. The uncertainty in angle parameters derived with neglect of shrinkage corrections (1.3.2) is also difficult to estimate, particularly when the system has a large-amplitude vibration. Where a slight deviation from linearity or planarity is indicated by an analysis which neglects shrinkage corrections or which does not explicitly take into account large-amplitude motions, the deviation may be spurious; in such a case an explicit remark is made to this effect.

The standard errors estimated in a least-squares analysis can be used as a measure of relative precision, i.e., to decide which parameters in the molecule under study can be determined more precisely than others. Nevertheless, their absolute magnitudes are always underestimated, since the essential parts of the systematic errors mentioned above are dissolved by adjustment of variable parameters and therefore overlooked. Such systematic errors can be discovered and corrected for only by a critical examination of measurements on a sample of precisely known structure made under analogous experimental conditions or by a comparison of the rotational constants calculated by use of the parameters obtained by electron diffraction with those determined by spectroscopy. Such a test has not always been made in the past, however. Even when the test is made, it never provides complete assurance that the data are free from all the systematic errors.

Thus many authors estimate total experimental uncertainties not only from the random standard errors obtained in their least-squares analyses but also from the systematic errors estimated somehow from their past experience, although their methods may differ appreciably from laboratory to laboratory. Accordingly, the uncertainties estimated in the following tables generally include *all* the supposedly possible systematic errors, i.e., estimated *total* errors rather than only *random* errors, unless indicated otherwise.

The styles of representation of the uncertainties also differ widely from laboratory to laboratory. Estimated standard deviations (e.s.d.) are often multiplied by a certain constant (2, 2.6 or 3). Several authors report "estimated limits of error" when they regard the probability of finding the geometric parameters *outside* the range as *negligible*. In general, the tables follow the styles reported by the authors and in each case define the meaning of the listed uncertainty, since it is considered to be inappropriate, if possible at all, to alter them into a uniform style.

The uncertainties have sometimes been re-estimated, and explicit remarks to this effect have then been made. Sometimes attention is drawn to a potential source of significant systematic error, such as the presence of sample impurity, implicit assumptions in the analysis such as the neglect of significant shrinkage effects, or strong correlation among the parameters. For some of the geometric parameters reported in the original papers, only their weighted average values are regarded as "well-defined" (1.2.2). In such cases, the average values and their uncertainties have been re-estimated and listed.

In summary, notwithstanding all possible caution in estimating a "reasonable" uncertainty in each of the geometric parameters obtained by electron diffraction, they can never be immune from hidden systematic errors. Accordingly, a warning has to be given that a minority of the listed uncertainties may have been underestimated and that, in the worst cases, even some of the authors' conclusions may be qualitatively incorrect.

1.5 Geometric structures of molecules in excited electronic states

In surveying the information on excited states given in these tables the reader may be struck by its limited extent and apparently haphazard distribution when compared with the analogous compilation for stable molecules in their ground states. The reasons are well known but are none the less worth briefly repeating, to be borne constantly in mind when trying to assess the meaning or reliability of the data in any particular case. There are two principal sources of difficulty, technical and theoretical.

Experimentally the source of information is almost exclusively spectroscopic. *Band-systems* may show *vibrational* and *rotational* structure whose analysis yields information on molecular geometry. Additional structure, such as electronic spin *fine-structure* or nuclear *hyperfine* structure, may be of great value in determining molecular electronic structure but rarely contribute much to knowledge of geometry. It may, however, greatly complicate the process of spectral analysis as a whole, making it difficult to extract the desired geometric parameters or limiting their accuracy: examples are to be found in NO_2 or triplet H_2CO.

Experimental limitations on the sources of primary information are usually chemical rather than instrumental. Thus chemically unstable species may be hard to prepare even in sufficient transient optical density or emitting concentration to yield a spectrum. More seriously, to obtain spectra of isotopic species requires usually the preparation of much larger samples than would be needed, e.g. in microwave spectroscopy, and in dominating concentration rather than as a minor constituent of a mixture or even in natural abundance. Thus in molecules with numerous geometric parameters to be determined, the technique of isotopic substitution has, with the exception of deuteration, been used only relatively rarely (see e.g. *s*-tetrazine). There are therefore in the literature many cases of molecules not listed here for which one or several rotational constants are known in excited states.

Instrumentally, limitations are rarely set by available resolving-powers. Spectra may be only partially resolved because of a combination of congestion and line-broadening: e.g. through Doppler- and pressure-broadening; and, more seriously, due to unimolecular lifetime-limiting non-radiative processes such as pre-dissociation, pre-ionization or, in large molecules, electronic relaxation into dense vibronic manifolds of lower-lying electronic states, arising from the limitations of the Born-Oppenheimer approximation. It is this last factor which is responsible for the apparent paradox that more seems to be known about the excited electronic states of chemically unstable molecules, such as free radicals, than about those of the common stable molecules; for, as a simple rule, stable closed-shell molecules have only high-lying excited states, at energies comparable to or above those of bond dissociation energies, and hence spectra that tend to be diffuse through predissociation, whereas open-shell molecules such as free radicals have low-lying electronic states with sharp spectra in experimentally ideal regions. When spectra are rich and well-resolved, rotational constants may be obtained which, although still not as good as those from microwave spectra, are of considerable accuracy, for the relatively low precision of individual line-frequencies is to a degree compensated by the large number of lines usually observed (see e.g. glyoxal). Where ground-state constants are known from microwave spectroscopy, excited-state constants may be obtainable with comparable accuracy, for it is usually possible to determine differences of rotational constants much more precisely than the constants themselves (see e.g. propynal).

Theoretically, the interpretation of geometric parameters tends to be hedged by qualifications. Most directly, the constants of *rotational analysis* may be interpreted in terms of average moments of inertia as in microwave spectroscopy except that the data tend to be much less extensive. From rotational constants A_v, B_v, C_v are calculated structures r_v which are effective averages over vibrational amplitudes in the level v. The level v is most often the zero-point level, and hence most of the structures quoted in these tables are the so-called "r_0-structures" (1.3.1). As in ground states, r_0-structures differ rather little from "true" r_e-structures in molecules that are relatively rigid; but in contrast, "non-rigid" molecules are much more common in excited states than in ground states. Many of the classical analyses involve quasi-linear or quasi-planar structures, and in these r_0-structures may differ considerably from r_e-structures. A striking example is to be found in the first excited quasi-planar singlet state of formaldehyde, in which the out-of-plane angles θ_0 and θ_e are about $20°$ and $36°$, respectively. Non-rigidity is often revealed by considerable inertial defects in planar molecules, but otherwise rotational analysis tends to be insensitive to the degree of non-rigidity and hence a poor way of determining "true" molecular structures in such cases.

Patchy but much more extensive information about potential surfaces as a whole may be obtainable from *vibrational analysis*. Electronic excitation usually leads to some changes of molecular geometry and these are reflected in the spectra by progressions of bands associated with transitions to a range of successive vibrational levels in a potential well to which their intensities are related through the *Franck-Condon Principle*. A knowledge of frequencies, assignments and intensities allows one therefore in principle to map the potential surface of one electronic state onto that of another, i.e. to determine structures relative to each other. In practice, there are great difficulties and structures determined with any precision and completeness by these methods are very rare. There exists however an enormous literature giving partial indications of molecular structures of widely varying reliability, probability or interest. The simplest examples are cases in which a minimum statement can be made on the basis of observed selection-rules alone: that a molecule has "*changed shape*" on electronic excitation. By this is meant that the point-group symmetry of the nuclear framework at potential turning-points of *stable* equilibrium has changed, as in e.g. linear-bent or planar-pyramidal transitions. Such behavior is now known to be rather common in polyatomic molecules and is usually regarded as interesting: the prototypes are in the first excitations of acetylene and formaldehyde. It also provides the cases in which the angular dependence of potential surfaces has been mapped out in greatest detail, with quite reliable estimates of equilibrium out-of-line or out-of-plane angles and barrier-heights between equivalent potential minima or conformers. Such structural information derived wholly from vibrational analysis has been included in these tables when regarded as of sufficient interest or reliability. The choice is necessarily arbitrary. Some idea of what structural information was available up to 1966 in other cases may be obtained from the tables in Herzberg's "Electronic Spectra of Polyatomic Molecules", General References I-1, which lists molecules with up to 12 atoms according to their known electronic states rather than geometries; a review of the azines (azabenzenes) by Innes et al. [40], and numerous articles devoted to the rotational analysis of complex electronic spectra by the technique of simulated computed band-contours mainly in "Journal of Molecular Spectroscopy" and "Molecular Physics". There is also a growing literature on the interpretation of the vibrational structure of photoelectron spectra which gives some information on the geometries of positive ions, but the results are only rarely definitive.

There has been great progress both in the experimental techniques of electronic molecular spectroscopy and in the theoretical techniques of analyzing spectra.

Experimentally, there have been several advances. The availability of narrow-line tunable lasers has transformed absorption spectroscopy. Instead of monitoring the dependence of absorption-coefficient on spectral frequency in terms of Beer-Lambert attenuation of incident white light intensity in the conventional way, the absorption is detected through the fluorescence it induces in the molecules excited. The sample is illuminated by a laser whose wavelength can be tuned continuosly through the molecular absorption-spectrum and the total undispersed fluorescence is detected. The *laser-induced fluorescence spectrum* ("LIFS") is thus simply related to the absorption spectrum except that the fluorescence intensities do not accurately reflect the absorption intensities since the fluorescence efficiencies of excited state levels depend on the extent to which these levels are mixed with high rovibronic levels of the ground state. The main advantages over conventional spectroscopy lie in sensitivity and resolution. Absorption of photons by molecules being a two-body process, the enormous radiation-intensity at the focus of a laser-beam will produce photon-absorption in a large proportion of the molecules present, even if few in number; and photoelectric detection of undispersed fluorescence can be taken down to single-photon counting levels. LIF spectroscopy can therefore be as much as ten orders of magnitude more sensitive than ordinary absorption spectroscopy in terms of either molecular number-density or sample-volume. The linewidths of tunable lasers can be made very narrow, sharper than the widths of absorption lines broadened by Doppler effect and pressure-broadening. Doppler-limited spectra are therefore replacing those which earlier were instrument-limited.

A further refinement uses countercurrent two-photon absorption which selects only molecules with negligible components of thermal velocity in the beam-direction, as in Lamb-dip spectroscopy. Such *sub-Doppler two-photon* LIF *spectra* have been obtained e.g. for benzene in its near ultraviolet spectrum down to within a factor of only four times the natural *homogeneous linewidth*, in this case 2 MHz, with an effective resolution of 1:2×10^8 [41]. Spectral bands seen previously only as contours of partially resolved rotational structure now have the simplicity of those of fully-resolved spectra of diatomic molecules. Rotational constants are improved by nearly two orders of magnitude, enough to pick up quite precisely an inertial defect in what, at equilibrium, is indisputably a planar symmetric rotor. But the problem of determining "the molecular geometry of the molecule" is taken little further, being merely projected into the same field of uncertainties and ambiguities revolving around r_v- versus r_e-structures familiar in microwave spectroscopy.

The above techniques are restricted to molecules which fluoresce. A more general technique is *multiphoton ionization* (MPI) in which a molecule absorbs several photons sufficient in energy to produce a molecular ion. This technique is very sensitive since single ions can be detected. The process may involve a single laser and several photons, or two (or more) lasers with various combinations of photons, e.g. 1+3, 2+2. One of the lasers can be adjusted so as to involve an intermediate excited state in which case the sensitivity is considerably enhanced and the process is known as *resonance enhanced multiphoton ionization* (REMPI). The detection of the resultant ion with a mass-spectrometer further refines the specificity of the method, and allows individual mass peaks, and isotopic species, to be monitored. The introduction of ZEKE (zero electron kinetic energy [42]) considerably increases the resolution which is beginning to approach the limit imposed by the widths of the laser.

The application of these techniques is considerably enhanced by the introduction of sample-seeded supersonic jets. Gas-phase spectra are obtained at effective temperatures close to the absolute zero and the problem of "Boltzmann congestion" is effectively overcome. Besides making the analysis of previously hopelessly congested spectra tractable it has revealed a new family of weakly-bound van der Waals dimers or clusters. Some of the analyses are limited to general conclusions, as e.g. the distinction between end-on and sideways-on orientation of diatomic iodine in a benzene-iodine complex. Such data are not included in the present compilation. Other analyses, however, yield accurate internuclear distances as in the benzene-rare gas complexes.

Theoretically, the advances in computational methods have made possible more direct approaches to structural analysis. In the past, analysis has been by inversion. Spectral data were reduced to parameters such as rotational and vibration constants. These in turn were inverted into other parameters that describe parts of molecular (Born-Oppenheimer) potential surfaces, such as the coordinates of their minima - the r_e-structures defining molecular geometry - and their curvatures around these minima - the "molecular force-field". These inversions involve many approximations, usually based on perturbation-theory, and tend to be ill-determined, leading to serious ambiguities. As examples, the separation of rotation and vibration was successful only in quasi-rigid molecules, as discussed at various places in the introduction above; and the treatment of molecular geometry in electronically degenerate or nearly-degenerate states confined to cases in which the vibronic interactions such as Jahn-Teller or Renner-Teller couplings were either very small or very large. Today the tendency is towards a more direct approach. A model potential surface is assumed, more and more frequently itself obtained from *ab initio* calculations (e.g. in H_3, q.v.), and rotational-vibrational energies calculated from it by direct diagonalization of large Hamiltonian matrices are compared with experimental values. The model is refined by iteration. Semi-rigid molecules can be successfully treated (see e.g. CH_2 and CH_2O), and Jahn-Teller effects have at last been convincingly analyzed in complex polydimensional systems such as $C_6F_6^+$ [43].

As previously, precise meaning of the data quoted in the present tables is likely to vary from case to case as are the uncertainties where indicated. The reader wishing to use them beyond the level of ordinary general purposes is therefore advised to return to the original sources and to evaluate these for himself.

1.6 Arrangement of the tables

1.6.1 General remarks on the content

All information on one molecule is listed together. Gross formula, name and symmetry are given as headline. The molecule is identified not only by its names (1.6.4) but also by a schematic structural formula or a figure. After the symbol for the experimental method used follow tables with data. In remarks and footnotes further information and/or comments are added before the references to the original papers used as source of the information. When results of two experimental methods are given, these are presented separately, designated each by the appropriate symbol (1.9.2). Molecules listed in Vols. II/7, II/15, II/21 and/or II/23 for which no geometric data are presented here, are cited in this volume at their proper place; their names are listed together with their substance numbers in the previous volumes.

1.6.2 Presentation of the data and comments

a) *Atoms* of the same kind are distinguished by numbers given in parentheses, such as C(1), C(2), or sometimes designated by primes, e.g., C(1), C(1'). Simple designations are used for hydrogen atoms; for example, H(s) and H(a) denote methyl hydrogen atoms located on a molecular symmetry plane and located out of plane, respectively. Non-equivalent hydrogen atoms in a methyl or methylene group are designated as H', H",... and defined in a footnote or identified in the figure.

b) An *internuclear distance* is represented by a solid line, like C(1)–H(s) for a pair of atoms directly bonded, and by a dotted line as N(1)$\cdots$N(2) for a nonbonded atom pair. A bond angle is represented as e.g. C(1)–C(2)–O(1); other angles such as a dihedral angle are defined in each case.

Nonbonded internuclear distances are tabulated in almost all the electron diffraction papers considered, but only those nonbonded distances that have been determined precisely and that are regarded as especially important are listed in the tables together with bond distances and angles.

Distances are given in Å (1 Å = 0.1 nm = 100 pm), and *angles* are given in degrees.

The *uncertainty* in a structural parameter (1.4), given in parentheses, applies to the last significant figure(s) of the parameter; e.g. 3.478(21) Å = (3.478±0.021) Å, 13.4(21) Å = (13.4±2.1) Å, 119.3(2) deg = (119.3±0.2) deg, and 119.3(20) deg = (119.3±2.0) deg.

c) *Point-group symmetry* of each molecule is also given in the tables. For most fairly rigid molecules this is the symmetry of the nuclear framework at stable equilibrium, i.e., of the minima of the interatomic potential surfaces. It is what is usually thought of as "the structure of the molecule" and acts as the origin of displacement coordinates used to describe the (small) internal motions of the atoms in the vibrating molecule. There are $3N-6$ such coordinates for a non-linear molecule and $3N-5$ for a linear molecule. The whole potential surface, in general, has several *minima* corresponding to what chemists distinguish as different isomers or conformers, depending on the topological relationship of local potential wells to each other. The symmetry of the whole potential surface is therefore always higher than that of its individual minima. If the potential barriers between minima are sufficiently low the internal molecular motions may become delocalized over several potential minima, either by passing classically over the barrier as in e.g. hindered internal rotations, or tunnelling through them quantum-mechanically. To describe such motions it often becomes convenient therefore to use the symmetry-properties of an enlarged portion of the potential surface encompassing the several minima accessible to the motion. The point-group symmetry of a potential *maximum* e.g. between *equivalent* minima, is often a useful symmetry for this purpose, then referred to as the "effective symmetry of the molecule". A classical example is ammonia which, in its electronic ground state "belongs to C_{3v}" because the equilibrium configuration of its hydrogen atoms about the nitrogen atom is a trigonal pyramid. But there are clearly two equivalent pyramidal potential minima disposed symmetrically about a planar trigonal potential maximum, a potential turning-point at which the molecule has, instantaneously or at unstable equilibrium, the symmetry D_{3h}. If the molecules were rigid and the hydrogen atoms distinguishably labeled, the two pyramidal isomers would be optically resolvable *d*- and *l*-isomers. As it is, the barrier is low, and the hydrogen atoms tunnel through it in a time which is short enough to make it necessary for some purposes to consider them as delocalized explicitly. For these purposes, mainly spectroscopic, the "effective symmetry of the molecule" is D_{3h}. For other purposes, such as a discussion of bulk dielectric properties reducible to an effective molecular electronic dipole moment, the effective symmetry remains C_{3V}. The point of these remarks is to stress again what is well known, that the selection of appropriate symmetry depends on the barrier-height which a molecule has to overcome in going from one equilibrium form to another (the criterion of *feasibility*), and the purpose to which the symmetry properties are to be applied. Conversely, in non-rigid molecules it may be quite difficult to establish the point-group symmetry at stable equilibrium – particularly when the potential barriers between minima do not even rise above the zero-point levels. A few cases of low barriers are annotated in footnotes.

Equilibrium symmetry is sometimes deduced by electron diffraction. Spectroscopy is often a better source of experimental information on symmetry. In other cases, where symmetry is simply *assumed*, explicit remarks are made to this effect; borderline cases are often encountered, however. For example electron diffraction data are often found to be "consistent" or "compatible" with a model of certain symmetry.

d) *Temperature* (with electron diffraction data): Since the geometric parameters determined by electron diffraction are thermal average values, they depend on the effective vibrational temperature of the sample molecules. Therefore, the temperatures of the experiment are listed explicitly whenever they are given in the original paper. The difference between the nozzle temperature and the effective temperature of the sample molecules depends on the experimental conditions such as the nozzle shape but it is usually not essential.

However, this temperature dependence is significant only when the experiment is done at very high temperature and/or when a property is examined which is very sensitive to the temperature, such as the relative abundance of isomers in a conformational equilibrium.

Mean amplitudes (from electron diffraction data): All current papers of electron diffraction report the observed or calculated mean amplitudes; no data are given in the tables.

e) *Wavenumbers* of the vibrational modes related to intermolecular bonds in weakly-bound complexes provide valuable information on their structure and bonding properties. Therefore, the wavenumbers and the force constants for the bond-stretching mode, denoted as v_s and k_s, respectively, are listed whenever accurate experimental values have been reported. In some cases the force constants for other intermolecular vibrational modes such as bending have also been determined.

1.6.3 Order of molecules

Inorganic molecules are arranged alphabetically according to their gross stoichiometric formulae also alphabetically ordered. All molecules containing one or more carbon atoms are listed as organic molecules and are arranged according to the Hill system.

1.6.4 Nomenclature

The names of molecules are mostly taken from the original papers, but an attempt is made to follow the usage in Chemical Abstracts and/or the rules of IUPAC (International Union of Pure and Applied Chemistry). Therefore many molecules have two or three names listed.

A complex or an addition compound consisting of *n*, *m*, ... atoms or molecules is indicated by the notation (*n*/*m*/...), e.g. argon-hydrogen bromide (1/1).

1.6.5 Figures and structural formulae

Nearly all molecules are represented by a schematic structural formula with chemical symbols or a figure with circles representing atoms, where the symbols of hydrogen atoms are sometimes omitted. In both structural formulae and figures, the special designations (numbers etc.) of individual atoms are assigned whenever necessary for discrimination. They are in general based on those given in the original papers, and are consistent with those used in the respective tables; in most cases they are also consistent with the international recommendations.

The lines shown connecting atoms in structural formulae need not necessarily represent correct bond orders or bond types, for outside the realm of organic chemistry a bond type in a molecule, particularly in the case of a free radical or a molecule in an electronically excited state, may be ambiguous, contentious, indefinable or even meaningless although perfectly characterizable experimentally in terms of an attractive potential function. Double and triple bonds etc. indicated on structural formulae and in data tables are not necessarily correct representations of their bond nature.

1.7 References for 1.1···1.6

1. Landolt-Börnstein, New Series, Vol. II/6 and II/14 (diatomic molecules); Tables de Constantes 17 (diatomic molecules).
2. Landolt-Börnstein, New Series, Vols. III/5, III/6, III/7, III/8, III/10 and III/14 (crystal structures); O. Kennard, D.G. Watson, Eds.: "Interatomic Distances 1960-65; Organic and Organometallic Crystal Structures". Vol. A 1, **1972**, and subsequent volumes.
3. Landolt-Börnstein, New Series. Vol. II/7 (polyatomic molecules): K.-H. and A.M. Hellwege, Eds., "Structure Data of Free Polyatomic Molecules", **1976**.
4. Landolt-Börnstein, New Series, Vol. II/15 (polyatomic molecules); K.-H. and A.M. Hellwege, Eds., "Structure Data of Free Polyatomic Molecules", **1987**.
5. Landolt-Börnstein, New Series, Vol. II/21 (polyatomic molecules); K. Kuchitsu, Ed., "Structure Data of Free Polyatomic Molecules", **1992**.
6. Landolt-Börnstein, New Series, Vol. II/23 (polyatomic molecules); K. Kuchitsu, Ed., "Structure Data of Free Polyatomic Molecules", **1995**.
7. L.E. Sutton, Ed.: "Tables of Interatomic Distances and Configuration in Molecules and Ions", special publication No. 11, London: The Chemical Society, **1958**, "Supplement 1956-1959", special publication No. 18, London: The Chemical Society, **1965**.
8. G. Guelachvili, M. Birk, Ch.J. Bordé, J.W. Brault, L.R. Brown, B. Carli, A.R.H. Cole, K.M. Evenson, A. Fayt, D. Hausamann, J.W.C. Johns, J. Kauppinen, Q. Kou, A.G. Maki, K. Narahari Rao, R.A. Toth, W. Urban, A. Valentin, J. Vergès, G. Wagner, M.H. Wappelhorst, J.S. Wells, B.P. Winnewisser, M. Winnewisser: Pure Appl. Chem. **68** (1996) 193; J. Mol. Spectrosc. **177** (1996) 164; Spectrochim. Acta **52A** (1996) 717.
9. See, for example, I.M. Mills: "Vibration-Rotation Structure in Asymmetric and Symmetric Top Molecules", in Molecular Spectroscopy: Modern Research, K.N. Rao, C.W. Mathews, Eds., New York: Academic Press **1972**. K. Kuchitsu, in: Reference G-2, Chap. 2.
 See also Reference I-3.
10. T. Oka: J. Phys. Soc. Jpn. **15** (1960) 2274; D.R. Herschbach, V.W. Laurie: J. Chem. Phys. **37** (1962) l668; M. Toyama, T. Oka, Y. Morino: J. Mol. Spectrosc. **13** (1964) 193.
11. J.K.G. Watson, in: "Vibrational Spectra and Structure", Chap. 1, Vol. 6, J.R. Durig, Ed., Amsterdam: Elsevier, **1977**.
12. See, for example, P.R. Bunker: J. Mol. Spectrosc. **46** (1973) 119; J.K.G. Watson: J. Mol. Spectrosc. **45** (1973) 99; P.R. Bunker: J. Mol. Spectrosc. **80** (1980) 411.
13. S. Yamamoto, M. Nakata, K. Kuchitsu: J. Mol. Spectrosc. **112** (1985) 173.

14	W.J. Hehre, L. Radom, P. von R. Schleyer, J.A. Pople: "Ab Initio Molecular Orbital Theory", New York: Wiley Interscience, 1986. J.E. Boggs, in: Reference E-8, Part B, Chap. 10.
15	L. Schäfer, J.D. Ewbank, K. Siam, N.-S. Chiu, H.L. Sellers, in: Reference E-8, Part A, Chap. 9.
16	H.J. Geise, W. Pyckhout, in: Reference E-8, Part A, Chap. 10.
17	N.L. Allinger, in: Reference G-2, Chap. 14.
18	P. Diehl, in: "NMR of Liquid Crystals" J.W. Emsley, Ed., Boston: Reidel, **1985**, Chap. 7; P. Diehl, in: Reference G-2. Chap. 12.
19	D.W.H. Rankin, in: Reference E-8, Chap. 14.
20	I. Hargittai, G. Schultz, J. Tremmel, N.D. Kagramanov, A.K. Maltsev, O.M. Nefedov: J. Am. Chem. Soc. **105** (1983) 2895; I. Hargittai, in: Reference E-8, Chap. 6.
21	C.C. Costain: J. Chem. Phys. **29** (1958) 864.
22	J. Kraitchman: Am. J. Phys. **21** (1953) 17.
23	A. Chutjian: J. Mol. Spectrosc. **14** (1964) 361; L. Nygaard: J. Mol. Spectrosc. **62** (1976) 292.
24	J.K.G. Watson: J. Mol. Spectrosc. **48** (1973) 479.
25	V.W. Laurie, D.R. Herschbach: J. Chem. Phys. **37** (1962) 1687.
26	K. Kuchitsu, T. Fukuyama, Y. Morino: J. Mol. Struct. **4** (1969) 41.
27	K. Kuchitsu, K. Oyanagi: Faraday Discuss. Chem. Soc. **62** (1977) 20.
28	E. Hirota: J. Mol. Struct. **146** (1986) 237.
29	M. Nakata, M. Sugie, H. Takeo, C. Matsumura, T. Fukuyama, K. Kuchitsu: J. Mol. Spectrosc. **86** (1981) 241.
30	M. Nakata, K. Kuchitsu, I.M. Mills: J. Phys. Chem. **88** (1984) 344.
31	M. Nakata, K. Kuchitsu: J. Mol. Struct. **320** (1994) 179.
32	M.D. Harmony, W.H. Taylor: J. Mol. Spectrosc. **118** (1986) 163.
33	M.D. Harmony, R.J. Berry, W.H. Taylor: J. Mol. Spectrosc. **127** (1988) 324.
34	R.J. Berry, M.D. Harmony: J. Mol. Spectrosc. **128** (1988) 176.
35	L.S. Bartell: J. Chem. Phys. **23** (1955) 1219.
36	K. Kuchitsu, L.S. Bartell: J. Chem. Phys. **35** (1961)1945; K. Kuchitsu: Bull. Chem. Soc. Jpn. **40** (1967) 498, 505.
37	Y. Morino, S.J. Cyvin, K. Kuchitsu, T. Iijima: J. Chem. Phys. **36** (1962)1109; see also Reference E-2.
38	Y. Morino, K. Kuchitsu, T. Oka: J. Chem. Phys. **36** (1962) 1108; K. Kuchitsu, T. Fukuyama, Y. Morino: J. Mol. Struct. **1** (1967-68) 463.
39	R.F. Curl Jr.: J. Comput. Phys. **6** (1970) 367.
40	K.K. Innes, J.P. Byrne, I.G. Ross: J. Mol. Spectrosc. **22** (1967) 125.
41	E. Riedle, H.J. Neusser: J. Chem. Phys. **80** (1984) 4686.
42	K. Müller-Dethlefs, E.W. Schlag: Ann. Rev. Phys. Chem. **42** (1991) 109.
43	T.A. Miller: J. Chem. Soc., Faraday Trans. II **82** (1986) 1123.

1.8 References to general literature

1.8.1 General

G-1	Critical Evaluation of Chemical and Physical Structural Information, D.R. Lide Jr., M.A. Paul, Eds., Washington, DC: National Academy of Sciences, **1974**.
G-2	Accurate Molecular Structures, Their Determination and Importance, A. Domenicano, I. Hargittai, Eds., Oxford: Oxford University Press, **1992**.
G-3	J. Demaison, G. Wlodarczak: Struct. Chem. **5** (1994) 57.

1.8.2 Microwave spectroscopy

Books

M-1	W. Gordy, W.V. Smith, R.F. Trambarulo: "Microwave Spectroscopy" New York: Dover, **1966**.
M-2	M.W.P. Strandberg: "Microwave Spectroscopy", London: Methuen, **1954**.
M-3	C.H. Townes, A.L. Schawlow: "Microwave Spectroscopy". New York: Dover, **1975**.
M-4	T.M. Sugden, C.N. Kenney: "Microwave Spectroscopy of Gases", London: Van Nostrand, **1965**.
M-5	J.E. Wollrab: "Rotational Spectra and Molecular Structure", New York: Academic Press, **1967**.
M-6	H.W. Kroto: "Molecular Rotation Spectra", New York: Wiley, **1984**.
M-7	W. Gordy, R.L. Cook: "Microwave Molecular Spectra", New York: Wiley, **1984**.
M-8	D.A. Ramsay, Ed.: "MTP International Review of Science, Physical Chemistry", Vol. 3 Spectroscopy, London: Butterworth, **1972**.
M-9	A. Carrington: "Microwave Spectroscopy of Free Radicals", New York: Academic Press, **1974**.
M-10	G.W. Chantry, Ed.: "Modern Aspects of Microwave Spectroscopy", London: Academic Press, **1979**.
M-11	K.N. Rao, C.W. Mathews, Eds.: "Molecular Spectroscopy: Modern Research", New York: Academic Press, Vol. 1 **1972**. Vol. 2 **1976**; Vol. 3 **1985**.
M-12	E. Hirota: "High-Resolution Spectroscopy of Transient Molecules", Heidelberg: Springer, **1985**.

Review papers

M-13	W.H. Flygare: Ann. Rev. Phys. Chem. **18** (1967) 325.
M-14	H. Dreizler: Fortschr. Chem. Forsch. **10** (1968) 59.
M-15	Y. Morino, E. Hirota: Ann. Rev. Phys. Chem. **20** (1969) 139.
M-16	H.D. Rudolph: Ann. Rev. Phys. Chem. **21** (1970) 733.
M-17	V.W. Laurie: Acc. Chem. Res. **3** (1970) 331.
M-18	R.J. Saykally, R.C. Woods: Ann. Rev. Phys. Chem. **32** (1981) 403.
M-19	B.P. van Eijck, Reference G-2, Chap. 3.
M-20	E. Hirota: Chem. Rev. **92** (1992) 141.
M-21	J. Demaison: "Accurate Structures of Non-Rigid Molecules by Microwave Spectroscopy" in: Structures and Conformations of non-rigid Molecules, Netherlands: Kluwer Academic Publishers, **1993**, p. 239.

Tables

M-22	E. Hirota: Ann. Rep. Sect. C. Phys. Chem. **1994**, p. 3.
M-23	Landolt-Börnstein, New Series, Vols. II/4, II/6, II/14a,b, II/19a-d, Berlin: Springer, **1967**, **1974**, **1982**, **1983**, **1992-1994** respectively, and references cited therein.
M-24	"Microwave Spectral Tables", Natl. Bur. Std. U.S. Monograph 70, Vols I-V, **1964-1969**.
M-25	A. Guarnieri, P. Favero: "Microwave Gas Spectroscopy Bibliography", Inst. Chimico G. Ciamician, Univ. di Bologna , **1968**.
M-26	M.D. Harmony, V.W. Laurie, R.L. Kuczkowski, R.H. Schwendeman, D.A. Ramsay, F.J. Lovas, W.J. Lafferty, A.G. Maki: J. Phys. Chem. Ref. Data **8** (1979) 619.

1.8.3 Infrared, Raman and electronic spectroscopy

Books

I-1   G. Herzberg: "Molecular Spectra and Molecular Structure. I. Spectra of Diatomic Molecules" Malabar, Fla.: Krieger **1989**. "II. Infrared and Raman Spectra of Polyatomic Molecules", ibid. **1991**, and "III. Electronic Spectra of Polyatomic Molecules", ibid. **1991**.

I-2   E.B. Wilson Jr., J.C. Decius, P.C. Cross: "Molecular Vibrations" New York: McGraw-Hill, **1955**.

I-3   H.H. Nielsen: Rev. Modern Phys. **23** (1951) 90; "Handbuch der Physik", S. Flügge, Ed., Vol. 37/1, Berlin: Springer, **1959**.

I-4   H.C. Allen Jr., P.C. Cross: "Molecular Vib-Rotors", New York: Wiley, **1963**.

I-5   H.G.M. Edwards, D.A. Long, in: Molecular Spectroscopy". Chap. 1, Vol. 1, London: The Chemical Society, **1973**.

I-6   H.G.M. Edwards, in: "Molecular Spectroscopy". Chap. 5, Vol. 3, London: The Chemical Society, **1975**.

I-7   J.M. Hollas: "High Resolution Spectroscopy", London: Butterworth, **1982**.

I-8   T.A. Miller, V.B. Bondybey, Eds.: "Molecular Ions, Spectroscopy, Structure and Chemistry Amsterdam: North Holland, **1983**.

I-9   K.N. Rao, C.W. Mathews, Eds.: "Molecular Spectroscopy: Modern Research", New York: Academic, **1972**; K. N. Rao, Ed.: "Molecular Spectroscopy: Modern Research", Vol. 2, New York: Academic, **1976**; Vol. 3, Orlando: Academic, **1985**.

I-10   E. Hirota, in: "Chemical and Biochemical Applications of Lasers", Chap. 2, Vol. V, C.B. Moore, Ed., New York: Academic, **1980**.

I-11   K.M. Evenson, R.J. Saykally, D.A. Jennings, R.F. Curl Jr., J.M. Brown, in: "Chemical and Biochemical Applications of Lasers", Chap. 3, Vol. V, C.B. Moore, Ed., New York: Academic, **1980**.

I-12   E. Hirota, in: "Vibrational Spectra and Structure", Chap. 1, Vol. 14, J. Durig, Ed., Amsterdam: Elsevier, **1985**.

I-13   D.A. Ramsay, in: "Vibrational Spectra and Structure", Chap. 2, Vol. 14, J.R. Durig, Ed., Amsterdam: Elsevier, **1985**.

I-14   J.M. Hollas: "Modern Spectroscopy", New York: Wiley, **1987**.

I-15   J.P. Maier, Ed.: "Ion and Cluster Ion Spectroscopy and Structure", Amsterdam: Elsevier, **1989**.

Review papers

I-16   C.S. Gudeman, R.J. Saykally: Ann. Rev. Phys. Chem. **35** (1984) 387.

I-17   E. Hirota, K. Kawaguchi: Ann. Rev. Phys. Chem. **36** (1985) 53.

I-18   E. Hirota, S. Saito: Revs. Chem. Intermed. **7** (1987) 353.

I-19   A. Carrington, B.A. Thrush, Eds.: "The Spectroscopy of Molecular Ions", London: The Royal Society, **1988**.

I-20   E. Hirota: Int. Revs. Phys. Chem. **8** (1989) 171.

I-21   P.F. Bernath: Ann. Rev. Phys. Chem. **41** (1990) 91.

I-22   G. Graner, Reference G-2, Chap. 4.

Tables

I-23   M.E. Jacox: J. Phys. Chem. Ref. Data **13** (1984) 945; **17** (1988) 269; **19** (1990) 1387.

1.8.4 Electron diffraction

Books

E-1   S.J. Cyvin: "Molecular Vibrations and Mean Square Amplitudes", Amsterdam: Elsevier, **1968**.
E-2   M. Davis: "Electron Diffraction in Gases", New York: Marcel Dekker, **1971**.
E-3   I. Hargittai, W.J. Orville-Thomas, Eds.: "Diffraction Studies on Non-Crystalline Substances", Budapest: Akademiai Kiado, and Amsterdam: Elsevier, **1981**.
E-4   L.V. Vilkov, V.S. Mastryukov, N.I. Sadova: "Determination of the Geometrical Structure of Free Molecules", Moscow: MIR Publishers, **1983**.
E-5   I. Buck, E. Maier, R. Mutter, U. Seiter, C. Spreter, B. Starck, I. Hargittai, O. Kennard, D.G. Watson, A. Lohr, T. Pirzadeh, H.G. Schirdewahn, Z. Majer: "Bibliography of Gas-Phase Electron Diffraction 1930-1979." Physik Daten/Physics Data Nr. 21-1, Karlsruhe: Fachinformationszentrum Energie, Physik, Mathematik GmbH, **1981**.
E-6   E. Maier, R. Mutter, U. Seiter, C. Spreter, B. Starck, I. Hargittai, D.G. Watson, A. Lohr: "Bibliography of Gas-Phase Electron Diffraction, Supplement 1980-1982". Physik Daten/Physics Data Nr. 21-2. Karlsruhe: Fachinformationszentrum Energie, Physik, Mathematik GmbH, **1985**.
E-7   P. Goodman, Ed.: "Fifty Years of Electron Diffraction" Dordrecht: Reidel, **1981**.
E-8   I. Hargittai, M. Hargittai, Eds.: "Stereochemical Application of Gas-Phase Electron Diffraction" Part A: The Electron Diffraction Technique. Part B: Structural Information for Selected Classes of Compounds. New York: VCH, **1988**.

Review papers

E-9   K. Kuchitsu, in: "MTP International Review of Science", G. Allen, Ed., Phys. Chem. Series 1, Vol. 2, Chap. 6, Oxford: Medical and Technical Publ. Co., **1972**.
E-10 K. Kuchitsu, in: "Molecular Structures and Vibrations". S.J. Cyvin, Ed., Chap. 12, Amsterdam: Elsevier, **1972**.
E-11 L.S. Bartell, K. Kuchitsu, H.M. Seip: "Guide for the Publication of Experimental Gas-Phase Electron Diffraction Data and Derived Structural Results in the Primary Literature", International Union of Crystallography, Acta Cryst. **A32** (1976) 1013.
E-12 K. Kuchitsu, in: Reference E-7, Part 3, Chap. 3.
E-13 I. Hargittai, in: "Topics in Current Chemistry", Vol. 96, Berlin: Springer, **1981**.
E-14 G. Gundersen, D.W.H. Rankin, in: "Spectroscopic Properties of Inorganic and Organometallic Compounds", Specialist Periodical Reports, G. Davidson, E.A.V. Ebsworth, Eds., London: The Chemical Society, **14** (1981) 389; **15** (1982) 374.
E-15 B. Beagley, in: "Problems in Molecular Structure", G.J. Bullen. M.G. Greenslade, Eds., Chap. 2.3, p.118, London: Pion, **1983**.
E-16 D.W.H. Rankin: Chem. Ber. **18** (1982) 426.
E-17 D.W.H. Rankin, H.E. Robertson, in: "Spectroscopic Properties of Inorganic and Organometallic Compounds", Specialist Periodical Reports. G. Davidson, E.A.V. Ebsworth, Eds., London: The Chemical Society **16** (1984) 350; **17** (1985) 381; **18** (1986) 449, **19** (1986) 452; **20** (1987) 475; D.G. Anderson, D.W.H. Rankin, ibid., **21** (1988) 488; D.W.H. Rankin, H.E. Robertson, ibid., **22** (1989) 462; **23** (1990) 471; **24** (1991) 470; **25** (1992) 433; **26** (1993) 463; **27** (1994) 438; **28** (1995) 428; **29** (1996) 418. D.W.H. Rankin, in: "Frontiers of Organosilicon Chemistry", A.R. Bassindale, P.P. Gasper, Eds., London: Royal Society of Chemistry, **1991**, 253.
E-18 I. Hargittai, in: "Static and Dynamic Implications of Precise Structural Information, Lecture Notes", A. Domenicano, I. Hargittai, P. Murray-Rust, Eds., Ettore Majorana Centre for Scientific Culture, International School of Crystallography, 11th Course, Erice, Italy, **1985**.

E-19   I. Hargittai, M. Hargittai: "Molecular Structures and Energetics"., J.F. Liebman, A. Greenberg, Eds., Deerfield Beach, Fla.: VCH Publ. **1986**.
E-20   K. Kuchitsu, in: Reference G-2, Chap. 2.
E-21   I. Hargittai, in: Reference E-8, Part A, Chap. 1.
E-22   M. Fink, D.A. Kohl, in: Reference E-8, Part A, Chap. 5.
E-23   J. Tremmel, I. Hargittai, in: Reference E-8, Part A, Chap. 6.
E-24   K. Kuchitsu, M. Nakata, S. Yamamoto, in: Reference E-8, Part A, Chap. 7.
E-25   V.P. Spiridonov, in: Reference E-8, Part A, Chap. 8.
E-26   L. Schäfer, J.D. Ewbank. K. Siam, N.-S. Chiu, H.L. Seller, in: Reference E-8, Part A, Chap. 9.
E-27   H.J. Geise, W. Pyckhout, in: Reference E-8, Part A, Chap. 10.
E-28   K. Hedberg, in: Reference E-8, Part A, Chap. 11.
F-29   A.H. Lowrey, in: Reference E-8, Part A, Chap. 12.
E-30   B. Beagley, in: Reference E-8, Part A, Chap. 13.
E-31   D.W.H. Rankin, in: Reference E-8, Part B, Chap. 1.
E-32   L.V. Vilkov, in: Reference E-8, Part B, Chap. 2.
E-33   V.S. Mastryukov, in: Reference E-8, Part B, Chap. 3.
E-34   H. Oberhammer, in: Reference E-8, Part B, Chap. 4.
E-35   L.K. Montgomery, in: Reference E-8, Part B, Chap. 5.
E-36   M. Traetteberg, in: Reference E-8, Part B, Chap. 6.
E-37   A. Domenicano, in: Reference E-8, Part B, Chap. 7.
E-38   A. Haaland, in: Reference E-8, Part B, Chap. 8.
E-39   M. Hargittai, in: Reference E-8, Part B, Chap. 9.
E-40   J.E. Boggs, in: Reference E-8, Part B, Chap. 10.
E-41   M. Hargittai, I. Hargittai: Int. J. Quantum Chem. **44** (1992) 1057.
E-42   A.A. Ischenko, J.D. Ewbank, L. Schäfer: J. Phys. Chem. **98** (1994) 4287.

1.9 Lists of symbols and abbreviations

1.9.1 List of symbols

A, B, C	Rotational constants (see 1.2.1)
$\tilde{X}, \tilde{A}, \tilde{B}, \tilde{C}\ldots$	Labels for electronic states, ground state conventionally labelled $\tilde{X}$
$\tilde{a}, \tilde{b}, \tilde{c}, \ldots$	Labels for excited electronic states of spin-multiplicity differing from that of the ground state $\tilde{X}$
r	Internuclear distance (X–Y = bond distance, X⋯Y = nonbonded distance)
r_e	Distance between equilibrium nuclear positions
r_{av}, r_z, r_α^0	Distance between average nuclear positions (ground vibrational state, $v_1 = v_2 \cdots = 0$)
r_α	Distance between average nuclear positions (thermal equilibrium)
r_0	Distance between effective nuclear positions derived from rotational constants of zero-point vibrational level ($v_1 = v_2 \cdots = 0$): that derived from rotational constants of a specified vibrational level v is denoted as r_v.
r_s	Distance between effective nuclear positions derived from isotopic differences in rotational constants
r_m	Distance between effective nuclear positions derived from the mass-dependence method of Watson
r_m^ρ	r_m obtained by a slightly modified method of Harmony et al.
r_g	Thermal average value of internuclear distance
r_a	Constant argument in the molecular term, Eq.(8), see Table 1.3.3
θ	Bond angle; for indexes, see r
	For example, θ_e, θ_z and θ_α represent angles defined by a set of three nuclear positions, equilibrium, average (ground vibrational state) and average (thermal equilibrium), respectively. Some electron diffraction papers report distances as r_g and angles as θ_α or θ_z.
k_s	Stretching force constant of an intermolecular bond of a weakly bound complex
v_s	Wavenumber of a stretching intermolecular vibration of a weakly bound complex

1.9.2 List of abbreviations

ED	Electron diffraction			
IR	Infrared spectroscopy		MPI	Multiphoton ionization
LIF	Laser induced fluorescence		CEI	Coulomb explosion imaging
LMR	Laser magnetic resonance		ac	anticlinal
MW	Microwave spectroscopy		ap	antiperiplanar
NMR	Nuclear magnetic resonance		ax	axial
PES	Photoelectron spectroscopy		br, b	bridge
Ra	Raman spectroscopy		cm	center of mass
REMPI	Resonance enhanced multiphoton ionization		eq	equatorial
			sc	synclinal
UV	Ultraviolet spectroscopy		sp	synperiplanar
			t	terminal

2 Inorganic molecules

AlB$_3$H$_{12}$	Aluminum triboron dodecahydride	II/7(**2**,1)
AlBr$_3$H$_3$N	Aluminum tribromide – ammonia (1/1)	II/15(**2**,1)
AlBr$_6$Sb	Aluminum tribromide – antimony tribromide (1/1)	II/7(**2**,3)
AlCl$_3$	Aluminum trichloride	II/7(**2**,4), II/15(**2**,2)
AlCl$_3$H$_3$N	Aluminum trichloride – ammonia (1/1)	II/7(**2**,6)
AlCl$_4$Cs	Cesium tetrachloroaluminate	II/15(**2**,3), II/21(**2**,1)
AlCl$_4$K	Potassium tetrachloroaluminate	II/7(**2**,7), II/15(**2**,4) II/21(**2**,2)
AlCl$_4$Na	Sodium tetrachloroaluminate	II/21(**2**,3)
AlCl$_4$Rb	Rubidium tetrachloroaluminate	II/15(**2**,5) II/21(**2**,4)
AlCsF$_4$	Cesium tetrafluoroaluminate	II/15(**2**,6), II/21(**2**,5)

ED   **AlF$_3$**   **Aluminum trifluoride**   **D$_{3h}$**
                                                                                                               AlF$_3$

	r_g	Å [a]
Al–F	1.630(3)	

A vibrational analysis of the ED intensity data provided firm evidence for the planarity of this molecule [1].
The nozzle temperature was 1300 K.

[a]) Estimated total error.

Hargittai, M., Kolonits, M., Tremmel, J., Fourquet, J.-L., Ferey, G.: Struct. Chem. **1** (1990) 75.
[1] Hargittai, M., Subbotina, N.Yu., Gershikov, A.G.: J. Mol. Struct. **245** (1991) 147.

                                                 II/7(**2**,8), II/15(**2**,7), II/21(**2**,6), II/23(**2**,1)

AlF$_4$K	Potassium tetrafluoroaluminate	II/15(**2**,8)
AlF$_4$Na	Sodium tetrafluoroaluminate	II/7(**2**,9)
AlF$_4$Rb	Rubidium tetrafluoroaluminate	II/15(**2**,9)
AlH$_2$	Aluminum dihydride	II/7(**2**,10), II/15(**2**,10)
AlI$_3$	Aluminum triiodide	II/7(**2**,11)
Al$_2$Br$_6$	Dialuminum hexabromide	II/7(**2**,2)
Al$_2$Cl$_6$	Dialuminum hexachloride	II/7(**2**,5)
Al$_2$O	Dialuminum monoxide	II/7(**2**,12), II/23(**2**,2)
ArBF$_3$	Argon – boron trifluoride (1/1)	II/15(**2**,11), II/23(**2**,3)
ArBrH	Argon – hydrogen bromide (1/1)	II/15(**2**,12), II/23(**2**,4)
ArClF	Argon – chlorine fluoride (1/1)	II/15(**2**,13)
ArClH	Argon – hydrogen chloride (1/1)	II/7(**2**,13), II/15(**2**,14), II/23(**2**,5)
ArFH	Argon – hydrogen fluoride (1/1)	II/15(**2**,15)
ArF$_3$P	Argon – phosphorus trifluoride (1/1) Argon – trifluorophosphine (1/1)	II/21(**2**,7), II/23(**2**,6)
ArHO	Argon – hydroxyl (1/1)	II/23(**2**,7)
ArH$_2$	Argon – hydrogen (1/1)	II/7(**2**,14)
ArH$_2$O	Argon – water (1/1)	II/21(**2**,8)

	ArH$_2$S	Argon – hydrogen sulfide (1/1)	II/21(**2**,9)
	ArH$_3^+$	Argon – trihydrogen(1+) (1/1)	II/21(**2**,10)

ArH$_3$N

MW

ArH$_3$N   **Argon – ammonia (1/1)**   **C$_s$**
(weakly bound complex)   (effective symmetry class)
Ar · NH$_3$

r_0	Ar–^{14}NH$_3$	Ar–^{15}NH$_3$	θ_0	deg
R_{cm} [Å] [a]	3.8358(50)	3.8334(50)	$\theta(1)$ [c]	84.5(55)
			$\theta(2)$ [c]	58.29(8)

	v_s [d] cm^{-1}	k_s [d] N m^{-1}
Ar–^{14}NH$_3$	34.6	0.840
Ar–^{15}NH$_3$	33.9	0.841

The Ar–NH$_3$ intermolecular potential is nearly isotropic.
The NH$_3$ subunit undergoes practically free internal rotation in each of its angular degrees of freedom.

[a]) Uncertainties are not listed in the original paper.
[b]) R_{cm} is the average of the distance between the center of mass of the NH$_3$ subunit and the Ar atom as calculated in the free rotor limit.
[c]) $\theta(1)$ is the average value of θ as determined by the inversion of the dipole moment data. $\theta(2)$ is the average value of θ as determined by inversion of the quadrupole coupling constant data. The supplements of $\theta(1)$ and $\theta(2)$ are equally valid estimates of $\langle\theta\rangle$.
[d]) v_s and k_s denote the wavenumber and the force constant, respectively, for the stretching vibration of the weak bond.

Nelson, D.D., Fraser, G.T., Peterson, K.I., Zhao, K., Klemperer, W., Lovas, F.J., Suenram, R.D.: J. Chem. Phys. **85** (1986) 5512.   II/21(**2**,11)

ArN$_2$O	Argon – dinitrogen monoxide (1/1)	II/15(**2**,16)
	Argon – nitrous oxide(1/1)	
ArO$_2$S	Argon – sulfur dioxide(1/1)	II/25(**2**,18), II/23(**2**,8)
ArO$_3$	Argon – ozone (1/1)	II/15(**2**,17)
ArO$_3$S	Argon – sulfur trioxide (1/1)	II/15(**2**,19)

MW

Ar$_2$ClH   **Diargon – hydrogen chloride (1/1)**   **C$_{2v}$**
Hydrogen chloride – argon (1/2)   (effective symmetry class)
(weakly bound complex)   Ar$_2$ · HCl

Isotopic species:	Ar$_2$···H^{35}Cl	Ar$_2$···H^{37}Cl
r_0 (Ar–Ar) [Å] [a]	3.8611(30)	3.8612(30)
R (Ar$_{2,cm}$···HCl$_{cm}$) [Å] [a]	3.4734(30)	3.4749(30)
R_0 (Ar$_{2,cm}$···Cl) [Å] [a]	3.5094(30)	3.5090(30)
d (Ar···HCl$_{cm}$) [Å] [a]	3.9739(30)	3.9752(30)
d_0 (Ar···Cl) [Å] [a]	4.0054(30)	4.0050(30)
β [b]) [deg]	7.32 [c]	7.32 [c]

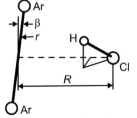

(*continued*)

IR	Isotopic species	$Ar_2\cdots H^{35}Cl$	$Ar_2\cdots H^{37}Cl$
	R_0 ($Ar_{2,cm}\cdots HCl_{cm}$) [Å]	3.4969(10)[d]	3.4975(10)[d]

The structures of the two monomers were supposed to be unchanged on complex formation.

[a]) Uncertainties were not estimated in the original paper.
[b]) Average Ar_2 torsional angle.
[c]) Assumed.
[d]) Uncertainties given in the original paper were multiplied by 10.

Elrod, M.J., Steyert, D.W., Saykally, R.J.: J. Chem. Phys. **94** (1991) 58.
Elrod, M.J., Steyert, D.W., Saykally, R.J.: J. Chem. Phys. **95** (1991) 3182.
Klots, T.D., Chuang, C., Ruoff, R.S., Emilsson, T., Gutowsky, H.S.: J. Chem. Phys. **86** (1987) 5315.
See also: Klots, T.D., Gutowsky, H.S.: J. Chem. Phys. **91** (1989) 63.   II/21(**2**,12), II/23(**2**,9)

Ar_2FH	Diargon – hydrogen fluoride (1/1)	II/23(**2**,10)
Ar_3ClH	Hydrogen chloride – triargon (1/1)	II/21(**2**,13), II/23(**2**,11)
	Hydrogen chloride – argon (1/3)	
Ar_3FH	Hydrogen fluoride – triargon (1/1)	II/21(**2**,14)
	Hydrogen fluoride – argon (1/3)	
Ar_4FH	Hydrogen fluoride – argon (1/4)	II/21(**2**,15)
$AsBr_3$	Arsenic tribromide	II/7(**2**,16), II/15(**2**,20), II/21(**2**,16)
$AsCl_2F_3$	Arsenic dichloride trifluoride	II/21(**2**,17)
	Dichlorotrifluoroarsenic(V)	
$AsCl_3$	Arsenic trichloride	II/7(**2**,17), II/15(**2**,21)
AsF_3	Arsenic trifluoride	II/7(**2**,18), II/15(**2**,22)
AsF_5	Arsenic pentafluoride	II/7(**2**,19)
AsH_2	Dihydrogen arsenic	II/7(**2**,20), II/15(**2**,23)
AsH_3	Arsine	II/7(**2**,21), II/15(**2**,24), II/21(**2**,18)
AsH_9Si_3	Trisilylarsane, Trisilylarsine	II/7(**2**,22), II/15(**2**,25)
AsI_3	Arsenic triiodide, Triiodoarsine	II/7(**2**,23)
As_4	Tetraarsenic	II/7(**2**,15)
Au_2F_{10}	Digold decafluoride	II/15(**2**,26)
Au_3F_{15}	Trigold pentadecafluoride	II/15(**2**,27)
$BBrS$	Bromosulfidoboron	II/23(**2**,12)
BBr_2H	Dibromoborane	II/7(**2**,26)
BBr_3	Boron tribromide, Tribromoborane	II/7(**2**,25)
$BClF_2$	Boron chloride difluoride, Chlorodifluoroborane	II/15(**2**,28)
$BClF_3H$	Hydrogen chloride – boron trifluoride (1/1)	II/21(**2**,19)
$BClO$	Boron chloride oxide, Chloro(oxo)boron	II/15(**2**,29)
$BClS$	Boron chloride sulfide, Chlorothioboron	II/15(**2**,30)
$BClSe$	Boron chloride selenide, Chloro(selenido)boron	II/23(**2**,13)
BCl_2H	Dichloroborane	II/7(**2**,30)
BCl_3	Boron trichloride, Trichloroborane	II/7(**2**,28), II/15(**2**,31)
$BCsO_2$	Cesium metaborate	II/15(**2**,33)
BFH^+	Fluoroborylium, Fluorohydroboron(1+)	II/21(**2**,20)
BFH_2O	cis–Fluorohydroxyborane	II/21(**2**,21)
BFH_2O_2	Fluorodihydroxyborane	II/15(**2**,34)

	BFO	Boron fluoride oxide, Oxoboryl fluoride	II/21(**2**,22)
	BFS	Boron fluoride sulfide, Fluorosulfidoboron	II/23(**2**,14)
	BF_2H	Difluoroborane	II/7(**2**,33)
	BF_2HO	Difluorohydroxyborane	II/7(**2**,37)
	BF_2H_2N	Aminodifluoroborane	II/7(**2**,34)
	BF_2H_4P	Difluorophosphine borane	II/7(**2**,38)
	$BF_2H_6NSi_2$	N–Difluoroboryldisilazane	II/7(**2**,36)
	BF_2O	Boron difluoride oxide	II/7(**2**,41), II/15(**2**,35)

BF_3   **Boron trifluoride** D_{3h}
Trifluoroborane BF_3

IR

r_e	Å
B–F	1.3070(1)

Zeisberger, E., Ruoff, A.: J. Mol. Spectrosc. **136** (1989) 295.
IR: Yamamoto, S., Kuwabara, R., Takami, M., Kuchitsu, K.: J. Mol. Spectrosc. **115** (1986) 333.
                                                                   II/7(**2**,32), II/15(**2**,36), II/21(**2**,23)

	BF_3H_3P	Phosphine – trifluoroborane (1/1)	II/7(**2**,39), II/15(**2**,37)
		Trifluorophosphine borane	
	BF_3Kr	Boron trifluoride – krypton (1/1)	II/21(**2**,24), II/23(**2**,15)
		Trifluoroborane – krypton (1/1)	
	BF_3N_2	Nitrogen-trifluoroborane (1/1)	II/15(**2**,38)
	BF_3Ne	Boron trifluoride – neon (1/1)	II/21(**2**,25), II/23(**2**,16)
	BF_5Si	Trifluorosilyldifluoroborane	II/7(**2**,42), II/15(**2**,38a)
	BF_7Si_2	(Pentafluorodisilyl)difluoroborane	II/7(**2**,43)
	$BGaH_6$	Di–μ–hydrido[dihydroboron(III)] [dihydridogallium(III)]	
		Gallaborane	II/23(**2**,17)
	BHO	Boron hydride oxide, Oxoboryl hydride	II/21(**2**,26)
		Hydroboriooxide	
	BHS	Boron hybride sulfide	II/7(**2**,53), II/15(**2**,39)

BH_2   **Boron dihydride**

UV BH_2

State	$\tilde{X}\,^2A_1$	$\tilde{A}\,^2B_1$ (II)
Symmetry	C_{2v}	$D_{\infty h}$
Energy [eV]	0	0.64 [a]
r_0 [Å] B–H	1.181 [b]	1.17 [c]
θ_0 [deg] H–B–H	131 [b]	180 [c]

Rotational analysis of electronic absorption spectrum $\tilde{A} \leftarrow \tilde{X}$.

[a]) Uncertain, based on long extrapolation and dependent on footnote [c]).
[b]) Accuracy limited by uncertainties due to zero-point motions: appreciable inertial defect.
[c]) Extrapolated values from levels $v_2 = 7 \cdots 11$. Shallow double potential minimum cannot be ruled out.

Herzberg, G., Johns, J.W.C.: Proc. Roy. Soc. (London) Ser. **A 298** (1967) 142.
                                                                                   II/7(**2**,44), II/15(**2**,40)

	BH_2N	Iminoborane, Hydroboronimide	II/21(**2**,27)

	BH₃	**Borane**	**D$_{3h}$**
IR		Boron trihydride	BH₃

r_0	Å
B–H	1.19001(20) [a]

The r_0 structure is derived only from the experimental ground state B_0 constants of the two isotopomers ^{10}BH₃ and ^{11}BH₃.

[a] Uncertainty given in the original paper was multiplied by 20.

Kawaguchi, K.: J. Chem. Phys. **96** (1992) 3411. II/23(**2**,18)

BH₃O	Borinic acid, Hydroxyborane	II/15(**2**,41)
BH₃O₂	cis, trans–Dihydroxyborane	II/21(**2**, 28)
BH₄Li	Lithium tetrahydroborate	II/23(**2**,19)
BH₄N	Aminoborane	II/15(**2**,42), II/21(**2**,29)

	BH₄Na	**Sodium tetrahydroborate**	**C$_{3v}$**
MW			NaBH₄

r_0	Å		θ_0	deg
Na···B	2.308(6)		H(b)–B–H(t)	111 [a]
B–H(b)	1.28(10)			
Δ(B–H) [b]	0.04 [a]			

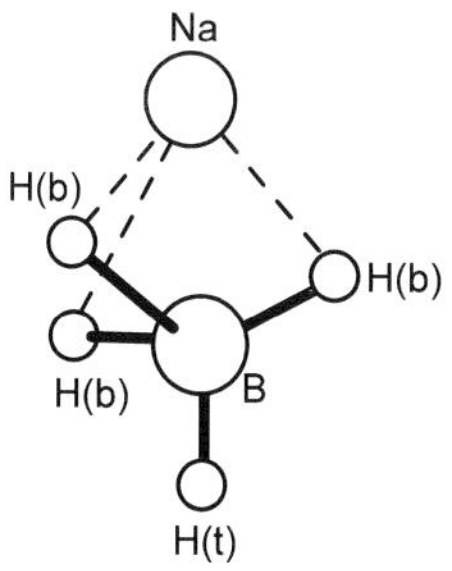

[a] Assumed.
[b] (B–H(b)) − (B–H(t)).

Kawashima, Y., Yamada, C., Hirota, E.: J. Chem. Phys. **94** (1991) 7707. II/23(**2**,20)

BH₅N₂	Diaminoborane	II/15(**2**,43)
BH₆N	Ammonia borane	II/15(**2**,44)
BH₆P	Phosphine borane	II/7(**2**,52)
BI₃	Boron triiodide, Triiodoborane	II/7(**2**,55)
BKO₂	Potassium metaborate	II/7(**2**,56), II/15(**2**,45)
BLiO₂	Lithium metaborate	II/7(**2**,57)
BNaO₂	Sodium metaborate	II/7(**2**,58)

	BO₂	**Boron dioxide**	**D$_{\infty h}$**
IR			O=B=O

r_0	Å
B=O	1.26485(5)

Structure calculated from the rotational constants of Kawaguchi et al., the average of those obtained from the ^{11}BO₂ and ^{10}BO₂ species.

Kawaguchi, K., Hirota, E., Yamada, C.: Mol. Phys. **44** (1981) 509.
Maki, A.G., Burkholder, J.B., Sinha, A., Howard, C.J.: J. Mol. Spectrosc. **130** (1988) 238.

(*continued*)

BO$_2$ (continued)

UV   Rotational and vibrational analysis

State	$\tilde{X}\ ^2\Pi_g$	$\tilde{A}\ ^2\Pi_g$	$\tilde{B}\ ^2\Sigma_u^+$
Symmetry	$D_{\infty h}$	$D_{\infty h}$	$D_{\infty h}$
Energy [eV]	0	2.268	3.039
r_0 [Å] B=O	1.2652	1.3025	1.2733
α_0 [deg] O=B=O	180	180	180

Johns, J.W.C.: Can. J. Phys. **39** (1961) 1738.   II/7(**2**,59), II/15(**2**,46), II/21(**2**,30)

BO$_2$Rb	Rubidium metaborate	II/7(**2**,61), II/15(**2**,47)
BO$_2$Tl	Thallium metaborate	II/15(**2**,48)
B$_2$BeH$_8$	Beryllium tetrahydroborate	II/7(**2**,24), II/15(**2**,32)
B$_2$BrH$_5$	1 – Bromodiborane(6)	II/7(**2**,27)
B$_2$ClH$_5$	1 – Chlorodiborane(6)	II/15(**2**,51), II/21(**2**,31)
B$_2$ClH$_7$	Diborane – hydrogen chloride (1/1)	II/23(**2**,21)
B$_2$Cl$_2$S$_3$	3,5–Dichloro–1,2,4–trithia–3,5–diborolane	II/7(**2**,31)
B$_2$Cl$_4$	Diboron tetrachloride	II/7(**2**,29)
B$_2$FH$_7$	Hydrogen fluoride – diborane (1/1)	II/21(**2**,32)
B$_2$F$_4$	Diboron tetrafluoride	II/15(**2**,49)
B$_2$F$_6$H$_4$P$_2$	Phosphorus trifluoride – diborane(4) (2/1)	II/7(**2**,40)
B$_2$GaH$_9$	Hydridogallium bis(tetrahydroborate)	II/15(**2**,50)
B$_2$H$_2$O$_3$	1,2,4,3,5–Trioxadiborolane	II/7(**2**,50)

IR   **B$_2$H$_6$**   **Diborane(6)** D_{2h}

r_e	Å
B–H(t)	1.184(3)
B–H(b)	1.314(3)
B⋯B	1.743 [a]

θ_e	deg
H(t)–B–H(t)	121.5 (5)
H(b)–B–H(b)	96.9 (5)

Structure calculated from ground state rotational constants of four isotopic species corrected to equilibrium constants using harmonic force filed calculations.

[a]) Dependent parameter.

Duncan, J.L., Harper, J.: Mol. Phys. **51** (1984) 371.   II/7(**2**,45), II/15(**2**,52)

B$_2$H$_7$N	Aminodiborane(6)	II/15(**2**,53)
B$_2$O$_3$	Diboron trioxide	II/7(**2**,60), II/15(**2**,54)
B$_2$S$_3$	Diboron trisulfide	II/7(**2**,62)
B$_3$F$_3$H$_3$N$_3$	2,4,6–Trifluoroborazine	II/7(**2**,35)
B$_3$GaH$_{10}$	2–Galla–*arachno*–tetraborane(10)	II/23(**2**,22)
B$_3$H$_3$O$_3$	Boroxin	II/7(**2**,51)
B$_3$H$_4$NO$_2$	3*H*–3–Azacylotriboroxane	II/21(**2**,33)
B$_3$H$_5$N$_2$O	3–Oxacyclotriborazane	II/21(**2**,34)
B$_3$H$_6$N$_3$	Borazine	II/7(**2**,48)

$B_3H_7N_4$	Aminoborazine	II/7(**2**,49)
$B_3H_{12}Ti$	Titanium tris(tetrahydroborate)	II/15(**2**,55), II/23(**2**,23)
B_4Cl_4	Tetrachloro–*tetrahedro*–tetraboron(4*B*–*B*)	II/21(**2**,35)
	Tetraboron tetrachloride	
B_4H_{10}	Tetraborane (10)	II/15(**2**,56)
$B_4H_{16}Zr$	Zirconium tetrakis(tetrahydroborate)	II/7(**2**,54)
B_5H_9	Pentaborane(9)	II/7(**2**,46), II/15(**2**,57), II/21(**2**,36)
B_5H_{11}	Pentaborane(11)	II/21(**2**,37)
$B_5H_{11}Si$	1–Silylpentaborane(9)	II/15(**2**,58)
$B_5H_{11}Si$	2–Silylpentaborane(9)	II/15(**2**,59)
B_6H_{10}	Hexaborane (10)	II/15(**2**,60)
B_6H_{12}	*arachno*–Hexaborane(12), Hexaborane(12)	II/21(**2**,38)
$B_{10}H_{14}$	Decaborane(14)	II/7(**2**,47), II/15(**2**,61)
$B_{11}H_{11}S$	1–Thia–*closo*–dodecaborane(11)	II/23(**2**,24)
BaHO	Monohydroxobarium, Barium monohydroxide	II/21(**2**,39)
BaI_2	Barium iodide	II/15(**2**,62)
BaO_4W	Barium tungstate	II/15(**2**,63)
BeF_2	Beryllium fluoride, Beryllium difluoride	II/23(**2**,25)
BeF_3K	Potassium trifluoroberyllate	II/7(**2**,63)
$Be_4N_6O_{19}$	Hexakis–μ–nitrato–(O,O')–μ_4–oxo–tetraberyllium	II/15(**2**,64)
$BiCl_3$	Bismuth trichloride	II/21(**2**,40)
$BrFO_2S$	Sulfuryl fluoride bromide	II/7(**2**,66)
	Sulfonyl fluoride bromide	
$BrFO_3$	Perbromyl fluoride	II/15(**2**,65)
BrF_2PS	Thiophosphoryl monobromide difluoride	II/15(**2**,66)
BrF_3	Bromine trifluoride	II/7(**2**,64), II/15(**2**,67)
BrF_3Si	Bromotrifluorosilane	II/7(**2**,68), II/21(**2**,41)
	Silicon trifluorobromide	
BrF_5	Bromine pentafluoride	II/7(**2**,65), II/15(**2**,68)
BrF_5S	Bromopentafluoro–λ^6–sulfane, Sulfur bromide pentafluoride,	
	Bromopentafluorosulfur(VI) II/7(**2**,67), II/15(**2**,68a), II/21(**2**,42)	
BrGeH	Germanium bromide hydride	II/23(**2**,26)
$BrGeH_3$	Bromogermane	II/7(**2**,70), II/15(**2**,69)
BrHKr	Krypton – hydrogen bromide (1/1)	II/15(**2**,70)
$BrHN_2$	Dinitrogen – hydrogen bromide (1/1)	II/23(**2**,27)
$BrHN_2O$	Hydrogen bromide – dinitrogen monoxide (1/1)	II/23(**2**,28)
BrHO	Hypobromous acid	II/23(**2**,29)
BrHSi	Bromosilylene	II/7(**2**,73), II/15(**2**,71)
	Silicon bromide hydride	
BrHXe	Xenon – hydrogen bromide (1/1)	II/15(**2**,71a)
BrH_3S	Hydrogen bromide – hydrogen sulfide (1/1)	II/21(**2**,43)
BrH_3Si	Bromosilane, Silyl bromide	II/7(**2**,72), II/21(**2**,44)
BrH_3Sn	Bromostannane, Stannyl bromide	II/7(**2**,75), II/23(**2**,30)
BrH_4N	Hydrogen bromide – ammonia (1/1)	II/21(**2**,45)
BrH_4P	Phosphine – hydrogen bromide (1/1)	II/15(**2**,72)
BrNO	Nitrosyl bromide	II/7(**2**,79)

Br$_2$Ca	Calcium dibromide	II/21(**2**,46)
Br$_2$Cd	Cadmium dibromide	II/15(**2**,73), II/21(**2**,47), II/23(**2**,31)
Br$_2$Co	Cobalt dibromide	II/21(**2**,48), II/23(**2**,32)
Br$_2$Cs$_2$	Dicesium dibromide	II/21(**2**,49)
Br$_2$Fe	Iron dibromide	II/15(**2**,74), II/23(**2**,33)
Br$_2$Ge	Germanium dibromide	II/15(**2**,75)
Br$_2$GeH$_2$	Dibromogermane	II/7(**2**,71)
Br$_2$Hg	Mercury dibromide	II/21(**2**,50)
Br$_2$K$_2$	Dipotassium dibromide	II/21(**2**,51)
Br$_2$Li$_2$	Dilithium dibromide	II/7(**2**,77)
Br$_2$Mn	Manganese dibromide Manganese(II) bromide	II/15(**2**,76), II/23(**2**,34)
Br$_2$MoO$_2$	Molybdenum dibromide dioxide	II/23(**2**,35)
Br$_2$Na$_2$	Disodium dibromide	II/7(**2**,80), II/21(**2**,52)
Br$_2$Ni	Nickel dibromide	II/15(**2**,77), II/23(**2**,36)
Br$_2$OS	Thionyl bromide	II/15(**2**,78)
Br$_2$Pb	Lead dibromide	II/15(**2**,79), II/21(**2**,53), II/23(**2**,37)
Br$_2$Rb$_2$	Dirubidium dibromide	II/21(**2**,54)
Br$_2$Si	Silicon dibromide	II/15(**2**,80), II/23(**2**,38)
Br$_2$Sn	Tin dibromide	II/15(**2**,81), II/21(**2**,55), II/23(**2**,39)
Br$_2$Sr	Strontium dibromide	II/23(**2**,40)
Br$_2$Zn	Zinc dibromide	II/21(**2**,56), II/23(**2**,41)
Br$_3$GaH$_3$N	Gallium tribromide – ammonia (1/1)	II/15(**2**,82)
Br$_3$Gd	Gadolinium tribromide	II/15(**2**,83)
Br$_3$HSi	Tribromosilane	II/7(**2**,74)
Br$_3$La	Lanthanum tribromide	II/7(**2**,76), II/15(**2**,84)
Br$_3$Lu	Lutetium tribromide	II/15(**2**,85)
Br$_3$Nd	Neodymium tribromide	II/7(**2**,82)
Br$_3$OP	Phosphoryl bromide	II/15(**2**,86)
Br$_3$P	Phosphorus tribromide Tribromophosphine	II/7(**2**,83), II/21(**2**,57)
Br$_3$PS	Thiophosphoryl bromide	II/15(**2**,87)
Br$_3$Sb	Tribromostibine, Antimony tribromide	II/7(**2**,85)
Br$_4$Co$_2$	Dicobalt tetrabromide Di–μ–bromo–bis[bromocobalt(II)]	II/21(**2**,58)
Br$_4$Fe$_2$	Diiron tetrabromide	II/15(**2**,88), II/23(**2**,42)
Br$_4$Ge	Tetrabromogermane, Germanium(IV) bromide	II/7(**2**,69)
Br$_4$H$_2$Si$_2$	1,1,2,2–Tetrabromodisilane	II/21(**2**,59)
Br$_4$Hf	Hafnium tetrabromide	II/15(**2**,89)
Br$_4$Mn$_2$	Dimanganese tetrabromide	II/15(**2**,90)
Br$_4$Mo	Molybdenum tetrabromide	II/7(**2**,78)
Br$_4$OW	Tungsten(VI) tetrabromide oxide Tetrabromooxotungsten(VI)	II/15(**2**,91), II/21(**2**,60)
Br$_4$SW	Tetrabromo(sulfido)tungsten(VI) Tungsten(VI) tetrabromide sulfide	II/23(**2**,43)
Br$_4$SeW	Tetrabromo(selenido)tungsten(VI) Tungsten(VI) tetrabromide selenide	II/23(**2**,44)
Br$_4$Si	Silicon tetrabromide	II/21(**2**,61)

	Br$_4$Th	Thorium tetrabromide	II/7(**2**,87)
	Br$_4$Ti	Titanium tetrabromide	II/15(**2**,92)
	Br$_4$U	Uranium tetrabromide	II/7(**2**,88), II/23(**2**,45)
	Br$_4$V	Vanadium tetrabromide	II/7(**2**,89), II/15(**2**,93)
	Br$_4$Zr	Zirconium tetrabromide	II/15(**2**,94)
	Br$_5$Nb	Niobium pentabromide	II/7(**2**,81)
	Br$_5$Ta	Tantalum pentabromide	II/7(**2**,86), II/15(**2**,95)
	Br$_9$Re$_3$	Trirhenium enneabromide	II/7(**2**,84)
	CaCl$_2$	Calcium dichloride	II/21(**2**,62)

CaHO　　　　**Monohydroxocalcium**　　　　$C_{\infty v}$
　　　　　　　　Calcium monohydroxide

UV　Rotational analysis of laser-induced fluorescence spectra　　　Ca–O–H

State	$\tilde{X}\ ^2\Sigma^+$	$\tilde{A}\ ^2\Pi$
Symmetry	$C_{\infty v}$	$C_{\infty v}$
Energy [eV]	0	1.984
r_0 [Å] Ca–O [a]	1.986	1.966
O–H [a]	0.901	0.897
r_e [Å] Ca–O [b]	1.976(4)	1.956(4)
O–H [b]	0.930(7)	0.923(7)

[a] Zero-point averages over large-amplitude bending motion. Experimental precision higher than the quoted significant figures might imply.
[b] Rotational constants B_e obtained from B_0 assuming values of rotation-vibration constants α_e transferred from other molecules. Uncertainties quoted set by the limits assumed in these transferred α's.

Hilborn, R.C., Zhu Quingshi, Harris, D.O.: J. Mol. Spectrosc. **97** (1983) 73.　　　II/15(**2**,96)

	CaDO	Calcium deuteriooxide	II/23(**2**,46)
	CaHS	Calcium hydrogensulfide	II/23(**2**,47)
	CaH$_2$N	Calcium monoamide	II/15(**2**,97)
	CaI$_2$	Calcium diiodide	II/15(**2**,98), II/21(**2**,63)
	CaN$_3$	Calcium monoazide	II/21(**2**,64)
	CdCl$_2$	Cadmium dichloride	II/23(**2**,48)

　　　　CdI$_2$　　　　　**Cadmium diiodide**　　　　$D_{\infty h}$
ED　　　　　　　　　　　Cadmium(II) iodide　　　　　CdI$_2$

r_g	Å [a]
Cd–I	2.582(5)

r_e [b]	Å [a]
Cd–I	2.570(6)

The nozzle temperature was 678 K.

[a] Estimated total error.
[b] Estimated by several model calculations.

Vogt, N., Hargittai, M., Kolonits, M., Hargittai, I.: Chem. Phys. Lett. **199** (1992) 441.

II/23(**2**,49)

CeF$_4$	Cerium tetrafluoride	II/15(**2**,99)
ClDNe	Neon – deuterium chloride (1/1)	II/15(**2**,112)
ClFH$_2$	Hydrogen chloride – hydrogen fluoride (1/1)	
		II/15(**2**,100), II/21(**2**,65)
ClFKr	Krypton – chlorine fluoride (1/1)	II/15(**2**,101)
ClFO$_2$	Chloryl fluoride	II/7(**2**,96), II/15(**2**,102)
ClFO$_2$S	Sulfuryl fluoride chloride	II/7(**2**,98)
ClFO$_3$	Perchloryl fluoride	II/7(**2**,97), II/23(**2**,50)
ClF$_2$H	Hydrogen fluoride – chlorine fluoride (1/1)	II/15(**2**,103)
ClF$_2$N	Chlorodifluoroamine, Nitrogen chloride fluoride	II/7(**2**,93)
ClF$_2$NOS	Sulfinyl chloroimide difluoride	II/7(**2**,94)
ClF$_2$NS	Sulfur chloride difluoride	II/7(**2**,95)
ClF$_2$P	Phosphorus chloride difluoride	II/7(**2**,99)
ClF$_2$PS	Difluorochlorophosphine sulfide	II/15(**2**,104)
	Thiophosphoryl chloride difluoride	
ClF$_3$	Chlorine trifluoride	II/7(**2**,91), II/15(**2**,105), II/23(**2**,51)
ClF$_3$Ge	Chlorotrifluorogermane	II/7(**2**,92)
	Germanium chloride trifluoride	
ClF$_3$O	Chlorine trifluoride oxide	II/15(**2**,106)
ClF$_3$Si	Chlorotrifluorosilane	II/7(**2**,101), II/21(**2**,66)
	Silicon chloride trifluoride	
ClF$_4$P	Phosphorus chloride tetrafluoride	II/21(**2**, 67)
	Chlorotetrafluorophosphorus(V)	
ClF$_5$	Chlorine pentafluoride	II/15(**2**,107)
ClF$_5$S	Sulfur chloride pentafluoride(VI)	II/7(**2**,100), II/15(**2**,108)
	Chloropentafluorosulfur	
ClF$_5$Te	Tellurium pentafluoride chloride	II/7(**2**,103)
	Chloropentafluorotellurium(VI)	
ClF$_5$W	Tungsten pentafluoride chloride	II/7(**2**,104)
	Chloropentafluorotungsten(VI)	
ClGeH	Germanium chloride hydride	II/21(**2**,68)
ClGeH$_3$	Germyl chloride	II/7(**2**,109), II/15(**2**,109), II/21(**2**,69)
	Chlorogermane	

ClHHg   **Mercury – hydrogen chloride (1/1)** $C_{\infty v}$
(weakly bound complex)   (large-amplitude motion)

MW

Isotopic species	R_{cm} Å [a]	θ_0 [b]) [c]) deg	γ_0 [c]) deg [a])	R (Hg···Cl) Å [a])
^{198}Hg · H^{35}Cl	4.0666(10)	31.22(10)	31.30(15)	4.0974(50)
^{199}Hg · H^{35}Cl	4.0666(10)	31.21(8)	31.29(12)	4.0974(50)
^{200}Hg · H^{35}Cl	4.0666(10)	31.23(2)	31.31(3)	4.0974(50)
^{201}Hg · H^{35}Cl	4.0666(10)	31.24	31.32	4.0974(50)
^{202}Hg · H^{35}Cl	4.0666(10)	31.25(4)	31.33(6)	4.0974(50)
^{204}Hg · H^{35}Cl	4.0665(10)	31.26(8)	31.34(12)	4.0973(50)
^{200}Hg · H^{37}Cl	4.0677(10)	31.16(32)	31.24(48)	4.0969(50)
^{202}Hg · H^{37}Cl	4.0677(10)	31.41(33)	31.49(50)	4.0968(50)

(*Table continued*)

(*Table continued*)

Isotopic species	R_{cm} Å [a]	θ_0 [b] [c] deg	γ_0 [c] deg [a]	R (Hg⋯Cl) Å [a]
^{198}Hg · D^{35}Cl	4.0490(10)	25.01(54)	25.14(81)	4.1123(50)
^{199}Hg · D^{35}Cl	4.0490(10)	25.26(1)	25.39(2)	4.1121(50)
^{200}Hg · D^{35}Cl	4.0490(10)	25.28(8)	25.41(12)	4.1121(50)
^{201}Hg · D^{35}Cl	4.0490(10)	25.30	25.43	4.1121(50)
^{202}Hg · D^{35}Cl	4.0490(10)	25.33(1)	25.46(2)	4.1121(50)
^{204}Hg · D^{35}Cl	4.0490(10)	25.31(5)	25.44(8)	4.1121(50)

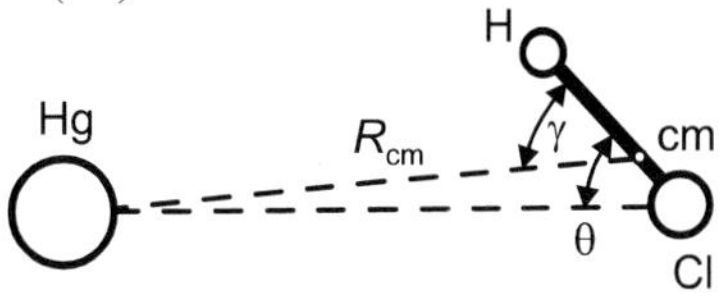

[a]) Uncertainties were not all estimated in the original paper.
[b]) Angle obtained from nuclear quadrupole coupling constant.
[c]) Average angle.

Shea, J.A., Campbell, E.J.: J. Chem. Phys. **81** (1984) 5326.   II/15(**2**,109a)

ClHKr	Krypton – hydrogen chloride (1/1)	II/15(**2**,110)
ClHN$_2$	Nitrogen – hydrogen chloride (1/1)	II/15(**2**,111), II/21(**2**,70)
ClHN$_2$O	Nitrous oxide – hydrogen chloride (1/1)	II/23(**2**,52)

ClHO **Hypochlorous acid** C_s

MW

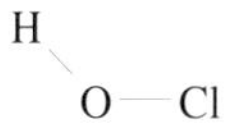

r_0	Å	θ_0	deg
O–H	0.964(75)	H–O–Cl	103(16)
O–Cl	1.695(27)		

r_s	Å	θ_s	deg
O–H	0.962(5)	H–O–Cl	102.4(3)
O–Cl	1.693(3)		

r_z	Å	θ_z	deg
O–H	0.9732(23)	H–O–Cl	102.45(42)
O–Cl	1.6974(7)		
δr_z(O–D)	−0.0028(19)		

r_e	Å	θ_e	deg
O–H	0.9636(25)	H–O–Cl	102.45(42)
O–Cl	1.6908(10)		

Anderson, W.D., Gerry, M.C.L., Davis, R.W.: J. Mol. Spectrosc. **115** (1986) 117.

IR

r_e	Å	θ_e	deg
O–H	0.9643(5)	H–O–Cl	102.96(8)
O–Cl	1.6891(2)		

Deeley, C.M.: J. Mol. Spectrosc. **122** (1988) 481.   II/7(**2**,114), II/21(**2**,71)

| ClHO$_2$S | Hydrogen chloride – sulfur dioxide (1/1) | II/21(**2**,72) |
| ClHO$_4$ | Perchloric acid | II/7(**2**,115) |

ClHSi   **Silicon chloride hydride** C_s
Monochlorosilylene

UV   Rotational and vibrational analysis

State		$\tilde{X}\ ^2A'$	$\tilde{A}\ ^2A''$ [a]
Symmetry		C_s	C_s
Energy [eV]		0	2.569
r_0 [Å]	H–Si	1.561	1.499
	Si–Cl	2.064	2.047
θ_0 [deg]	Cl–Si–H	102.8	116.1

[a] See Herzberg (1966), p. 270 (see also II/7(**2**,73)); Hougen et al.: previously assigned tentatively to $\tilde{a}\ ^3A''$.

Herzberg, G., Verma, R.D., Can. J. Phys. **42** (1964) 395.
Herzberg, G.: Molecular Spectra and Molecular Structure Vol. **III** (1966).
Hougen, J.T., Watson, J.K.G.: Can. J. Phys. **43** (1965) 298; Billingsley, J.: Can. J. Phys. **50** (1972) 531.

State		$\tilde{A}\ ^1A''$
r_e [Å]	H–Si	1.510(10)
	Si–Cl	2.0465(14)
θ_e [deg]	Cl–Si–H	116.1(8)

Rotational-vibrational energy levels fitted to a quadratic-cum-Lorentzian model potential of cylindrical symmetry about the linear unstable equilibrium configuration.
Barrier to inversion in the molecular plane 1.54(37) eV.

Gilchrist, W.A., Reyna, E., Coon, J.B.: J. Mol. Spectrosc. **74** (1979) 345.

                                                                         II/7(**2**,116), II/15(**2**,113)

ClHXe	Xenon – hydrogen chloride (1/1)	II/15(**2**,114)
ClH_2^+	Chloronium cation	II/21(**2**,73)
ClH_2N	Monochloroamine	II/7(**2**,112)
ClH_3O	Water – hydrogen chloride (1/1)	II/15(**2**,114a)
ClH_3S	Hydrogen sulfide – hydrogen chloride (1/1)	II/15(**2**,114b)

ClH_3Si   **Chlorosilane** C_{3v}
Silyl chloride

IR, MW SiH_3Cl

r_0	Å	θ_0	deg
Si–H	1.47496(11) [a]	H–Si–Cl	108.295(12)
Si–Cl	2.05057(6)		

Improved structure results from obtaining an A_0 rotational constant for the $SiHD_2Cl$ isotopic species. All available microwave data were included in the fitting.

[a] The assumption was made that δr_0 (SiH–SiD) = 0.0023 Å.

Duncan, J.L., Harvie, J.L., McKean, D.C., Cradock, S.: J. Mol. Struct. **145** (1986) 225.
                                                                                      II/7(**2**,117), II/21(**2**,74)

ClH_3Sn	Chlorostannane, Stannyl chloride	II/23(**2**,53)

	ClH₄N	**Hydrogen chloride – ammonia (1/1)**	**C₃ᵥ**
MW		(weakly bound complex)	(effective symmetry class)
			NH₃ · HCl

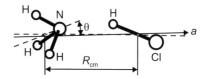

Isotopic species	$\langle R_{cm}^2 \rangle^{1/2}$ Å	$r_0(N \cdots Cl)$ Å	k_s Nm^{-1}	ν_s cm^{-1}
^{14}NH₃···H^{35}Cl	3.1654(2)	3.1364(7)	17.6(3)	161(2)
^{14}NH₃···H^{37}Cl	3.1673(2)	3.1363(6)	18.1(2)	162(1)
^{15}NH₃···H^{35}Cl	3.1614(2)	3.1358(7)	18.2(2)	160(1)
^{15}NH₃···H^{37}Cl	3.1632(2)	3.1358(7)	18.1(2)	159(1)
^{14}NH₃···H^{35}Cl	3.1367(2)	3.1410(11)		

The angle θ is assumed to be in the range of 15(3)°.

Howard, N.W., Legon, A.C.: J. Chem. Phys. **88** (1988) 4694.
See also: Goodwin, E.J., Howard, N.W., Legon, A.C.: Chem. Phys. Lett. **131** (1986) 319.

II/21(**2**,75)

ClH₄P	Phosphine – hydrogen chloride (1/1)	II/15(**2**,115)
ClNO	Nitrosyl chloride	II/7(**2**,130), II/15(**2**,116)
ClNOS	Thionyl chloroimide	II/7(**2**,132)
ClNO₂	Nitryl chloride	II/7(**2**,131), II/15(**2**,117)
ClNO₂	Chlorine nitrite	II/15(**2**,118)
ClNS	Thiazyl chloride	II/7(**2**,134), II/15(**2**,118a)
ClN₃	Chlorine azide	II/7(**2**,128)
ClO₂⁻	Chlorite ion, Dioxachlorate(1–) ion	II/23(**2**,54)
ClO₂	Chlorine dioxide	II/7(**2**,138), II/15(**2**,119), II/21(**2**,76)
ClO₃Re	Rhenium chloride trioxide	II/7(**2**,142)
Cl₂Co	Cobalt dichloride	II/7(**2**,90), II/23(**2**,55)
Cl₂Cr	Chromium dichloride	II/21(**2**,77)
Cl₂CrO₂	Chromyl dichloride	II/15(**2**,120)
	Chromium(VI) dichloride dioxide	
Cl₂Cs₂	Dicesium dichloride	II/21(**2**,78)
Cl₂Eu	Europium dichloride	II/23(**2**,56)
Cl₂FH	Chlorine – hydrogen fluoride (1/1)	II/15(**2**,121)
Cl₂F₃P	Phosphorus dichloride trifluoride	II/21(**2**,79)
	Dichlorotrifluorophosphorus(V)	
Cl₂F₈N₂S₂	Di–μ–chloroimido–bis(tetrafluorosulfur)	II/15(**2**,122)
Cl₂Fe	Iron dichloride	II/7(**2**,105), II/15(**2**,123), II/23(**2**,57)
Cl₂Ga₂H₄	Di–μ–chloro–bis[dihydridogallium(III)]	II/21(**2**,80)
	Digallium dichloride tetrahydride	
Cl₂Ge	Germanium dichloride	II/15(**2**,124)
Cl₂GeH₂	Dichlorogermane	II/7(**2**,110)
Cl₂HN	Dichloroamine	II/7(**2**,113)
Cl₂HNO₄S₂	Imidobis(sulfonyl chloride)	II/15(**2**,125)

	Cl$_2$H$_2$		**Hydrogen chloride dimer**		**C$_s$**
Far-IR					(HCl)$_2$

r_0	Å		θ_0	deg
R_{cm}	3.81(2)		θ_{av}	46.8(20)
			θ_1	70···80
			θ_2	0···10
			ϕ	< 10

Blake, G.A., Busarow, K.L., Cohen, R.C., Laughlin, K.B., Lee, Y.T., Saykally, R.J.: J. Chem. Phys. **89** (1988) 6577.

II/21(**2**,81)

Cl$_2$H$_2$Si	Dichlorosilane	II/15(**2**,126)
Cl$_2$Hg	Mercury dichloride	II/7(**2**,120), II/15(**2**,127)
Cl$_2$K$_2$	Dipotassium dichloride	II/21(**2**,82)
Cl$_2$Li$_2$	Dilithium dichloride	II/7(**2**,124)
Cl$_2$Mg	Magnesium chloride	II/15(**2**,128)
Cl$_2$Mn	Manganese dichloride	II/15(**2**,129), II/23(**2**,58)
Cl$_2$MoO$_2$	Molybdenum dichloride dioxide	II/15(**2**,130), II/23(**2**,59)
Cl$_2$Na$_2$	Disodium dichloride	II/7(**2**,135), II/21(**2**,83)
Cl$_2$Ni	Nickel dichloride	II/23(**2**,60)
Cl$_2$O	Dichlorine monoxide, Oxygen dichloride	II/7(**2**,139), II/15(**2**,131), II/21(**2**,84)
Cl$_2$OS	Thionyl chloride, Sulfinyl chloride	II/7(**2**,143), II/15(**2**,132)
Cl$_2$OSe	Selenyl chloride	II/15(**2**,133)

	Cl$_2$O$_2$		**Dioxygen dichloride**		**C$_2$**
MW					ClOOCl

r_0	Å		θ_0	deg
Cl–O	1.7044(10)		Cl–O–O	110.07(1)
O–O	1.4259(21)		ϕ [a])	81.03(8)

[a]) Dihedral angle ClOOCl.

Birk, M., Friedl, R.R., Cohen, E.A., Pickett, H.M., Sander, S.P.: J. Chem. Phys. **91** (1989) 6588.

II/21(**2**,85)

Cl$_2$O$_2$S	Sulfuryl chloride, Sulfonyl chloride	II/7(**2**,144), II/15(**2**,134)
Cl$_2$O$_7$	Dichlorine heptaoxide	II/7(**2**,140)
Cl$_2$Pb	Lead dichloride	II/15(**2**,135), II/21(**2**,86), II/23(**2**,61)
Cl$_2$Rb$_2$	Dirubidium dichloride	II/21(**2**,87)
Cl$_2$S	Sulfur dichloride	II/7(**2**,151), II/15(**2**,136)
Cl$_2$S$_2$	Disulfur dichloride	II/7(**2**,152), II/15(**2**,137)
Cl$_2$Se	Selenium dichloride	II/15(**2**,138)
Cl$_2$Si	Silicon dichloride	II/15(**2**,139), II/23(**2**,62)
Cl$_2$Sm	Samarium dichloride	II/23(**2**,63)
Cl$_2$Sn	Tin dichloride	II/21(**2**,88), II/23(**2**,64)
Cl$_2$Te	Tellurium dichloride	II/21(**2**,89)
Cl$_2$V	Vanadium dichloride	II/21(**2**,90)

Cl₂Yb	Ytterbium dichloride, Ytterbium(II) chloride	II/23(**2**,65)
Cl₂Zn	Zinc dichloride	II/21(**2**,91)
Cl₃FSi	Fluorotrichlorosilane	II/7(**2**,102)
Cl₃F₂P	Phosphorus trichloride difluoride Trichlorodifluorophosphorus(V)	II/21(**2**,92)
Cl₃Fe	Iron trichloride	II/7(**2**,106)
Cl₃Ga	Gallium trichloride	II/23(**2**,66)
Cl₃GaH₃N	Gallium trichloride – ammonia (1/1)	II/15(**2**,140)
Cl₃Gd	Gadolinium trichloride	II/15(**2**,141)
Cl₃GeH	Trichlorogermane	II/7(**2**,111)
Cl₃HSi	Trichlorosilane	II/7(**2**,118), II/15(**2**,142)
Cl₃Ho	Holmium trichloride	II/15(**2**,143)
Cl₃In	Indium trichloride	II/23(**2**,67)
Cl₃La	Lanthanum trichloride	II/7(**2**,123), II/21(**2**,93), II/23(**2**,68)
Cl₃Lu	Lutetium trichloride	II/15(**2**,144)
Cl₃N	Nitrogen trichloride, Trichloroamine	II/7(**2**,129)
Cl₃Nd	Neodymium trichloride	II/7(**2**,137)
Cl₃OP	Phosphoryl trichloride	II/7(**2**,141), II/21(**2**,94)
Cl₃OV	Vanadium(V) trichloride oxide	II/7(**2**,146)
Cl₃P	Phosphorus trichloride, Trichlorophosphine	II/7(**2**,148)
Cl₃PS	Thiophosphoryl trichloride	II/7(**2**,150)
Cl₃Pr	Praseodymium trichloride	II/15(**2**,145)
Cl₃Sb	Antimony trichloride Trichlorostibine	II/7(**2**,153), II/15(**2**,146), II/21(**2**,95)
Cl₃Tb	Terbium trichloride	II/15(**2**,147)

Cl₃Ti	**Titanium trichloride**	**D₃ₕ**
ED and vibrational spectroscopy	Titanium(III) chloride	TiCl₃

r_g	Å [a]) [1]
Ti–Cl	2.203(5)

r_e	Å [a])
Ti–Cl	2.178(5)

The ED intensity data of [1] were reanalyzed using the force field. The nozzle temperature was 705(20) °C.

[a]) 2.5 times the estimated standard error including a scale error.

Giricheva, N.I., Girichev, G.V., Shlykov, S.A.: Zh. Strukt. Khim. **32** No. 4 (1991) 165; Russ. J. Struct. Chem. (Engl. Transl.) **32** (1991) 602.

[1] Girichev, G.V., Shlykov, S.A., Petrova, V.N., Subbotina, N.Yu., Lapshina, S.B., Danilova, T.G.: Izv. Vyssh. Uchebn. Zaved., Khim. Khim. Tekhnol. **31** No.8 (1988) 46.

II/21(**2**,96), II/23(**2**,69)

Cl₃U	Uranium trichloride	II/23(**2**,70)
Cl₃W	Tungsten trichloride	II/21(**2**,97), II/23(**2**,71)
Cl₄Cr₂	Dichromium tetrachloride Di–μ–chloro–bis[chlorochromium(II)]	II/21(**2**,98)

	Cl₄FP	Phosphorus tetrachloride fluoride	II/21(**2**,99)
		Tetrachlorofluorophosphorus(V)	
	Cl₄Ge	Germanium tetrachloride, Tetrachlorogermane	II/7(**2**,108)
	Cl₄Hf	Hafnium tetrachloride	II/7(**2**,119), II/15(**2**,148)
	Cl₄InTl	Indium(III) thallium tetrachloride	II/7(**2**,121)
	Cl₄KY	Potassium yttrium tetrachloride	II/7(**2**,122)
	Cl₄Mo	Molybdenum tetrachloride	II/7(**2**,125)
	Cl₄MoO	Molybdenum(VI) tetrachloride oxide	II/7(**2**,127), II/15(**2**,149)
	Cl₄NV	Vanadium(V) trichloride chloroimide	II/15(**2**,150)
	Cl₄OOs	Osmium(VI) tetrachloride oxide	II/21(**2**,100)
		Tetrachlorooxoosmium(VI)	
	Cl₄ORe	Rhenium(VI) tetrachloride oxide	II/21(**2**,101)
		Tetrachlorooxorhenium(VI)	
	Cl₄OW	Tungsten(VI) tetrachloride oxide	II/7(**2**,147), II/15(**2**,151)
	Cl₄Pb	Lead tetrachloride	II/23(**2**,72)
	Cl₄SW	Tunsten(VI) tetrachloride sulfide	II/15(**2**,152)
		Tetrachloro(thio)tungsten(VI)	
	Cl₄SeW	Tungsten(VI) tetrachloride selenide	II/15(**2**,153)
		Tetrachloro(selenido)tungsten(VI)	
	Cl₄Si	Tetrachlorosilane, Silicon tetrachloride	II/7(**2**,154)
	Cl₄Sn	Tetrachlorostannane, Tin(IV) tetrachloride	II/7(**2**,156)
	Cl₄Th	Thorium tetrachloride	II/7(**2**,158), II/23(**2**,73)
	Cl₄Ti	Titanium tetrachloride	II/7(**2**,159)
	Cl₄U	Uranium tetrachloride	II/7(**2**,160), II/21(**2**,102)
	Cl₄V	Vanadium tetrachloride	II/7(**2**,161)
	Cl₄V₂	Divanadium tetrachloride	II/21(**2**,103)
		Di–μ–chloro–bis[chlorovanadium(II)]	
	Cl₄W	Tungsten tetrachloride	II/15(**2**,154), II/23(**2**,74)
	Cl₄Zr	Zirconium tetrachloride	II/7(**2**,163), II/21(**2**,104)
	Cl₅Mo	Molybdenum pentachloride	II/7(**2**,126), II/15(**2**,155), II/23(**2**,75)
	Cl₅Nb	Niobium pentachloride	II/7(**2**,136), II/15(**2**,156)
	Cl₅P	Phosphorus pentachloride	II/7(**2**,149), II/15(**2**,157)
	Cl₅Sb	Antimony pentachloride	II/15(**2**,158)
	Cl₅Ta	Tantalum pentachloride	II/7(**2**,157), II/15(**2**,159)
	Cl₅W	Tungsten pentachloride	II/7(**2**,162), II/15(**2**,160)
	Cl₆Fe₂	Diiron hexachloride	II/7(**2**,107), II/15(**2**,161)

	Cl₆Ga₂	**Digallium hexachloride**	**D₂ₕ**
ED		Di–μ–chloro–bis[dichlorogallium(III)]	Cl₆Ga₂

r_a	Å [a])	θ_a	deg [a])
Ga–Cl(t)	2.093(5)	Cl(t)–Ga–Cl(t)	124.5(1)
Ga–Cl(b)	2.298(6)	Cl(b)–Ga–Cl(b)	90(1)
Ga...Ga	3.250(8)		

(*continued*)

The vapor was found to contain dimeric and monomeric molecules of GaCl₃ in amounts 79 and 21 mol%, respectively. It was assumed that r_α(Ga–Cl) in GaCl₃ was equal to r_α(Ga–Cl(t)) in Ga₂Cl₆.

The nozzle temperature was 49(3) °C.

[a]) 2.5 times the estimated standard errors including an experimental scale error.

Petrov, V.M., Giricheva, N.I., Girichev, G.V., Titov, V.A., Chusova, T.P.: Zh. Strukt. Khim. **32** No. 4 (1991) 56; Russ. J. Struct. Chem. (Engl. Transl.) **32** (1991) 498.

II/23(**2**,76)

Cl₆In₂	Diindium hexachloride	II/23(**2**,77)
Cl₆N₃P₃	Phosphorus dichloride nitride trimer	II/7(**2**,133)
Cl₆OSi₂	Hexachlorodisiloxane	II/7(**2**,145)
Cl₆Si₂	Hexachlorodisilane	II/7(**2**,155)
Cl₆W₂	Ditungsten hexachloride	II/21(**2**,105), II/23(**2**,78)
Cl₈Si₃	Octachlorotrisilane	II/15(**2**,162)
CoF₂	Cobalt difluoride	II/21(**2**,106), II/23(**2**,79)
CrCs₂O₄	Cesium chromate	II/15(**2**,163)
CrF₂	Chromium difluoride	II/21(**2**,107), II/23(**2**,80)
CrF₂O₂	Chromyl fluoride	II/7(**2**,164), II/15(**2**,164)
CrF₃	Chromium trifluoride	II/21(**2**,108), II/23(**2**,81)
CrF₄	Chromium tetrafluoride	II/21(**2**,109)
CrF₄O	Chromium(VI) tetrafluoride oxide Tetrafluorooxochromium(VI)	II/21(**2**,110)
CrK₂O₄	Potassium chromate	II/7(**2**,165)

ED **CrN₂O₈** **Chromyl bis(nitrate)** C_2
Bis(nitrato–*O*)dioxochromium CrO₂(NO₃)₂

r_g	Å [a])	θ_α	deg [a])
Cr=O(2,3)	1.586(2)	O(2,3)=Cr=O(2,3)	112.6(35)
Cr–O(4,8)	1.957(5)	O(4,8)–Cr–O(4,8)	140.4(33)
Cr...O	2.254(20)	O(2)=Cr–O(4)	97.2(18)
N–O(4)	1.341(4)	O(2)=Cr–O(8)	104.5(9)
N=O(6)	1.254(4)	O(2,3)=Cr...O(6)	83.7(34)
N=O(7)	1.193(4)	Cr–O(4,8)–N	97.5(5)
		O(4)–N=O(6)	112.2(71)
		O(4)–N=O(7)	119.7(40)
		O(6)=N=O(7)	128.1(36)
		τ(Cr–O) [b])	144.7(39)
		τ(CrO₂) [c])	4.6(11)

The Cr–O–NO₂ group is only slightly nonplanar with the dihedral angle of the planes CrON and NO₂ equal to 16(3)°. It appears that there is relatively little torsional motion around either of the single bonds in the –ONO₂ groups.

(*continued*)

CrN$_2$O$_8$ (*continued*)

The nozzle temperature was 44...51 °C.

[a]) Twice the estimated standard errors.
[b]) Torsional angle about the Cr–O bond; $\tau = 0°$ when the N–O–Cr–O–N chain is planar *trans-trans*; a positive value corresponds to clockwise rotation of the N–O bond looking from O to Cr along the Cr–O bond.
[c]) Torsional angle of the OCrO plane about the C$_2$ axis; $\tau(\text{CrO}_2) = 0°$ when the O=Cr=O fragment is perpendicular to the OCrO plane.

Marsden, C.J., Hedberg, K., Ludwig, M.M., Gard, G.L.: Inorg. Chem. **30** (1991) 4761.

II/23(**2**,82)

CrN$_4$O$_4$	Tetranitrosylchromium(0)	II/21(**2**,111)
CrNa$_2$O$_4$	Sodium chromate	II/7(**2**,166)
Cr$_4$O$_{12}$	Tetrachromium dodecaoxide	II/15(**2**,165)
CsHO	Cesium hydroxide	II/7(**2**,167)
CsNO$_2$	Cesium nitrite	II/15(**2**,166)
CsNO$_3$	Cesium nitrate	II/15(**2**,167)
CsO$_3$P	Cesium metaphosphate	II/15(**2**,168)
CsO$_4$Re	Cesium perrhenate	II/15(**2**,169)
Cs$_2$F$_2$	Dicesium difluoride	II/21(**2**,112)
Cs$_2$H$_2$O$_2$	Dicesium dihydroxide	II/21(**2**,113)
Cs$_2$I$_2$	Dicesium diiodide	II/21(**2**,114)
Cs$_2$MoO$_4$	Cesium molybdate	II/7(**2**,168)
Cs$_2$O$_4$S	Cesium sulfate	II/7(**2**,169), II/15(**2**,170)
Cs$_2$O$_4$W	Cesium tungstate	II/7(**2**,170)
CuF$_2$	Copper difluoride	II/21(**2**,115), II/23(**2**,83)

CuHO   **Copper monohydroxide radical**   C$_s$

LIF

Cu–O–H

State	$\tilde{X}\ ^1A'$	$\tilde{A}\ ^1A'$	$\tilde{B}\ ^1A''$
Energy [eV]	0.0	1.973	2.284
Reference	[1]	[2]	[1]
r_s [Å] Cu–O	1.7689(2) [a]	1.775(3) [a]	1.7841(5)
O–H	0.952(5)	1.035(4)	0.951(3)
θ_s [deg] Cu–O–H	110.24(8)	111.0(16)	117.67(10)

Rotational analysis of bands of ^{63}CuOH, ^{65}CuOH, ^{63}CuOD and ^{65}CuOD.

[a]) Error limits are 1σ.

Jarman, C.N., Fernando, W.T.M., Bernath, P.F.: J. Mol. Spectrosc. [1] **144** (1990) 286, [2] **145** (1991) 151.

II/15(**2**,171), II/21(**2**,116), II/23(**2**,84)

CuN$_2$O$_6$	Copper(II) nitrate	II/7(**2**,171), II/15(**2**,172)

	FGeH$_3$		**Germyl fluoride**		**D$_{3v}$**
MW			Fluorogermane		GeH$_3$F

	r_e	Å		θ_e	deg
	Ge–F	1.73095(40)		F–Ge–H	106.071(17)
	Ge–H	1.51453(13)			
	r_0	Å		θ_0	deg
	Ge–F	1.734026(67)		F–Ge–H	106.370(28)
	Ge–H	1.52427(23)			
	r_0 [a])	Å		θ_0 [a])	deg
	Ge–F	1.73008(12)		F–Ge–H	105.466(35)
	Ge–H	1.52529(51)			

[a]) Assuming ε to be common to all isotopic species, where ε is defined by $I_0 = I_e + \varepsilon$. This structure is often referred as $r_{\varepsilon,I}$.

Le Guennec, M., Chen, W., Wlodarczak, G., Demaison, J., Eujen, R., Bürger, H.: J. Mol. Spectrosc. **150** (1991) 493.　　　　　　　II/7(**2**,173), II/15(**2**,173), II/23(**2**,85)

	FHKr		Krypton – hydrogen fluoride (1/1)		II/15(**2**,174)

	FHN		**Aminylene fluoride**		**C$_s$**
UV			Monofluoroaminyl radical		

State	$\tilde{X}\ ^2A''$	$\tilde{A}\ ^2A'$
Symmetry	C$_s$	C$_s$
Energy [eV]	0	2.497

r_0 [Å]	N–H	1.09(1)	1.01(1)
	N–F	1.364(7)	1.343(3)
θ_0 [deg]	H–N–F	109(1)	124(1)

From the rotational analysis of bands of HNF and DNF. Error limits are 1σ.

Chen, J., Dagdigian, P.J.: J. Chem. Phys. **96** (1992) 7333.　II/7(**2**,176), II/15(**2**,175), II/23(**2**,86)

FHN$_2$	Nitrogen – hydrogen fluoride (1/1)	II/15(**2**,176)
FHN$_2$O	Nitrous oxide – hydrogen fluoride (1/1)	
	Dinitrogen monoxide – hydrogen fluoride (1/1)	
		II/15(**2**,177), II/21(**2**,118), II/23(**2**,87)
FHO	Hypofluorous acid	II/7(2,179), II/23(**2**,88)
FHO$_2$S	Hydrogen fluoride – sulfur dioxide (1/1)	II/21(**2**,119)

	FHSi		**Fluorosilylene**			**C$_s$**
LIF			Silicon fluoride hydride			HSiF

State	$\tilde{X}\ ^1A'$		$\tilde{A}\ ^1A''$	
Energy [eV]	0.00		2.884	
Reference	[1]	[2]	[1]	[2]

r [Å]	Si–H	1.530 [a])	1.534	1.484 [a])	1.543
	Si–F	1.605	1.604	1.609	1.599
θ [deg]	H–Si–F	97	97.6 [b])	111	115.3 [b])

(continued)

FHSi (*continued*)

Rotational analysis of laser-excited fluorescence spectrum.

[a]) Assumed value based on related molecules.
[b]) Fixed at *ab initio* values [3]. The value of $\Delta\theta_0$ agrees well with the Franck-Condon distribution of band intensities.

[1] Suzuki, T., Hakuta, K., Saito, S., Hirota, E.: J. Chem. Phys. **82** (1985) 3580.
[2] Dixon, R.N., Wright, N.G.: Chem. Phys. Lett. **117** (1985) 280.
[3] Colvin, M.E., Grev, R.S., Schaefer III, H.F.: Chem. Phys. Lett. **99** (1983) 399.

II/15(**2**,178), II/21(**2**,120)

FHXe	Xenon – hydrogen fluoride (1/1)	II/15(**2**,179)
FH_2^+	Fluoronium ion	II/15(**2**,179a)
FH_2I	Hydrogen fluoride – hydrogen iodide (1/1)	II/21(**2**,121)
FH_2N	Monofluoroamine	II/21(**2**,122)
FH_2NO_2S	Sulfonyl amide fluoride	II/15(**2**,180)
FH_3	Hydrogen fluoride – dihydrogen (1/1)	II/21(**2**,123)
	Hydrogen – hydrogen fluoride (1/1)	
FH_3O	Water-hydrogen fluoride (1/1)	II/15(**2**,181)
FH_3S	Hydrogen sulfide – hydrogen fluoride (1/1)	II/15(**2**,181a)

FH_3Si   **Fluorosilane** C_{3v}
Silyl fluoride

IR, MW SiH_3F

r_0	Å
Si–H	1.47608(19) [a])
Si–F	1.59450(13)

θ_0	deg
H–Si–F	108.269(21)

Improved structure results from obtaining an A_0 rotational constant for the $SiHD_2F$ isotopic species. All available microwave data were included in the fitting.

[a]) The assumption was made that δr_0(SiH–SiD) = 0.0023 Å.

Duncan, J.L., Harvie, J.L., McKean, D.C., Cradock, S.: J. Mol. Struct. **145** (1986) 225.
Robiette, A.G., Georghiou, C., Baker, J.G.: J. Mol. Spectrosc. **63** (1976) 391.

II/7(**2**,186), II/15(**2**,182), II/21(**2**,124)

FH_4P	Phosphine – hydrogen fluoride (1/1)	II/15(**2**,182a)
FH_5Si_2	Disilanyl fluoride, Fluorodisilane	II/7(**2**,187)
$FMnO_3$	Manganese(VII) fluoride trioxide	II/7(**2**,196)
FNO	Nitrosyl fluoride	II/7(**2**,202), II/15(**2**,182b)
FNO_2	Nitryl fluoride	II/7(**2**,203)
FNS	Thiazyl fluoride	II/7(**2**,206), II/15(**2**,183)
FN_3	Fluorine azide	II/21(**2**,125)

2 Inorganic molecules

	FOS		**Fluorooxosulfur radical**		C_s
MW					FSO

	r_0	Å		θ_0	deg
	F–F	1.602(3)		F–S–O	108.32(6)
	S–O	1.452(3)			

Endo, Y., Saito, S. Hirota, E.: J. Chem. Phys. **74** (1981) 1568.

II/15(**2**,184)

	FO₂		**Dioxygen fluoride**		C_s
IR			Peroxofluorine radical		O–O
					F

r_e	Å		θ_e	deg
O–O	1.200		F–O–O	111.2
F–O	1.649			

Structure obtained from rotational constants and force field calculations.

Yamada, C., Hirota, E.: J. Chem. Phys. **80** (1984) 4694.

II/15(**2**,185)

FO₃Re		Rhenium(VII) fluoride trioxide	II/7(**2**,218)

	FO₃S		**Sulfur fluoride trioxide**		
UV, ED			Fluorosulfate radical		SO₃F

State [a]		$\tilde{X}\,^2A_2$	$^2A_1$	$^2E\,(1)$	$^2E\,(2)$
Symmetry		C_{3v}			C_{3v}
Energy [eV]		0			2.403
Reference		[1] [3] [4]	[1]	[1]	[1] [2] [4]
r [Å]	S–F	1.64			1.64
	S=O	1.46			1.49
θ [deg]	O=S–F	109	(97)	(105)	106
Notes		[b] [c]	[d]	[d]	[c]

[a] Assignments based on the theoretical calculations, consistent with spectra observed.
[b] Electron diffraction: quoted in ref. [4].
[c] Rotational analysis of $\tilde{X}\,^2A_2 - {}^2E\,(2)$ system, 5160 Å.
[d] Electronic *ab initio* calculations.

[1] King, G.W., Santry, D.P., Warren, C.H.: J. Mol. Spectrosc. **32** (1969) 108.
[2] King, G.W., Warren, C.H.: J. Mol. Spectrosc. **32** (1969) 121.
[3] Bauer, S.H., Hencher, J.L.: quoted in ref. [4].
[4] King, G.W., Warren, C.H.: J. Mol. Spectrosc. **32** (1969) 138.

II/7(**2**,219), II/15(**2**,186)

F₂Fe	Iron difluoride	II/21(**2**,126), II/23(**2**,89)
F₂Ge	Germanium difluoride	II/7(**2**,172)
F₂GeH₃PS	Difluoro(germylthio)phosphine	II/15(**2**,187)
F₂Ge₂H₆P₂	1,1–Difluoro–2,2–digermylphosphane	II/15(**2**,188)

	F₂H⁻	**Difluorohydrogenate(1–) ion**	**D∞h**
IR		Hydrogendifluoride(1–) ion	F–H–F⁻

r_0	Å
F...F	2.304432(52)

r_e	Å
F...F	2.27771(7)

Kawaguchi, K., Hirota, E.: J. Chem. Phys. **84** (1986) 2953.   II/21(**2**,127)

F₂HN	Difluoroamine	II/7(**2**,177)
F₂HNOS	Sulfinyl difluoride imide	II/15(**2**,189)
F₂HOP	Difluorophosphine oxide	II/7(**2**,180)
	Phosphoryl difluoride hydride	
F₂HP	Difluorophosphine	II/7(**2**,181)
F₂HPS	Thiophosphoryl difluoride monohydride	
	Difluorophosphine sulfide	
		II/7(**2**,184), II/15(**2**,190), II/21(**2**,128)
F₂HPSe	Difluorophosphine selenide	II/15(**2**,191)

	F₂H₂	**Hydrogen fluoride dimer**	**C₂**
MW			(large-amplitude motion)

r_0	Å [a])	θ_0 [b])	deg
F···F	2.78	θ_1^0	63(6)
		θ_2^0	10(2)
		α(HFHF)	1.01(50)
		α(HFDF)	1.15(50)
		α(DFDF)	1.92(50)
		β(HFHF)	0.34(50)
		β(HFDF)	0.43(50)
		β(DFDF)	0.62(50)

[a]) Uncertainties were partially estimated in the original paper.
[b]) See figure for definition of angles. Average angles.

Howard, B.J., Dyke, T.R., Klemperer, W.: J. Chem. Phys. **81** (1984) 5417.
IR: Pine, A.S., Lafferty, W.J., Howard, B.J.: J. Chem. Phys. **81** (1984) 2939.

II/7(**2**,175), II/15(**2**,191a)

F₂H₂NP	Aminodifluorophosphine	II/7(**2**,178)
F₂H₂P₂	Phosphinodifluorophosphine	II/7(**2**,182)
F₂H₂Si	Difluorosilane	II/7(**2**,188), II/15(**2**,192)
F₂H₃P	λ^5–Difluorophosphorane, Difluorophosphane	II/23(**2**,90)
F₂H₄NPSi	(Difluorophosphino)(silyl)amine	II/23(**2**,91)
F₂H₅N₂P	Diaminodifluorophosphorane	II/15(**2**,193)
F₂H₆NPSi₂	Difluorophosphino(disilyl)amine	II/15(**2**,194)
F₂IPS	Difluoroiodophosphine sulfide	II/15(**2**,195)
	Thiophosphoryl difluoride iodide	
F₂K₂	Dipotassium difluoride	II/21(**2**,129)

F$_2$Kr	Krypton difluoride		II/7(**2**,193)
F$_2$Li$_2$	Dilithium difluoride		II/7(**2**,195), II/15(**2**,196), II/23(**2**,92)
	Di–μ–fluoro–dilithium		
F$_2$Mg	Magnesium difluoride		II/15(**2**,197)
F$_2$Mn	Manganese difluoride		II/15(**2**,198), II/21(**2**,130), II/23(**2**,93)
F$_2$N	Nitrogen difluoride, Difluoroaminyl radical		II/7(**2**,198)
F$_2$N$_2$	Difluorodiazine		II/7(**2**,199)
F$_2$Na$_2$	Disodium difluoride		II/7(**2**,209), II/21(**2**,131)
F$_2$Ni	Nickel difluoride		II/15(**2**,199), II/21(**2**,132), II/23(**2**,94)
F$_2$O	Oxygen difluoride		II/7(**2**,214), II/23(**2**,95)
F$_2$OS	Thionyl difluoride, Sulfinyl difluoride		II/7(**2**,220)
F$_2$OSe	Seleninyl oxyfluoride		II/7(**2**,225)
	Selenium difluoride oxide		
F$_2$O$_2$	Dioxygen difluoride		II/7(**2**,215), II/21(**2**,133)
	Difluorine dioxide		
F$_2$O$_2$S	Sulfuryl difluoride		II/7(**2**,221), II/15(**2**,200)
	Sulfonyl difluoride		
F$_2$O$_2$Se	Selenonyl difluoride		II/15(**2**,201)
F$_2$O$_5$S$_2$	Disulfuryl difluoride		II/7(**2**,222)
F$_2$O$_8$S$_3$	Trisulfuryl difluoride		II/7(**2**,223)
F$_2$P	Phosphorus difluoride		II/21(**2**,134)
F$_2$Pb	Lead difluoride		II/15(**2**,202), II/21(**2**,135), II/23(**2**,96)
F$_2$Rb$_2$	Dirubidium difluoride		II/21(**2**,136)

F$_2$S   **Sulfur difluoride**   **C$_{2v}$**

MW   SF$_2$

r_e	Å	θ_e	deg
S–F	1.58745(12)	F–S–F	98.048(13)

Endo, Y., Saito, S., Hirota, E., Chikaraishi, T.: J. Mol. Spectrosc. **77** (1979) 222.

II/7(**2**,237), II/15(**2**,203)

F$_2$S$_2$	1,1–Difluorodisulfane	II/7(**2**,238), II/21(**2**,137)
	1,1–Difluorodisulfur, Disulfur 1,1–difluoride	
F$_2$S$_2$	1,2–Difluorodisulfane	II/7(**2**,239), II/21(**2**,138), II/23(**2**,97)
	1,2–Difluorodisulfur, Disulfur 1,2–difluoride	
F$_2$Si	Silicon difluoride	II/7(**2**,243), II/15(**2**,204)
F$_2$Tl$_2$	Dithallium difluoride	II/15(**2**,205), II/23(**2**,98)
	Di–μ–fluoro–dithallium	

F$_2$Xe   **Xenon difluoride**   **D$_{\infty h}$**

IR   F–Xe–F

r_0	Å
F–Xe	1.977965(15) [a]

r_e	Å
F–Xe	1.974355(30) [a]

(*continued*)

F$_2$Xe (*continued*)

The ground state structure is obtained by fitting the B_0 values of six isotopomers. Experimental α_i values for the ^{136}Xe species only lead to B_e and to the equilibrium structure.

[a]) Uncertainties have been increased from the original paper.

Bürger, H., Ma, S.: J. Mol. Spectrosc. **157** (1993) 536. II/7(**2**,252), II/23(**2**,99)

F$_2$Zn	Zinc difluoride	II/21(**2**,139), II/23(**2**,100)
F$_3$Fe	Iron trifluoride	II/23(**2**,101)
F$_3$Ga	Gallium trifluoride	II/21(**2**,140)
F$_3$Gd	Gadolinium trifluoride	II/21(**2**,141)
F$_3$HSi	Trifluorosilane	II/7(**2**,189)
F$_3$H$_2$OP	Water – phosphorus trifluoride (1/1)	II/23(**2**,102)
F$_3$H$_2$P	Trifluoro–λ^5–phosphane	II/21(**2**,142)
	Trifluorodihydridophosphorus, Trifluorodihydridophosphorane	
F$_3$H$_2$PSi	(Trifluorosilyl)phosphine	II/7(**2**,185)
F$_3$H$_3$Si$_2$	1,1,1–Trifluorodisilane	II/15(**2**,206)
F$_3$H$_4$N$_2$P	Diaminotrifluorophosphorane	II/15(**2**,207)
F$_3$Ho	Holmium trifluoride	II/21(**2**,143)
F$_3$ISi	Trifluoroiodosilane	II/21(**2**,144)
F$_3$KrP	Krypton – trifluorophosphine (1/1)	II/21(**2**,145)
F$_3$La	Lanthanum trifluoride	II/7(**2**,194)

F$_3$N   **Nitrogen trifluoride**   **C$_{3v}$**
Trifluoroamine   NF$_3$

MW

r_e	Å		θ_e	deg
N–F	1.3648(20)		F–N–F	102.37(3)

Otake, M., Matsumura, C., Morino, Y.: J. Mol. Spectrosc. **28** (1968) 316. II/7(**2**,200)

F$_3$NO	Trifluoroamine oxide, Nitrosyl trifluoride	II/7(**2**,204)
F$_3$NO$_2$S	Sulfonyl difluoroamide fluoride	II/21(**2**,146)
	Sulfuryl difluoroamide fluoride	
F$_3$NS	Sulfur trifluoride nitride	II/7(**2**,207), II/15(**2**,208)

F$_3$N$_3$S$_3$ **1,3,5–Trifluoro–1λ^4,3λ^4,5λ^4–cyclotriazatrithia–2,4,6–triene**
ED, MW   Trithiatriazine trifluoride   **C$_{3v}$**
Thionitrosyl fluoride trimer

r [a])	Å [b])	θ [a])	deg [b])
S–F	1.624(7)	N–S–N [c])	112.7(12)
S–N	1.582(4)	S–N–S [c])	124.3(6)
S...S	2.798(4)	N–S–F	100.9(9)
N...N	2.635(12)	δ(S–N–S–N [c])[e])	24.2(43)
Δ(SSS–NNN) [c])[d])	0.19(4)		

(*continued*)

The nozzle temperature was 30 °C.

[a]) Not specified, possibly r_z and θ_z.
[b]) Three times the estimated standard errors.
[c]) Dependent parameter.
[d]) Distance between the SSS and NNN planes.
[e]) Dihedral angle.

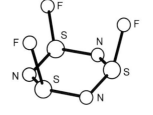

Jaudas-Prezel, E., Maggiulli, R., Mews, R., Oberhammer, H., Stohrer, W.-D.: Chem. Ber. **123** (1990) 2117.

ED

r_a	Å [a])		θ_a	deg [a])
S–F	1.619(4)		N–S–N	113.3(2)
S–N	1.592(2)		S–N–S	123.5(2)
			N–S–F	101.8(2)
			SF dip [b])	68.1(4)
			SF rock [c])	90 [e])
			pucker [d])	13.4(4)

The nozzle temperature was 293 K.

[a]) Estimated standard errors including systematic errors.
[b]) Angle between the S–F bond and the NSN plane.
[c]) Angle made by the NSN plane with the plane containing the NSN bisector and the S–F bond.
[d]) Angle between the NSN and NNN planes.
[e]) Assumed.

Downs, A.J., Efiong, A.B., McGrady, G.S., Rankin, D.W.H., Robertson, H.E.: J. Mol. Struct. **216** (1990) 201. II/23(**2**,103)

F$_3$Nd	Neodymium trifluoride		II/7(**2**,211)
F$_3$NeP	Neon – trifluorophosphine (1/1)		II/21(**2**,147), II/23(**2**,104)
F$_3$OP	Phosphoryl fluoride		II/7(**2**,216), II/15(**2**,209)
	Phosphorus trifluoride oxide		
F$_3$OV	Vanadium(V) trifluoride oxide		II/15(**2**,210)

F$_3$P	**Phosphorus trifluoride**		**C$_{3v}$**
	Trifluorophosphine		PF$_3$

MW

r_e	Å		θ_e	deg
P–F	1.561(1)		F–P–F	97.7(2)

r_z	Å		θ_z	deg
P–F	1.565(1)		F–P–F	97.6(2)

Kawashima, Y., Cox, A.P.: J. Mol. Spectrosc. **65** (1977) 319. II/7(**2**,229), II/15(**2**,211)

F$_3$PS	Thiophosphoryl trifluoride	II/7(**2**,233), II/15(**2**,212)
F$_3$Pr	Praseodymium trifluoride	II/21(**2**,148)
F$_3$Sc	Scandium trifluoride	II/7(**2**,241), II/15(**2**,213), II/21(**2**,149)

F₃V	Vanadium trifluoride	II/21(**2**,150)
F₄GeH₃NP₂	Bis(difluorophosphino)germylamine	II/15(**2**,214)
F₄HNP₂	Bis(difluorophosphino)amine	II/15(**2**,215)
F₄HP	Tetrafluoro–λ^5–phosphane	II/7(**2**,183), II/21(**2**,151)
	Tetrafluorohydridophosphorus, Tetrafluorohydridophosphorane	
F₄H₃NP₂Si	Bis(difluorophosphino)silylamine	II/15(**2**,216)
F₄H₃NSi	Tetrafluorosilane – ammonia (1/1)	II/23(**2**,105)
F₄Hf	Hafnium tetrafluoride	II/15(**2**,217)
F₄MoO	Molybdenum tetrafluoride oxide	II/15(**2**,218)
F₄N₂	Tetrafluorohydrazine	II/7(**2**,201)
F₄OOs	Osmium(VI) tetrafluoride oxide	II/15(**2**,219)
F₄OP₂	Bis[difluorophosphorus(1+)]oxide	II/7(**2**,217), II/15(**2**,220)
F₄OP₂S₂	Bis(thiophosphoryl) tetrafluoride oxide	II/15(**2**,221)
F₄ORe	Rhenium(VI) tretrafluoride oxide	II/15(**2**,222)
F₄OS	Thionyl tetrafluoride	II/7(**2**,224), II/15(**2**,223)
	Sulfur tetrafluoride oxide	
F₄OW	Tungsten(VI) terafluoride oxide	II/15(**2**,224)
F₄OXe	Xenon tetrafluoride oxide	II/7(**2**,227)
F₄P₂	Tetrafluorodiphosphane	II/7(**2**,230)
F₄P₂S	Bis[difluorophosphorus(1+)] sulfide	II/15(**2**,225)
	Diphosphorus tetrafluoride sulfide	
F₄P₂Se	Bis[difluorophosphorus(1+)] selenide	II/15(**2**,226)
F₄S	Sulfur tetrafluoride	II/7(**2**,240)
F₄SW	Tetrafluoro(thio)tungsten(VI)	II/15(**2**,227)
	Tungsten(VI) tetrafluoride sulfide	
F₄S₂	Disulfur tetrafluoride	II/15(**2**,228)
F₄Se	Selenium tetrafluoride	II/7(**2**,242)
F₄SeW	Tungsten(VI) tetrafluoride selenide	II/21(**2**,152)
	Tetrafluoro(seleno) tungsten(VI)	

F₄Si   **Silicon tetrafluoride** T_d
IR, MW double resonance   SiF₄

r_0	Å
Si–F	1.5540423(17)

Bond distance computed from the rotational constant given by Takami and Kuze.

Takami, M., Kuze, H.: J. Chem. Phys. **78** (1983) 2204.
ED: Bartell, L.S., Stanton, J.F.: J. Chem. Phys. **81** (1984) 37.   II/7(**2**,244), II/15(**2**,229)

F₄Si⁺	Silicon tetrafluoride(1+) ion	II/21(**2**,153)
F₄Th	Thorium tetrafluoride	II/7(**2**,247)
F₄Ti	Titanium tetrafluoride	II/15(**2**,230)
F₄U	Uranium tetrafluoride	II/7(**2**,248), II/15(**2**,231), II/23(**2**,106)
F₄Xe	Xenon tetrafluoride	II/7(**2**,253), II/21(**2**,154)
F₄Zr	Zirconium tetrafluoride	II/7(**2**,255), II/15(**2**,232)
F₅I	Iodine pentafluoride	II/7(**2**,190), II/15(**2**,233)
F₅IO	Iodine pentafluoride oxide	II/15(**2**,234)

	F₅NS	Tetrafluoro(fluoroimido) sulfur(VI) Sulfur tetrafluoride fluoroimide	II/15(**2**,235)
	F₅Nb	Niobium(V) fluoride Niobium pentafluoride	II/7(**2**,210), II/21(**2**,155)
	F₅ORe	Rhenium(VII) oxide pentafluoride	II/15(**2**,236)
	F₅P	Phosphorus pentafluoride II/7(**2**,231), II/15(**2**,237), II/21(2,156), II/23(**2**,107)	
	F₅Ta	Tantalum pentafluoride	II/7(**2**,245), II/21(**2**,157)
	F₅V	Vanadium pentafluoride	II/7(**2**,250), II/15(**2**,238)
	F₆Ir	Iridium hexafluoride	II/7(**2**,192)

F₆Mo **Molybdenum hexafluoride** O_h

ED

	r_g	Å ^{a)}
	Mo–F	1.820(3)

^{a)}) Estimated standard error.

Seip, H.M., Seip, R.: Acta Chem. Scand. **20** (1966) 2698.
Seip, H.M., in: "Selected Topics in Structure Chemistry", Eds.: P. Andersen, O. Bastiansen,
  S. Furberg: Oslo, Universitetsforlaget, 1967, p. 25.   II/7(**2**,197)

F₆NP₃	Tris(difluorophosphino)amine	II/15(**2**,239)
F₆N₃P₃	Phosphonitrilic fluoride trimer Phosphorus nitride difluoride trimer	II/7(**2**,205)
F₆Np	Neptunium hexafluoride	II/7(**2**,213)

F₆OS **Pentafluoro[oxofluorato(1–)] sulfur** C_s
ED, MW   Pentafluoro(hypofluorito)sulfur   SF₅OF

r_a	Å ^{a)}	θ_a	deg ^{a)}
S–F (mean)	1.555(3)	S–O–F	108.3(11)
Δ(SF) ^{b)}	0.0 ^{c)}	F(eq)–S–F(ax)	90.1(8)
S–O	1.671(7)	tilt ^{d)}	2.1(13)
O–F	1.408(9)		

The O–F bond is staggered with respect to the equatorial F atoms.
Local C_{4v} symmetry of the SF₅ group was assumed.
The nozzle was at room temperature.

^{a)}) Three times the estimated standard errors including the experimental scale error.
^{b)}) Δ(SF) = (SF(eq) – SF(ax)).
^{c)}) Assumed.
^{d)}) Tilt angle between the C₄ axis and the S–O bond.

Jaudas-Prezel, E., Christen, D., Oberhammer, H., Mallela, S.P., Shreeve, J.M.: J. Mol. Struct.
  248 (1991) 415.   II/23(**2**,108)

F₆OSi₂	Hexafluorodisiloxane	II/7(**2**,226)
F₆O₂V₂	Divanadium(V) hexafluoride dioxide	II/15(**2**,240)
F₆Os	Osmium hexafluoride	II/7(**2**,228)
F₆Pu	Plutonium hexafluoride	II/7(**2**,234)
F₆Re	Rhenium hexafluoride	II/7(**2**,235), II/21(**2**,158)

	F₆S	**Sulfur hexafluoride**	**O_h**
IR

r_0	Å
S–F	1.560722(7)

Bond distance computed from the rotational constant determined by Patterson et al.

Patterson, C.W., Herlemont, F., Azizi, M., Lemaire, J.: J. Mol. Spectrosc. **108** (1984) 31.

ED   (counting method)

r_a	Å [a])
S–F	1.5622(7)

The measurements were made at room temperature.

[a]) Twice the estimated standard error.

Miller, B.R., Fink, M.: J. Chem. Phys. **75** (1981) 5326.
See also: Miller, J.D., Fink, M.: J. Chem. Phys. **97** (1992) 8197.
Bartell, L.S., Doun, S.K.: J. Mol. Struct. **43** (1978) 245.
Bartell, L.S., Doun, S.K., Goates, S.R.: J. Chem. Phys. **70** (1979) 4585.
Bartell, L.S., Kacner, M.A.: J. Chem. Phys. **81** (1984) 280.
Bartell, L.S., Stanton, J.F.: J. Chem. Phys. **81** (1984) 3792.   II/15(**2**,241), II/23(**2**,109)

F₆Se	Selenium hexafluoride	II/15(**2**,242)

	F₆Si₂	**Hexafluorodisilane**	**D_{3d}**
		Disilicon hexafluoride	F₃Si–SiF₃

ED

r_α^0	Å [a])		θ_α^0	deg [a])
Si–Si	2.317(6)		F–Si–F	108.6(3)
Si–F	1.564(2)			

The barrier to internal rotation was found to be between 0.51(10) and 0.73(14) kcal mol⁻¹, depending on different assumptions of temperature drop due to gas expansion in the nozzle. The nozzle temperature was 15 °C; the temperature of the reservoir was –55 °C.

[a]) Three times the estimated standard errors.

Oberhammer, H.: J. Mol. Struct. **31** (1976) 237.
See also: Rankin, D.W.H., Robertson, A.: J. Mol. Struct. **27** (1975) 438.   II/15(**2**,243)

F₆Te	Tellurium hexafluoride	II/7(**2**,246), II/15(**2**,244)

	F₆U	**Uranium hexafluoride**	**O_h**
IR | | | UF₆

r_0	Å
U–F	1.9962(7)

Aldridge, J.P., Brock, E.G., Filip, H., Flicker, H., Fox, K., Galbraith, H.W., Holland, R.F., Kim, K.C., Krohn, B.J., Magnuson, D.W., Maier, W.B., McDowell, R.S., Patterson, C.W., Person, W.B., Smith, D.F., Werner, G.K.: J. Chem. Phys. **83** (1985) 34.

II/7(**2**,249), II/15(**2**,245)

F₆W	Tungsten hexafluoride	II/7(**2**,251)

F₆Xe	Xenon hexafluoride	II/7(**2**,254)
F₇I	Iodine heptafluoride	II/7(**2**,191)
F₇NS	Sulfur pentafluoride difluoroamide	II/7(**2**,208)
F₇NS₂	Pentafluorosulfurionitrilosulfur difluoride	II/21(**2**,159)
	(Difluorosulfurioimidato)pentafluorosulfur	
F₇Re	Rhenium heptafluoride	II/7(**2**,236)
F₈O₂Se₂	Di–μ–oxo–bis[tetrafluoroselenium(VI)]	II/15(**2**,246)
F₈O₂Te₂	Di–μ–oxo–bis[tetrafluorotellurium(VI)]	II/15(**2**,247)
F₉NOP₃Rh	Nitrosyltris(trifluorophosphine)rhodium(0)	II/15(**2**,248)
F₁₀HNS₂	Bis(pentafluoro–λ⁶–sulfanyl)amine	II/21(**2**,160)
F₁₀OS₂	Bis(pentaflourosulfur) oxide	II/15(**2**,249)
F₁₀OSe₂	Bis(pentafluoroselenium) oxide	II/15(**2**,250)
F₁₀OTe₂	Bis(pentafluorotellurium) oxide	II/15(**2**,251)
F₁₀O₂S₂	μ–Peroxo–bis[pentafluorosulfur(VI)]	II/23(**2**,110)
	Bis(pentafluorosulfur) peroxide	
F₁₀O₂Se₂	μ–Peroxo–bis[pentafluoroselenium(VI)]	II/23(**2**,111)
	Bis(pentafluoroselenium) peroxide	
F₁₀O₂Te₂	μ–Peroxo–bis[pentafluorotellurium(VI)]	II/23(**2**,112)
	Bis(pentafluorotellurium) peroxide	
F₁₀S₂	Disulfur decafluoride	II/21(**2**,161)
	Bis(pentafluoro–λ⁶–sulfane)	
F₁₁NS₂	Fluorobis(penta–λ⁶–sulfanyl)amine	II/21(**2**,162)
F₁₂NiP₄	Tetrakis(trifluorophosphine)nickel(0)	II/7(**2**,212)
F₁₂P₄Pt	Tetrakis(trifluorophosphine)platinum(0)	
		II/7(**2**,232), II/15(**2**,252)
F₁₅Mo₃	Trimolybdenum pentadecafluoride	II/15(**2**,253)
F₁₅Nb₃	cyclo–Tri–μ–fluoro–tris[tetrafluoroniobium(V)]	
	Triniobium pentadecafluoride	II/15(**2**,254), II/23(**2**,113)
F₁₅Sb₃	Triantimony pentadecafluoride	II/15(**2**,255)
F₁₅Ta₃	Tritantalum pentadecafluoride	II/15(**2**,256)
(FH)ₙ	Hydrogen fluoride polymer	II/7(**2**,174)
GaI₃	Gallium triiodide	II/7(**2**,256), II/15(**2**,257)
Ga₂H₆	Di–μ–hydrido–tetrahydridodigallium	II/23(**2**,114)
	Di–μ–hydrido–bis[dihydridogallium(III)]	
Ga₂O	Digallium nonoxide	II/7(**2**,257), II/15(**2**,258), II/21(**2**,163)
GdI₃	Gadolinium triiodide	II/15(**2**,259)
GeH₃I	Germyl iodide, Iodogermane	II/15(**2**,260)
GeH₃N₃	Germyl azide	II/7(**2**,260), II/15(**2**,261), II/21(**2**,164)
	Azidogermane	

	GeH₄	**Germane**	**T_d**
MW			GeH₄
	r_e Å		
Ge–H	1.5143(6)		

Third-order anharmonicity constant: $f_{rrr} = -12.3$ mdyn Å^{-2}.

Ohno, K., Matsuura, H., Endo, Y., Hirota, E.: J. Mol. Spectrosc. **118** (1986) 1.

II/7(**2**,258), II/15(**2**,262), II/21(**2**,165)

GeH$_6$Si	Germylsilane, Silylgermane	II/7(**2**,266)
GeH$_{12}$Si$_4$	Tetrasilylgermane	II/23(**2**,115)
GeI$_4$	Germanium tetraiodide	II/21(**2**,166)
Ge$_2$H$_6$	Digermane	II/7(**2**,259)
Ge$_2$H$_6$O	Digermyl oxide	II/7(**2**,262)
Ge$_2$H$_6$S	Digermyl sulfide	II/7(**2**,264)
Ge$_2$H$_6$Se	Digermyl selenide	II/7(**2**,265)
Ge$_3$H$_9$N	Trigermylamine	II/7(**2**,261)
Ge$_3$H$_9$P	Trigermylphosphine	II/7(**2**,263), II/15(**2**,263)
HISi	Monoiodosilylene, Silicon hydride iodide	II/7(**2**,267), II/15(**2**,264)

HKO **Potassium hydroxide** $C_{\infty v}$

MW KOH

r_0	Å [a]
K–O	2.212(5)
O–H	0.912(10)

r_e	Å [a]
K–O	2.196(3)
O–H	0.960(10)

[a]) Uncertainties were not estimated in the original paper.

Pearson, E.F., Winnewisser, B.P., Trueblood, M.B.: Z. Naturforsch. **31a** (1976) 1259.

II/7(**2**,269), II/15(**2**,265)

HNO **Nitrosyl hydride** C_s

UV Rotational and vibrational analysis

H–N=O

State	$\tilde{X}\ ^1A'$	$\tilde{A}\ ^1A''$	
Symmetry	C_s	C_s	
Energy [eV]	0	1.631	
Reference	[1]	[1] [2] [3]	
r_0 [Å] N–H	1.063(2) [a]	1.0360(5) [a]	[1]
r_e [Å] N–H		1.020(20)	[3]
r_0 [Å] N–O	1.212(1) [a]	1.241(1) [a]	[1]
r_e [Å] N–O		1.2389(50)	[3]
θ_0 [deg]	108.6(2) [a]	116.25(4)	[1]
θ_e [deg]		114.42(200)	[3]

[a]) Mean values of HNO and DNO; the uncertainties quoted spanning the differences due to zero point motion.

[1] Dalby, F.W.: Can. J. Phys. **36** (1958) 1336.
[2] Clement, M.J.Y., Ramsay, D.A.: Can. J. Phys. **39** (1961) 205.
[3] Bancroft, J.L., Hollas, J.M., Ramsay, D.A.: Can. J. Phys. **40** (1962) 322.
   Petersen, J.C.: J. Mol. Spectrosc. **110** (1985) 277, and earlier references therein.

II/7(**2**,276), II/15(**2**,266)

HNOS	Thionylimine	II/7(**2**,281), II/15(**2**,268)

	HNO_2	Nitrous acid	II/7(**2**,277)

HNO_3 Nitric acid C_s

MW

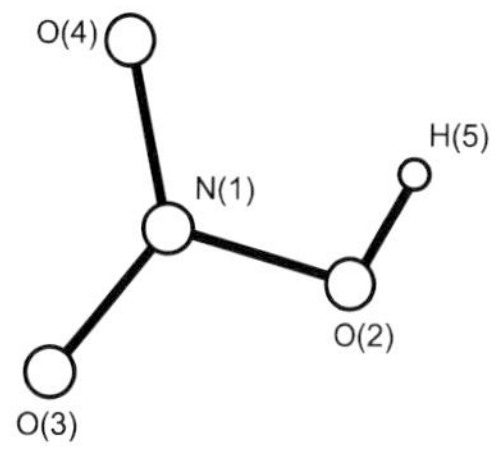

r_0	Å [a]	θ_0	deg [a]
N–O(2)	1.406(3)	O(2)–N=O(4)	113.9(3)
N=O(4)	1.203(3)	O(2)–N=O(3)	116.1(3)
N=O(3)	1.210(3)	N–O(2)–H	101.9(5)
O(2)–H	0.959(5)		

[a]) Uncertainties were not estimated in the original paper.

Ghosh, P.N., Blom, C.E., Bauder, A.: J. Mol. Spectrosc. **89** (1981) 159.

II/7(**2**,278), II/15(**2**,267)

HNO_4	Peroxonitric acid	II/21(**2**,167)
HN_2^+	Diazynium ion, Diazenylium	II/15(**2**,269), II/21(**2**,168)
HN_3	Hydrogen azide	II/7(**2**,271), II/15(**2**,270)

HNaO   Sodium hydroxide $C_{\infty v}$
NaOH

MW

r_0	Å [a]
Na–O	1.95(2)

[a]) Uncertainty was not estimated in the original paper.

Kuijpers, P., Törring, T., Dymanus, A.: Chem. Phys. **15** (1976) 457.

II/15(**2**,271)

HNeO	Hydroxyl – neon (1/1)	II/23(**2**,116)
HOP	Phosphoryl hydride	II/7(**2**,290), II/15(**2**,272)
HORb	Rubidium hydroxide	II/7(**2**,291)

HOS   Thionyl hydride radical C_s

MW, LMR,
UV   Rotational analysis

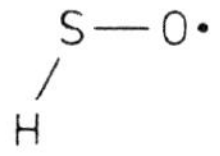

State	$\tilde{X}\,^2A''$	$\tilde{A}\,^2A'$
Symmetry	C_s	C_s
Energy [eV]	0	1.781 [2]
Reference	[1]	[1]

r_{000} [Å]	S–O	1.494(5)	
r_{003}			1.661(10)
r_{000}	S–H	1.389(5)	
r_{003}			1.342(8)
θ_{000} [deg]	H–S–O	106.6(5)	
θ_{003}			95.7(21)

[1] Obhashi, N., Kakimoto, M., Saito, S., Hirota, E.: J. Mol. Spectrosc. **84** (1980) 204.
[2] Schurath, U., Weber, M., Becker, K.H.: J. Chem. Phys. **67** (1977) 110.

MW: Endo, Y., Sito, S., Hirota, E.: J. Chem. Phys. **75** (1981) 4379.
LMR: Sears, T.J., McKellar, A.R.W.: Mol. Phys. **49** (1983) 25.

II/15(**2**,273)

	HOSi+	Hydroxosilicon(1+) ion	II/21(**2**,169), II/23(**2**,117)
	HOSr	Strontium hydroxide radical	II/15(**2**,274)

	HO$_2$	**Hydrogenperoxyl**	**C$_s$**
MW, IR, EPR, LMR		Perhydroxyl radical	H–O–O

r_0	Å [a])		θ_0	deg [a])
O–H	0.9774(30)		H–O–O	104.15(30)
O–O	1.3339(10)			

r_e	Å		θ_e	deg
O–H	0.9708(21)		H–O–O	104.30(39)
O–O	1.33051(94)			

[a]) Uncertainties were not estimated in the original paper.

r_0: Barnes, C.E., Brown, J.M. Radford, H.E.: J. Mol. Spectrosc. **84** (1980) 179.
r_e: Uehara, H., Kawaguchi, K., Hirota, E.: J. Chem. Phys. **83** (1985) 5479.

MW: Beers, Y., Howard, C.J.: J. Chem. Phys. **64** (1976) 1541.
IR: Lubic, K.G., Amano, T., Uehara, H., Kawaguchi, K., Hirota, E.: J. Chem. Phys. **81** (1984) 4826.

near-IR (electronic) Rotational analysis

State		$\tilde{X}\,^2A''$	$\tilde{A}\,^2A'$
Symmetry		C$_s$	C$_s$
Energy [eV]		0	0.8716 [a])
Reference		[1]	[2]
r_0 [Å]	O–O	1.3339(10)	1.393(6)
	O–H	0.9774(30)	0.966(2)
θ_0 [deg]	H–O–O	104.15	102.7(2)

[1] See above.
[2] Tucket, R.P., Freedman, P.A., Jones, W.J.: Mol. Phys. **37** (1979) 379, 403.   II/15(**2**,275)

	H$_2$I$_4$Si$_2$	1,1,2,2–Tetraiododisilane	II/23(**2**,118)
	H$_2$K$_2$O$_2$	Dipotassium dihydroxide	II/15(**2**,276), II/21(**2**,170)
		Potassium hydroxide dimer	
	H$_2$Kr	Hydrogen – krypton (1/1)	II/7(**2**,270)

	H$_2$N$^-$	**Amide ion**	**C$_{2v}$**
IR		Azanide ion	H$_2$N$^-$

r_0	Å		θ_0	deg
N–H	1.037(15)		H–N–H	102.0(33)

r_e	Å		θ_e	deg
N–H	1.028		H–N–H	101.9

In order to obtain the equilibrium structure it was necessary to make the assumption that the α_2 constants are the same as for NH$_2$.

Tack, L.M., Rosenbaum, N.H., Owrutsky, J.C., Saykally, R.J.: J. Chem. Phys. **85** (1986) 4222.
II/21(**2**,171)

| | **H₂N** | | **Aminyl radical** | | C_{2v} |
IR

r_e	Å		θ_e	deg
N–H	1.025(1)		H–N–H	102.9(3)

NH₂

Structure was calculated using the ground state constants and α_2 constants reported by [1], and the α_1 and α_3 constants given by [2].

[1] Burkholder, J.B., Howard, C.J., McKellar, A.R.W.: J. Mol. Spectrosc. **127** (1988) 415.
[2] Amano, T., Bernath, P.F., McKellar, A.R.W.: J. Mol. Spectrosc. **94** (1982) 100.

UV

State	$\tilde{X}\,^2B_1$	$\tilde{A}\,^2A_1\,(\Pi)$
Energy [eV]	0	1.271
Reference	[3]	[4]
r_0 [Å] N–H	1.024(5)	0.976 [a]
θ_0 [deg] H–N–H	103.36(50)	144(5) [a]

[a] From vibrational analysis fitting observed levels to a harmonic-cum-Gaussian double minimum potential combined with rotational analysis. Height of barrier = 777 ± 1000 cm⁻¹; lowest vibrational level is 164 cm⁻¹ below potential maximum, all others above.

[3] Dressler, K., Ramsay, D.A.: Phil. Trans. Roy. Soc. (London) Ser. **A 251** (1959) 553.
[4] Dixon, R.N.: Mol. Phys. **9** (1965) 357.

II/7(**2**,272), II/15(**2**,277), II/21(**2**,172)

| | **H₂N⁺** | | **Aminyl cation** | | C_{2v} or $D_{\infty h}$ |
| | | | Aminylium ion | | (quasilinear) |
IR | | | | | | NH₂⁺

r_{eff}	Å		θ_{eff}	deg
N–H	1.029		H–N–H	165

The molecule is quasilinear with a very small barrier to linearity. Because of this vibrational averaging plays a large role, and the bond distances cited must be considered an approximate zero order estimate. *Ab initio* calculations [Jensen *et al.*] obtained r_e(N–H) = 1.034 Å and θ_e(H–N–H) = 153.2°.

Okumura, M., Rehfuss, B.D., Dinelli, B.M., McLean, A.D.: Chem. Phys. Lett. **90** (1989) 5918.
See also: Jensen, P., Bunker, P.R., McLean, A.D.: Chem. Phys. Lett. **141** (1987) 53.

II/21(**2**,173)

| H₂NO | Dihydronitrosyl radical | II/23(**2**,119) |

	H$_2$N$_2$	**Diimide**			**C$_{2h}$**
		Diazene			
UV	Rotational and Franck-Condon analysis				

State	$\tilde{X}\,^1A_g$	$\tilde{A}\,^1B_g$	$\tilde{B}\,^1B_u$ (R, 3p)	$\tilde{X}\,^2A_g$, N$_2$H$_2^+$
Symmetry	C$_{2h}$	C$_{2h}$	C$_{2h}$	C$_{2h}$
Energy [eV]	0	2.55 [a]	7.182	10.02
Reference	[4]	[1] [2]	[2]	[2] [3]
r_0 [Å] N=N	1.252(2) [c]	1.340 [b]	1.167	1.222 [b]
N–H	1.028(5) [c]	1.022 [b]	1.028	1.045 [b]
θ_0 [deg] H–N=N	106.85(47) [c]	123 [b]	127.6	127 [b]

[a]) The transition $\tilde{A} - \tilde{X}$ is electronically forbidden and the absorption spectrum is based on a Herzberg-Teller vibronic false origin. $(T_{00} + v'_5) = 2.963$ eV; precise value of v_5 is not known.
[b]) Relative to the ground state from Franck-Condon factors guided by *ab initio* calculations.
[c]) Derived from the A_0, B_0 and C_0 values of N$_2$H$_2$ and N$_2$D$_2$.

[1] Back, R.A., Willis, C., Ramsay, D.A.: Can. J. Chem. **52** (1974) 1006; **56** (1978) 1575.
[2] Neudorfl, P.S., Back, R.A., Douglas, A.E.: Can. J. Chem. **59** (1981) 506.
[3] Frost, D.C., Lee S.T., McDowell, C.A., Greenwood, N.P.C.: J. Chem. Phys. **64** (1976) 4719.
[4] IR: Carlotti, M., Johns, J.W.C., Trombetti, A.: Can. J. Phys. **52** (1974) 340.

II/7(**2**,273), II/15(**2**,278)

H$_2$N$_2$O	Dinitrogen – water (1/1)	II/21(**2**,174)
H$_2$N$_2$O$_2$	Nitrylamide	II/7(**2**,279)
H$_2$N$_2$O$_2$	Dinitrogen oxide – water (1/1)	II/21(**2**,175)
	Nitrous oxide – water (1/1)	
H$_2$Ne	Dihydrogen – neon (1/1)	II/7(**2**,286)

	H$_2$O	**Water**		**C$_{2v}$**
IR				H$_2$O

r_e	Å		θ_e	deg
O–H	0.95843(1) [a]		H–O–H	104.43976 [b]

Observed experimental energy levels for six isotopomers were fitted using a Morse oscillator-rigid bender model with 19 parameters including r_e but θ_e was fixed at *ab initio* value.

[a]) One standard error.
[b]) Constrained at *ab initio* value given by [1].

Jensen, P.: J. Mol. Spectrosc. **133** (1989) 438.
[1] Bartlett, R., Cole, S.J., Purvis, G.D., Ermler, W.C., Hsieh, H.C., Shavitt, I.: J. Chem. Phys. **87** (1987) 6579.

(*continued*)

UV, MPI

State		$\tilde{C}\,^1B_1$	$\tilde{D}\,^1B_1$
Energy [eV]		9.996	10.469
Reference		[2]	[3]
r_0 [Å]	H–O	0.998(3) [a]	0.973
θ_0 [deg]	H–O–H	109.1(4) [a]	114

[a]) From constants corrected for *l*-uncoupling.

[2] Johns, J.W.C.: Can. J. Phys. **49** (1971) 944.
[3] Ashfold, M.N.R., Bayley, J.M., Dixon, R.N.: J. Chem. Phys. **79** (1983) 4080; Can. J. Phys. **62** (1984) 1806.
$\qquad$ II/7(**2**,287), II/15(**2**,279), II/23(**2**,120)

H₂O⁺   **Dihydrogenoxygen(1+) ion** C_{2v}

CEI [a]   Oxoniumyl
    Water cation

State		$\tilde{X}\,^2B_1$
r_e [Å]	O–H	1.00(4)
θ_e [deg]	H–O–H	108.4(5)

$\qquad\qquad\qquad\qquad$ O⁺
$\qquad\qquad\quad$ H $\qquad$ H

These values agree well with earlier spectroscopic values, see II/21(**2**,176A).

[a]) Coulomb Explosion Imaging.

Zajfman, D., Belkacem, A., Graber, T., Kanter, E.P., Mitchell, R.E., Naaman, R., Vager, Z., Zabransky, B.J.: J. Chem. Phys. **94** (1991) 2543.

IR

r_e	Å		θ_e	deg
O–H	0.9992(6)		H–O–H	109.30(10)

The equilibrium rotational constants were calculated using experimental α_i constants. The structure was deduced from the values of B_e and C_e only.

Huet, T.R., Pursell, C.J., Ho, W.C., Dinelli, B.M., Oka, T.: J. Chem. Phys. **97** (1992) 5977.
See also: Kauppi, E., Halonen, L.: Chem. Phys. Lett. **169** (1990) 393.
$\qquad$ II/7(**2**,288), II/15(**2**,280), II/21(**2**,176A), II/23(**2**,121)

D₂O⁺   **Deuterated water cation** C_{2v}
    Dideuteriumoxygen(1+) ion

UV $\qquad\qquad\qquad\qquad\qquad\qquad\qquad\qquad\qquad\qquad\qquad$ D₂O⁺

State		$\tilde{X}\,^2B_1$
r_0 [Å]	O–D	0.9987(2) [a]
θ_0 [deg]	D–O–D	110.17(2)

Rotational analysis of the bands of the $\tilde{A}\,^2A_1 - \tilde{X}\,^2B_1$ system. No structural data are given for the excited state since the low vibrational levels have not been observed.

[a]) Agrees closely with the values for H₂O⁺, viz. $r_0 = 0.9988$ Å, $\theta_0 = 110.46°$ [1].

Lew, H., Groleau, R.: Can. J. Phys. **65** (1987) 739.
[1] Lew, H.: Can. J. Phys. **54** (1976) 2028.
$\qquad\qquad\qquad\qquad\qquad\qquad\qquad\qquad\qquad\qquad\qquad$ II/21(**2**,176B)

H_2O_2 Hydrogen peroxide C_2

MW, IR

r_0	Å
O–H	0.950(5) [a]
O–O	1.475(4)

θ_0	deg
H–O–O	94.8(20)
[b]	120.0(5)
[c]	116.1(5)

H\
　O — O\
　　　　H

The r_0 (O–O) and ∡ H–O–O values were obtained from the IR data of [1] using the assumption that r_0 (O–H) = 0.950 Å. Using these parameters which are consistent with the MW spectrum, the values of the dihedral angles were obtained in [2].

[a]) Assumed.
[b]) Dihedral angle for the lower level of the ground-state doublet.
[c]) Dihedral angle for the upper level of the ground-state doublet.

[1] Redington, R.L., Olson, W.B., Cross, P.C.: J. Chem. Phys. **36** (1962) 1311.
[2] Oelfke, W.C., Gordy, W.: J. Chem. Phys. **51** (1969) 5336.
IR: Hunt, R.H., Leacock, R.A.: J. Chem. Phys. **45** (1966) 3141.
IR: Hunt, R.H., Leacock, R.A., Peters, C.W., Hecht, K.T.: J. Chem. Phys. **42** (1965) 1931.

II/7(**2**,289)

$H_2O_2Rb_2$	Dirubidium dihydroxide	II/23(**2**,122)
	Di–μ–hydroxo–dirubidium	
$H_2O_2S_2$	Hydrogen sulfide – sulfur dioxide (1/1)	II/21(**2**,177), II/23(**2**,123)

H_2O_3S Sulfur dioxide – water (1/1) C_s

MW

(weakly bound complex)　　　(effective symmetry class)

$SO_2 \cdot H_2O$

r_0	Å
R_{cm}	2.962(5)
O⋯S	2.824(16) [a]
H⋯O	3.33(15) [a]

θ_0	deg
θ_1	69.7(10)
θ_2	66.3(14)
ϕ_p [b]	134(2) [a]

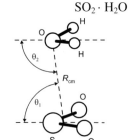

[a]) Derived parameter.
[b]) Angle between monomer planes, 90° is the parallel configuration and 0° has the H atoms directed toward SO_2.

Matsumura, K., Lovas, F.J., Suenram, R.D.: J. Chem. Phys. **91** (1989) 5887.　　II/21(**2**,178)

H_2O_4 Ozone – water (1/1) C_s

MW

(weakly bound complex)　　　(effective symmetry class)

$O_3 \cdot H_2O$

Species	$O_3 \cdot H_2O$, $O_3 \cdot H_2^{18}O$	$O_3 \cdot H_2^{16}O$, $O_3 \cdot H_2^{18}O$, $O_3 \cdot HDO$	$O_3 \cdot H_2^{16}O$, $O_3 \cdot H_2^{18}O$, $O_3 \cdot D_2O$
r_0	Å	Å	Å
R_{cm}	2.958(1)	2.957(2)	2.956(1)
O⋯O [a])[d]	3.208(1)	3.174(7)	3.172(5)
H⋯O [a])[e]	2.836(18)	3.145(45)	2.505(17)

(continued)

Species	$O_3 \cdot H_2O$, $O_3 \cdot H_2^{18}O$	$O_3 \cdot H_2^{16}O$, $O_3 \cdot H_2^{18}O$, $O_3 \cdot HDO$	$O_3 \cdot H_2^{16}O$, $O_3 \cdot H_2^{18}O$, $O_3 \cdot D_2O$
θ_0	deg	deg	deg
θ_1 [b])	120.7(2)	125.0(14)	118.4(7)
θ_2 [c])	54.2(12)	32.6(31)	78.7(14)
DO–a-axis [a])	74	84	50

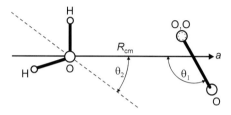

[a]) Derived properties.
[b]) Angle between the C_2 axis of ozone and R_{cm}.
[c]) Angle between R_{cm} and the C_2 axis of water.
[d]) O of H_2O to a terminal O of O_3.
[e]) H of H_2O (closer to O_3) to a terminal O of O_3.

Gillies, J.Z., Gillies, C.W., Suenram, R.D., Lovas, F.J., Schmidt, T., Cremer, D.: J. Mol. Spectrosc. **146** (1991) 493.

II/23(**2**,124)

H_2O_4S	Sulfuric acid	II/15(**2**,281)

H_2P λ^5–**Phosphane** C_{2v}
Phosphino radical
Dihydrogen phosphorus radical

UV Rotational and vibrational analysis

State	$\tilde{X}\ ^2B_1$	$\tilde{A}\ ^2A_1$
Energy [eV]	0	2.266
r_0 [Å] P–H	1.429	1.401
θ_0 [deg] H–P–H	91.67	123.07

Dixon, R.N., Duxbury, G., Ramsay, D.A.: Proc. Roy. Soc. (London) Ser. A **296** (1967) 137.

II/7(**2**,294), II/15(**2**,282)

H_2S   **Hydrogen sulfide** C_{2v}
MW H_2S

r_s	Å [a])	θ_s	deg [a])
H–S	1.3376(50)	H–S–H	91.6(5)
D–S	1.3362(50)	D–S–D	92.2(5)

r_{av}	Å [a])	θ_{av}	deg [a])
H–S	1.3518(30)	H–S–H	92.13(30)
D–S	1.3474(30)	D–S–D	92.11(30)

r_e	Å [a])	θ_e	deg [a])
H–S	1.3356(30)	H–S–H	92.11(30)
D–S	1.3362(30)	D–S–D	92.06(30)

[a]) Uncertainties were not estimated in the original paper.

Cook, R.L., DeLucia, F.C., Helminger, P.: J. Mol. Struct. **28** (1975) 237.

(*continued*)

H₂S (continued)

UV Rotational analysis

State		$^1B_1$ [a]
r [Å]	S–H	1.35(15)
θ [deg]	H–S–H	93(1)

[a] Rydberg state at 8.92 eV (71895 cm^{-1}) assigned to the configuration ··(3p, b$_1$)(5s, a$_1$).

Gallo, A.R., Innes, K.K.: J. Mol. Spectrosc. **54** (1974) 472.   II/7(**2**,299), II/15(**2**,283)

H₂S⁺	**Hydrogen sulfide cation**	**C$_{2v}$**
	Dihydrogensulfur(1+) ion	H₂S⁺

UV Rotational and vibrational analysis

State		$\tilde{X}\,^2B_1$	$\tilde{A}\,^2A_1$ [a]
Energy [eV]		0	2.296
r_0 [Å]	S–H	1.358	1.366
θ_0 [deg]	H–S–H	91.97	127

[a] Barrier to linearity estimated from vibrational spacings to be about 4400 cm^{-1}.

Duxbury, G., Horani, M., Rostas, J.: Proc. Roy. Soc. (London) Ser. **A 331** (1972) 109.

II/7(**2**,300), II/15(**2**,284)

H₂S₂	Dihydrogendisulfide, Disulfane	II/7(**2**,301)

H₂Se	**Hydrogen selenide**	**C$_{2v}$**

UV, MW Rotational and vibrational analysis

H–Se–H

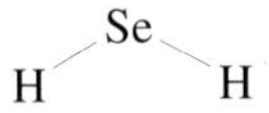

State		$\tilde{X}\,^1A_1$	$^1B_1$ (R, 5sa$_1$)
Energy [eV]		0	8.362
Reference		[1]	[2]
r_0 [Å]	Se–H	1.460	1.460
θ_0 [deg]	H–Se–H	90.9	91.4

[1] Helminger, P., DeLucia, F.C.: J. Mol. Spectrosc. **58** (1975) 375.
[2] Hollas, J.M., Lemanczyk, Z.R.: J. Mol. Spectrosc. **66** (1977) 79.

II/7(**2**,305), II/15(**2**,285)

H₂Si	**Silicon dihydride**	**C$_{2v}$**
	λ^2-Silane, Silylene	

IR

SiH₂

r_0	Å		θ_0	deg
Si–H	1.525(6)		H–Si–H	91.8(10)
r_e	Å		θ_e	deg
Si–H	1.514 [a]		H–Si–H	92.1 [a]

(continued)

a) The equilibrium parameters are estimated by scaling α constants from H_2S.

Yamada, C., Kanamori, H., Hirota, E., Nishiwaki, N., Itabashi, N., Kato, K., Goto, T.: J. Chem. Phys. **91** (1989) 4582. II/7(**2**,307), II/15(**2**,286), II/21(**2**, 179)

	H_2Si_2	**Di–µ–hydrogendisilane**	C_{2v}
MW		Di–µ–hydrido–disilicon	

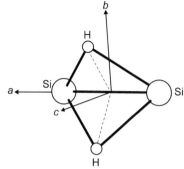

r_0	Å [a]		θ_0	deg [a]
Si–Si	2.2079(30)		η [b]	103.18(30)
Si–H	1.6839(30)			

[a]) Uncertainties were not estimated in the original paper.
[b]) Dihedral angle between two Si_2H planes.

Bogey, M., Bolvin, H., Demuynck, C., Destombes, J.L.: Phys. Rev. Lett. **66** (1991) 413.
II/23(**2**,125)

H_2Te	Hydrogen telluride	II/7(**2**,311)
H_2Xe	Dihydrogen – xenon (1/1)	II/7(**2**,312)

	H_3	**Trihydrogen**				D_{3h}
UV	Rotational and vibrational analysis					

		$\tilde{X}$	$\tilde{A}$	$\tilde{B}$	$\tilde{C}$	$\tilde{D}$
State [a]		(2p) $1^2E'$	(2s) $1^2A'_1$	(2p) $1^2A''_2$	(3p)$2^2E'$	(3s)$2^2A'_1$
Symmetry		—[b]	D_{3h}	D_{3h}	D_{3h} [c]	D_{3h}
Energy [eV]		[−5.641] [e]	0	0.123	1.731	2.182
Reference		—	[1, 2]	[1, 3]	[2, 3]	[1, 3]
r_0 [Å]	H_3 [f]		0.845(1)	0.8665(3)	0.889(2)	0.8702(2)
	D_3 [f]		0.852(1)	0.87014(2)	0.8895(1)	0.8723(4)

		$\tilde{C}'$	$\tilde{E}$	$\tilde{E}'$	$\tilde{E}''$	···	H_3^+ $\tilde{X}$
State		(3p) $2^2A''_2$	(3d) $3^2E'$	(3d)$1^2E''$	(3d)$^2A''_1$		$^1A'_1$
Symmetry				D_{3h} [d]			D_{3h}
Energy [eV]		2.206	2.236	2.282	2.295	···	[3.717]
Reference		[4]	[4]	[4]	[4]		[7]
r_0 [Å]	H_3 [f]	0.8396(2)		0.882(3)		···	0.87631(2) [g]
	D_3 [f]	0.8581(12)		0.8778(5)			
r_e [Å]						···	[0.8770] [h]

(*continued*)

H_3 (*continued*)

a) Labelling of states in conventional spectroscopic notation adopted here. Stable states $\tilde{A}, \tilde{B} \cdots$ are all Rydberg states with one electron loosely bound to the ionic core, H_3^+ $\tilde{X}^1A_1'$. Labels (nl) refer to the united-atom electron configuration; numbers that follow them enumerate states in ascending order of energies within each electronic symmetric-species [7].
b) Repulsive, unstable with respect to $H_2 + H$.
c) Small asymmetric distortion of the adiabatic potential minima through a weak Jahn-Teller effect; the stabilization-energy away from trigonal symmetry (*ca.* 0.013 eV) is considerably less than the zero-point energy of the distorting vibration [4].
d) The electronic energy-separation of these three states are small compared with rotational energies of even a few quanta. Rotation mixes the states so that their analysis proceeds as that of a single 5-fold degenerate electronic state.
e) Energy of zero-point of $H_2 + H$ relative to the zero-point of $\tilde{A}$, calculated from the *ab initio* theoretical dissociation energy and zero-point energy of H_3^+ ($D_0^0 = 4.488$ eV [6] and the *ab initio* ionization energy of $\tilde{A}$, assuming its zero-point energy to be the same as that of H_3^+ (IP_0^0) = 3.737 eV [7]).
f) From rotational constants B_0. Values of these constants obtained from several bands often differ by more than the analytical uncertainties estimated in any one of them, indicating residual correlated errors in what are complicated multiparametric analyses. Uncertainties quoted here encompass such discrepancies in those cases in which two references are cited, and may therefore greatly exceed individual estimates (as quoted, for instance, in [5]). In any case, the differences between r_0 for the isotopes H_3 and D_3 are greater still, indicating that the experimental uncertainties are smaller than the uncertainties inherent in the r_0 approximation. Precisions may differ considerably between H_3 and D_3 because of differences in line-broadening due to predissociations.
g) Ref. [8].
h) *Ab initio* calculation, ref. [6].

[1] Dabrowski, I., Herzberg, G.: Can. J. Phys. **58** (1980) 1238.
[2] Herzberg, G., Watson, J.K.G.: Can. J. Phys. **58** (1980) 1250.
[3] Herzberg, G., Lew, H., Sloan, J.J., Watzon, J.K.G.: Can. J. Phys. **59** (1981) 428.
[4] Herzberg, G., Hougen, J.T., Watson, J.K.G.: Can. J. Phys. **60** (1982) 1261.
[5] Herzberg, G.: J. Mol. Struct. **113** (1984) 1.
[6] Carney, G.D., Porter, R.N.: J. Chem. Phys. **60** (1974) 4251.
[7] King, H.F., Morokuma, K.: J. Chem. Phys. **71** (1979) 3213.
[8] Watson, J.K.G., Foster, S.C., McKellar, A.R.W., Bernath, P., Amano, T., Pan, F.S., Crofton, M.W., Altman, R.S., Oka, T.: Can. J. Phys. **62** (1984) 1875.   II/15(**2**,287)

H_3^+	**Trihydrogen cation**	D_{3h}
IR		H_3^+

	r_e	Å
H–H	0.877 a)	

a) Bond distance given is the average obtained from the D_2H^+ and H_2D^+ isotopic species.

Foster, S.C., McKellar, A.R.W., Watson, J.K.G.: J. Chem. Phys. **85** (1986) 664.
Foster, S.C., McKellar, A.R.W., Peterkin, I.R., Watson, J.K.G., Pan, F.S., Crofton, M.W., Altman, R.S., Oka, T.: J. Chem. Phys. **84** (1986) 91.   II/15(**2**,288), II/21(**2**,180)

	H₃ISi	**Iodosilane**			**C₃ᵥ**
IR, MW		Silyl iodide			SiH₃I

	r_0	Å		θ_0	deg
	Si–H	1.4741(14) [a]		H–Si–I	108.16(17)
	Si–I	2.43835(59)			

Improved structure results from obtaining an A_0 rotational constant for the SiHD₂I isotopic species. All available microwave data were included in the fitting.

[a] The assumption was made that δr_0 (SiH–SiD) = 0.0023 Å.

Duncan, J.L., Harvie, J.L., McKean, D.C., Cradock, S.: J. Mol. Struct. **145** (1986) 225.

II/7(**2**,268), II/21(**2**,181)

H₃ISn	Iodostannane, Stannyl iodide	II/23(**2**,126)

	H₃N	**Ammonia**	**C₃ᵥ**
MW			NH₃

	r_s [Å] [a]	θ_s [deg] [a]
Isotopic species [b]	N–H	H–N–H
¹⁴NH₃ : ¹⁵NH₃, ¹⁴ND₃	1.0138(10)	107.23(20)
¹⁵NH₃ : ¹⁴NH₃, ¹⁵ND₃	1.0138(10)	107.23(20)
¹⁴ND₃ : ¹⁵ND₃, ¹⁴NH₃	1.0136(10)	107.07(20)
¹⁵ND₃ : ¹⁴ND₃, ¹⁵NH₃	1.0137(10)	107.07(20)
¹⁴NH₃ : ¹⁵NH₃, ¹⁴NT₃	1.0132(10)	107.22(20)
¹⁴ND₃ : ¹⁵ND₃, ¹⁴NT₃	1.0128(10)	107.03(20)

[a] Uncertainties were not estimated in the original paper.

[b] The first species is the parent, and the differences in moments of inertia between the second and first species and also between the third and first species are used to calculate the r_s parameters.

Helminger, P., DeLucia, F.C., Gordy, W., Morgan, H.W., Staats, P.A.: Phys. Rev. **A 9** (1974) 12.

See also: Cohen, E.A., Pikett, H.M.: J. Mol. Spectrosc. **93** (1982) 83 (r_0 structures for NH₂D and NHD₂).

UV   Rotational and vibrational analysis

State	$\tilde{A}$ ¹A″₂	$\tilde{B}$ ¹E″	$\tilde{C}'$ ¹A₁′ [a]
Symmetry	D₃ₕ	D₃ₕ	D₃ₕ
Energy [eV]	5.720	7.343	7.919
References	[1] [5]	[2]	[3]
r_0 [Å] N–H	1.08	1.027	1.027
θ_0 [deg] H–N–H	120	120	120

[a] Some partial structural information is available on further eight electronic states [4], members of Rydberg series having essentially the planar geometry of the ionic core, NH₃⁺.

[1] Douglas, A.E.: Discuss. Faraday Soc. **35** (1963) 158.
[2] Douglas, A.E., Hollas, J.M.: Can. J. Phys. **39** (1961) 479.

(*continued*)

H$_3$N (*continued*)

[3] Nieman, G.C., Colson, S.D.: J. Chem. Phys. **71** (1979) 571.
[4] Glownia, J.H., Riley, S.J., Colson, S.D., Nieman, G.C.: J. Chem. Phys. **73** (1980) 4296.
[5] Ziegler, L.D.: J. Chem. Phys. **82** (1985) 664.   II/7(**2**,274), II/15(**2**,289)

IR | **H$_3$N$^+$** | **Ammoniumyl ion** | **D$_{3h}$** |
Ammonia cation | NH$_3^+$

r_0 [a])	Å
N–H	1.027

r_z	Å
N–H	1.0304

r_e [b])	Å
N–H	1.014

[a]) The r_0 distance is the average obtained from B_0 and C_0.
[b]) The r_e distance was obtained using several assumptions.

Bawendi, M.G., Rehfuss, B.D., Dinelli, B.M., Okumura, M., Oka, T.: J. Chem. Phys. **90** (1989) 5910.   II/21(**2**,182)

H$_3$NO	Hydroxylamine	II/7(**2**,280)
H$_3$NOSSi	Silylsulfinylamine	II/15(**2**,290)
H$_3$NS	Thiohydroxylamine	II/15(**2**,291)
H$_3$N$_3$O	Ammonia – nitrous oxide (1/1)	II/21(**2**,183)
H$_3$N$_3$Si	Azidosilane, Silyl azide	II/7(**2**,282)

IR | **H$_3$O$^+$** | **Oxonium cation** | **C$_{3v}$**

r_e	Å		θ_e	deg
O–H	0.9758		H–O–H	111.3

Sears, T.J., Bunker, P.R., Davies, P.B., Johnson, S.A., Špirko, V.: J. Chem. Phys. **83** (1985) 2676.   II/15(**2**,292)

MW | **H$_3$P** | **Phosphine** | **C$_{3v}$**
 | | | PH$_3$

r_e	Å		θ_e	deg
P–H	1.413(2)		H–P–H	93.45(9)

r_z	Å		θ_z	deg
P–H	1.42774(9)		H–P–H	93.286(12)
P–D	1.42373(13)		D–P–D	93.332(18)

McRae, G.A., Gerry, M.C.L., Cohen, E.A.: J. Mol. Spectrosc. **116** (1986) 58.
    II/7(**2**,295), II/15(**2**,293), II/21(**2**,184)

	H₃S⁺	**Sulfonium(1+) ion**		**C₃ᵥ**
IR				H₃S⁺

r_0	Å	θ_0	deg
S–H	1.3585	H–S–H	94.113

Nakanaga, T., Amano, T.: J. Mol. Spectrosc. **131** (1989) 201.   II/21(**2**,185)

H₃Sb	Stibine	II/7(**2**,303)
H₃Si⁺	Silyl(1+) ion	II/23(**2**,127)
H₄IP	Phosphine – hydrogen iodide (1/1)	II/23(**2**,128)
H₄I₂Si₂	1,2–Diiododisilane	II/23(**2**,129)

	H₄N	**Ammonium radical**	**T_d**
UV	Rotational analysis		

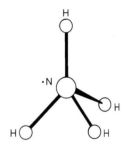

State	$\tilde{A}\ ^2A_1\ (3s)$ [a]	$\tilde{B}\ ^2F_2\ (3p)$ [a]
Symmetry	T_d	T_d
Energy [eV]	e [b]	$e + 1.838$
Reference	[1] [2]	[1] [2]
$r_{0,e}$ [Å] N–H	(1.0160) [c]	1.0028 [d]

[a] Stable states of NH₄ are well represented in terms of an ionic core NH₄⁺ with one Rydberg electron whose description in the united-atom limit (n, l) is indicated in brackets.

[b] Energy relative to NH₃ + H, the dissociation-products of the unstable ground-state of NH₄. Energy relative to the ground-state of NH₄⁺ calculated *ab initio* [3] to be ≈ -3.989 eV.

[c] Value calculated *ab initio* [3]. The known spectra are compatible with this value, but are of such a form that they do not allow it to be derived independently.

[d] From $r\ (3s^2A_1,\ ab\ initio)$ and $(r' - r'')$ determined from the spectrum.

[1] Watson, J.K.G.: J. Mol. Spectrosc. **107** (1984) 124.
[2] Alberti, F., Huber, K.P., Watson, J.K.G.: J. Mol. Spectrosc. **107** (1984) 133.
[3] Havriliak, S., King, H.F.: J. Am. Chem. Soc. **105** (1983) 4.   II/15(**2**,294)

	H₄N⁺	**Ammonium ion**	**T_d**
IR			NH₄⁺

r_0	Å
N–H	1.02874(2)

r_e	Å
N–H	1.021(2) [a]

[a] The equilibrium bond distance has been estimated.

Crofton, M.W., Oka, T.: J. Chem. Phys. **86** (1987) 5983.   II/21(**2**,186)

H_4N_2 Hydrazine C_2 [a]

MW

r_0	Å	θ_0	deg
N–H	1.008(8)	H–N–H	113.3(30)
N–N	1.447(5)	H–N–N	109.2(8)
		τ [b]	88.9(15)

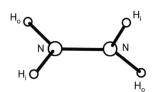

[a]) Effective symmetry is higher.
[b]) Dihedral angle.

Tsunekawa, S.: J. Phys. Soc. Jpn. **41** (1976) 2077.

ED, MW

r_{av}	Å [a]	θ_{av}	deg [a]
N–H [b]	1.015(2)	N–N–H_o	106(2)
N–D	1.447(2)	N–N–H_i	112(2)
		τ [c]	91(2)
		H–N–H	106.6 [d]

The nozzle temperature and the temperature of the reservoir were 60 °C.

[a]) Estimated limits of error.
[b]) Average of N–H_i and N–H_o. It was assumed that (N–H_i)–(N–H_o) = 0.003 Å. See figure for definition of H_o and H_i.
[c]) The dihedral angle between the planes containing the N–N bond and the bisectors of the H–N–H angles.
[d]) Assumed.

Kohata, K., Fukuyama, T., Kuchitsu, K.: J. Phys. Chem. **86** (1982) 602.

II/7(**2**,275), II/15(**2**,295)

H_4O_2 Water dimer C_s
(large-amplitude motion)

MW

r_0	Å [a]	θ_0	deg [a]
O···O	2.976 (30)	θ (1)	–51 (10)
		θ (2)	57 (10)
		χ_a [b]	6 (20)

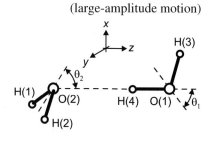

[a]) Uncertainties were not estimated in the original paper.
[b]) The monomer orientation is expressed by the Eulerian angles, ϕ, θ, χ (ϕ_a = 0 arbitrarily); χ_a represents the twisting of the acceptor H_2O.

Odutola, J.A., Dyke, T.R.: J. Chem. Phys. **72** (1980) 5062.

II/15(**2**,296)

H_4P_2 Diphosphane

II/7(**2**,296)

H_4Si Silane T_d

MW SiH_4

r_e	Å
Si–H	1.4707(6)

(continued)

Third-order anharmonicity constant: $f_{rrr} = -13.0$ mdyn Å^{-2}.

Ohno, K., Matsuura, H., Endo, Y., Hirota, E.: J. Mol. Spectrosc. **118** (1986) 1.

II/7(**2**,308), II/15(**2**,297), II/21(**2**,187)

MW | **H₄Sn** | **Stannane** | **T_d** |
   | | | SnH₄ |

r_e	Å
Sn–H	1.6909(24)

Third-order anharmonicity constant: $f_{rrr} = -9.06$ mdyn Å^{-2}.

Ohno, K., Matsuura, H., Endo, Y., Hirota, E.: J. Mol. Spectrosc. **118** (1986) 1.

IR

r_0	Å
Sn–H	1.7029344(11)

Bond distance calculated from the B_0 rotational of ^{116}SnH₄.

Brunet, F., Pierre, G., Bürger, H.: J. Mol. Spectrosc. **140** (1990) 237.

II/7(**2**,310), II/21(**2**,188)

| H₅ISi₂ | Iododisilane | II/15(**2**,198) |

MW | **H₅NO** | **Ammonia – water (1/1)** | **C_s** |
   | | (weakly bound comlex) | (effective symmetry class) |
   | | | NH₃ · H₂O |

r_0	Å ᵃ⁾	θ_0	deg ᵃ⁾
N···H–O	2.983(3)	θ_N (NH₃ · H₂O)	23.1(2)
N···D–O	2.979(3)	θ_N (ND₃ · D₂O)	20.4(5)
		θ_O	56(2)
		χ_O	≈ 0

$v_s = 168$ cm^{-1}
$k_s = 14.5$ N m^{-1}

ᵃ⁾ Uncertainties were not all estimated in the original paper.

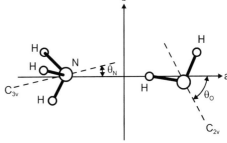

Herbine, P., Dyke, T.R.: J. Chem. Phys. **83** (1985) 3768.

II/21(**2**,189)

	H₅NS	**Ammonia – hydrogen sulfide (1/1)**	**C$_s$**
MW		(weakly bound complex)	(effective symmetry class)
			NH$_3$ · H$_2$S

r_0	Å		θ_0	deg
N···S	3.6393(35)		θ_N	24.63(18)
			θ_S	40.5(15)
			χ_S	0(38)

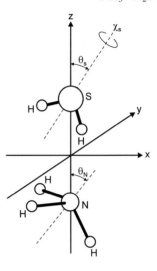

The figure shows the coordinate system for NH$_3$ · H$_2$S. The Euler angle rotations are defined from a reference configuration whose origin is at the sulfur atom. The H$_2$S is in the xz plane with the hydrogens pointing in the negative z direction. For NH$_3$, the nitrogen is at the origin. One N–H bond is in the xz plane with positive x and negative z coordinates. The symmetry axes of both molecules are initially along the z axis.

Herbine, P., Hu, T.A., Johnson, G., Dyke, T.R.: J. Chem. Phys. **93** (1990) 5485.   II/23(**2**,130)

H₅PSi	Silylphosphane, Phosphinosilane	II/7(**2**,297), II/15(**2**,299)

	H₆N₂	**Ammonia dimer**	**C$_1$**
MW		(weakly bound complex)	(effective symmetry class)

Isotopic species	R_{cm} [a]) Å	θ_1 deg	θ_2 deg
¹⁴NH₃···¹⁴NH₃	3.337	48.6(1)	64.5(1)
¹⁴NH₃···¹⁵NH₃	3.336	48.7(1)	
¹⁵NH₃···¹⁴NH₃	3.335		64.3(1)
¹⁴ND₃···¹⁴ND₃	3.325	49.6(2)	62.6(2)
¹⁴ND₃···¹⁴ND₂H	3.325 [b])	45.3(2)	65.8(2)

[a]) Experimental uncertainties in R_{cm} are less than 1 in the last decimal place. Systematic uncertainties in the reported values for R_{cm} due to the neglect of internal rotation effects could be on the order of 0.01 Å.

[b]) The value of R_{cm} for this isotopomer in the rigid motor approximation depends on the value of χ of the ND$_2$H subunit and on whether that subunit is described by θ_1 or by θ_2. The possible values of R_{cm} range between 3.322 and 3.327 Å.

Nelson, D.D., Klemperer, W., Fraser, G.T., Lovas, F.J., Suenram, R.D.: J. Chem. Phys. **87** (1987) 6364.   II/21(**2**,190)

H₆OSi₂	Disilyl oxide, Disiloxane	II/7(**2**,292)

H₆O₃ (D₆O₃) **Water trimer** C_1
IR
(D₂O)₃

r_0	Å		θ_0	deg
O(1)...O(2)	2.97(3)		D(1)–O(1)...O(2)	28(2) [a]
O(2)...O(3)	2.94(3)		D(3)–O(2)...O(3)	30(2) [a]
O(3)...O(1)	2.97(3)		D(5)–O(3)...O(1)	27(2) [a]

This structure is compatible with the experimental ground state rotational constant. It is assumed that the hydrogen-bonded D atoms lie in the plane of the three O atoms and that the out-of-plane D atoms lie perpendicular to this plane. The D on O(1) is below the plane whereas those on O(2) and O(3) point above the plane. Such a structure is supported by *ab initio* calculations.

[a] Uncertainties were not given in the original paper.

Pugliano, N., Saykally, R.J.: Science **257** (1992) 1937.
See also: Honegger, E., Leutwyler, S.: J. Chem. Phys. **88** (1988) 2582.   II/23(**2**,131)

| H₆SSi₂ | Disilyl sulfide | II/7(**2**,302), II/15(**2**,300) |
| H₆SeSi₂ | Disilyl selenide | II/7(**2**,306) |

H₆Si₂   **Disilane** D_{3d}
IR
H_3SiSiH_3

r_0	Å		θ_0	deg
Si–H	1.4874(17) [a]		H–Si–Si	110.66(16)
Si–Si	2.3317(15)			

Improved structure results from obtaining an A_0 rotational constant for the Si₂HD₅ isotopic species.

[a] The assumption was made that δr_0(SiH–SiD) = 0.0023 Å.

Duncan, J.L., Harvie, J.L., McKean, D.C., Cradock, S.: J. Mol. Struct. **145** (1986) 225.
II/7(**2**,309), II/21(**2**,191)

H₇NSi₂	Disilazane	II/7(**2**,283)
H₈O₄Si₄	Tetrameric silicon dihydride oxide Cyclotetrasiloxane	II/7(**2**,293)
H₉NSi₃	Trisilylamine	II/7(**2**,284)
H₉PSi₃	Trisilylphosphane	II/7(**2**,298), II/15(**2**,301)
H₉SbSi₃	Trisilylstibine	II/7(**2**,304), II/15(**2**,302)
H₁₀Si₅	Cyclopentasilane	II/15(**2**,303)
H₁₂N₂Si₄	Tetrasilylhydrazine	II/7(**2**,285)
H₁₂Si₆	Cyclohexasilane	II/15(**2**,304)
HfI₄	Hafnium tetraiodide	II/7(**2**,313), II/15(**2**,305)
HgI₂	Mercury diiodide	II/15(**2**,306)
I₂K₂	Dipotassium diiodide	II/21(**2**,192)
I₂Li₂	Dilithium diiodide	II/7(**2**,315)
I₂Na₂	Disodium diiodide	II/7(**2**,316), II/21(**2**,193)
I₂Pb	Lead diiodide	II/15(**2**,307), II/21(**2**,194), II/23(**2**,132)

I$_2$Rb$_2$	Dirubidium diiodide	II/21(**2**,195)
I$_2$Sm	Samarium diiodide	II/15(**2**,308)
I$_2$Sn	Tin diiodide	II/15(**2**,309), II/21(**2**,196), II/23(**2**,133)
I$_2$Sr	Strontium diiodide	II/15(**2**,310)
I$_2$Zn	Zinc diiodide	II/21(**2**,197)
I$_3$In	Indium triiodide	II/21(**2**,198)
I$_3$La	Lanthanum triiodide	II/7(**2**,314)
I$_3$Lu	Lutetium triiodide	II/15(**2**,311)
I$_3$NbO	Triiodo(oxo)niobium(V) Niobium(V) triiodide oxide	II/23(**2**,134)
I$_3$Nd	Neodymium triiodide	II/7(**2**,317), II/15(**2**,312)
I$_3$Pr	Praseodymium triiodide	II/15(**2**,313)
I$_3$Sb	Antimony triiodide	II/7(**2**,318)
I$_3$Ti	Titanium triiodide	II/23(**2**,135)
I$_3$U	Uranium triiodide	II/23(**2**,136)
I$_4$In$_2$	Diindium tetraiodide	II/21(**2**,199)
I$_4$Nb	Niobium tetraiodide	II/23(**2**,137)
I$_4$Ti	Titanium tetraiodide	II/15(**2**,314)
I$_4$Zr	Zirconium tetraiodide	II/7(**2**,319), II/15(**2**,315)
I$_6$In$_2$	Diindium hexaiodide Di−μ−iodo−bis[diiodoindium(III)]	II/21(**2**,200)
In$_2$MoO$_4$	Diindium molybdenum tetroxide Indium(I) molybdate	II/7(**2**,320)
In$_2$O	Indium(I) oxide Diindium monooxide	II/7(**2**,321), II/15(**2**,316), II/21(**2**,201)
In$_2$Se	Indium(I) selenide	II/15(**2**,317)
In$_2$Te	Indium(I) telluride	II/15(**2**,318)
KNO$_3$	Potassium nitrate	II/23(**2**,138)
KO$_3$P	Potassium metaphosphate	II/21(**2**,202)
KO$_4$Re	Potassium perrhenate	II/7(**2**,322)
K$_2$O$_4$S	Potassium sulfate	II/7(**2**,323)
LiNO$_3$	Lithium nitrate	II/7(**2**,324)
Li$_2$O	Lithium oxide	II/7(**2**,325)
MoO$_4$Rb$_2$	Rubidium molybdate	II/15(**2**,319)
MoO$_4$Tl$_2$	Thallium molybdate	II/7(**2**,327)
Mo$_3$O$_9$	Trimolybdenum nonaoxide	II/7(**2**,326)
NNaO$_3$	Sodium nitrate	II/7(**2**,329)
NO$_2^-$	Nitrite ion	II/21(**2**,203)

NO$_2$ — **Nitrogen dioxide** — **C$_{2v}$**

IR, MW

r_e	Å	θ_e	deg
N=O	1.19455(3)	O=N=O	133.851(2)

Morino, Y., Tanimoto, M.: Can. J. Phys. **62** (1984) 1315.

(*continued*)

UV, IR Vibrational and rotational analysis

State	$\tilde{X}\ ^2A_1$	$\tilde{A}\ (1)\ ^2B_2$ [a]	$(1)\ ^2B_1\ (^2\Pi_u)$ [b]	$\tilde{B}\ (2)\ ^2B_2$	$\tilde{E}\ ^2\Sigma^+\ (R)$
Symmetry	C_{2v}	C_{2v}	$D_{\infty h}$	C_{2v}	$D_{\infty h}$
Energy [eV]	0	1.21(9) [c]	1.828 [e]	4.975	7.23
Reference	[1]	[1] [2]	[6] [7]	[8] [9]	[10]
r_0 [Å] N=O	1.196376(6)	1.26 [d]	1.23 [e]	1.314	1.13 [f]
θ_0 [deg] O=N=O	134.250(2)	102	180	120.9	180

[a]) The label $\tilde{A}$ was used by Herzberg (1966) for what was thought to be a single excited electronic state reached in the visible absorption spectrum of NO_2. It was transferred to the lower of the two separate electronic states subsequently identified in this spectrum [1, 2], responsible for most of the intensity.

[b]) The first state to be fully characterized by a rotational analysis [6]. Becomes degenerate as one component of a $^2\Pi_u$ state with the $\tilde{X}\ ^2A_1$ ground state as the latter is opened into a linear configuration: the two states form a Renner-Teller pair. The labels (1), (2) ... are an alternative to $\tilde{X}$, $\tilde{A}$ ... and are enumerated in each symmetry-class separately.

[c]) The identification of a level at 9750 cm^{-1} as the zero-point level could be in error by one quantum of the bending-mode, to which the uncertainty quoted refers [3]. This value agrees almost exactly with one obtained from *ab initio* calculations [5].

[d]) From rotational constants obtained after extensive and somewhat uncertain deperturbation and extrapolation in a very complex spectrum. *Ab initio* calculations [4] give 1.252 Å and 102°.

[e]) Extrapolated from levels $8 \leq v_2 \leq 12$ for three isotopes, including ^{15}N and ^{18}O.

[f]) From the fact that $\tilde{E} - \tilde{X}$ bands are not shaded and hence $B' \simeq B''$.

[1] Stevens, C.R., Zare, R.N.: J. Mol. Spectrosc. **56** (1975) 167.
[2] Brand, J.C.D., Chan. W.H., Hardwick, J.L.: J. Mol. Spectrosc. **56** (1975) 309.
[3] Merer, A.J., Hallin, K.-E.J.: Can. J. Phys. **56** (1978) 838.
[4] Gillispie, G.D., Khan, A.J., Wahl, A.C., Hosteny, R.P., Kraus, M.: J. Chem. Phys. **63** (1975) 3425.
[5] Gillispie, G.D., Khan, A.J.: J. Chem. Phys. **65** (1976) 1624.
[6] Douglas, A.E., Huber, K.P.: Can. J. Phys. **43** (1965) 74.
[7] Hardwick, J.L., Brand, J.C.D.: Chem. Phys. Lett. **21** (1973) 458.
[8] Ritchie, R.W., Walsh, A.D., Warsop P.A.: Proc. Roy. Soc. (London) Ser. **A 266** (1962) 257.
[9] Hallin, K.-E.J., Merer, A.J.: Can. J. Phys. **54** (1976) 1157.
[10] Ritchie, R.W., Walsh, A.D.: Proc. Roy. Soc. (London) Ser. **A 267** (1962) 395.

II/7(**2**,330), II/15(**2**,320)

NO_2^+	Nitryl cation	II/23(**2**,139)
NO_2Rb	Rubidium nitrite	II/15(**2**,321)
NO_3	Nitrogen trioxide radical	II/15(**2**,322)
NO_3Rb	Rubidium nitrate	II/15(**2**,323)
NO_3Tl	Thallium(I) nitrate	II/7(**2**,336), II/15(**2**,324)

	N₂O	**Nitrous oxide**	**C**$_{∞v}$
IR		Dinitrogen monoxide	N=N=O

r_e	Å
N=N	1.127292(37) [a]
N=O	1.185089(37) [a]

Values reported for r(N=N) and r(N=O) correspond to the minimum of residuals achieved from a global fit of the equilibrium structure and sextic force field to a set of molecular band constants, belonging to six isotopomers.

[a]) Uncertainties, not given in the original paper, are obtained from standard deviations of the sum r(N=N) + r(N=O) (4 · 10⁻⁶ Å) and the difference r(N=O) − r(N=N) (7 · 10⁻⁵ Å).

Teffo, J.L., Chédin, A.: J. Mol. Spectrosc. **135** (1989) 389.

II/7(**2**,331), II/23(**2**,140)

	N₂O⁺	**Nitrous oxide cation**	**C**$_{∞v}$
UV		Dinitrogen monoxide cation	N=N=O

State		$\tilde{X}\ ^2\Pi_i$	$\tilde{A}\ ^2\Sigma^+$
Symmetry		C$_{∞v}$	C$_{∞v}$
Energy [eV]		0	3.492
r_0 [Å]	N=N	1.155(9)	1.140(6)
	N=O	1.185(9)	1.142(6)
θ_0 [deg]	N=N=O	180	180

Callomon, J.H., Creutzberg, F.: Phil. Trans. Roy. Soc. (London) Ser. A **277** (1974) 157.

II/7(**2**,332), II/15(**2**,325)

N₂O₂	Dinitrogen dioxide, Nitric oxide dimer	II/15(**2**,326)
N₂O₂S	Dinitrogen - sulfur dioxide (1/1)	II/23(**2**,141)
N₂O₃	Dinitrogen trioxide	II/7(**2**.333)
N₂O₄	Dinitrogen tetroxide	II/7(**2**,334), II/23(**2**,142)
	Nitrogen hemitetroxide, Nitrous oxide dimer	
N₂O₅	Dinitrogen pentaoxide	II/7(**2**,335), II/15(**2**,327)
N₂S	Dinitrogen sulfide	II/23(**2**,143)
N₃⁻	Azide(1-) ion	II/21(**2**,204)

	N₃	**Azide radical**	**D**$_{∞h}$
UV			N₃

State	$\tilde{X}\ ^2\Pi_g$	$\tilde{B}\ ^2\Sigma_u^+$
Symmetry	D$_{∞h}$	D$_{∞h}$
Energy [eV]	0	4.555
r_0 [Å]	1.1815	1.1799
θ_0 [deg]	180	180

Douglas, A.E., Jones W.J.: Can. J. Phys. **43** (1965) 2216.

II/7(**2**,328), II/15(**2**,328)

N₃Sr	Strontium monoazide	II/21(**2**,205)

	N₄O	Dinitrogen – dinitrogen monoxide (1/1)	II/21(**2**,206)
		Nitrogen molecule – nitrous oxide (1/1)	
	N₄O₂	Dinitrogen monoxide dimer	II/21(**2**,207)
	N₄S₄	Tetranitrogen tetrasulfide	II/21(**2**,208)
	NaO₃P	Sodium metaphosphate	II/15(**2**,329)
	Na₃	Trisodium	II/23(**2**,144)
	NbO₂	Niobium dioxide	II/15(**2**,330)

OS₂ **Disulfur monoxide** C_s

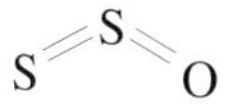

IR, MW

S=S=O

r_0	Å	θ_0	deg
S=S	1.887(12)	S=S=O	118.01(43)
S=O	1.457(15)		

r_s	Å	θ_s	deg
S=S	1.8852(22)	S=S=O	117.91(17)
S=O	1.4586(19)		

r_m^ρ	Å	θ_m^ρ	deg
S=S	1.8840(2)	S=S=O	117.89(1)
S=O	1.4554(3)		

r_e	Å	θ_e	deg
S=S	1.88424(11)	S=S=O	117.876(4)
S=O	1.45621(13)		

Lindenmayer, J., Rudolph, H.D., Jones, H.: J. Mol. Spectrosc. **119** (1986) 56.
Lindenmayer, J.: J. Mol. Spectrosc. **116** (1986) 315.
Tiemann, E., Hoeft, J., Lovas, F.J., Johnson, D.R.: J. Chem. Phys. **60** (1974) 5000.
Harmony, M.D., Berry, R.J., Taylor, W.H.: J. Mol. Spectrosc. **127** (1988) 324.

II/7(**2**,343), II/15(**2**,331), II/21(**2**,209)

OTl₂	Thallium(I) oxide, Dithallium monoxide	
	II/7(**2**,351), II/15(**2**,332), II/21(**2**,210), II/23(**2**,145)	
O₂P	Phosphorus dioxide	II/21(**2**,211)

O₂S **Sulfur dioxide** C_{2v}

MW

SO₂

r_e	Å	θ_e	deg
S=O	1.43080(1)	O=S=O	119.329(2)

Morino, Y., Tanimoto, M., Saito, S.: Acta Chem. Scand. Ser. **A 42** (1988) 346.

ED (counting method)

r_e	Å [a]	θ_e	deg [a]
S=O	1.4313(6) [b]	O=S=O	119.5(3)

The measurements were made at room temperature. The temperature of the sample was about 23 °C lower than the nozzle temperature.

(*continued*)

O_2S (*continued*)

[a]) Twice the estimated standard errors.
[b]) Calculated from r_a using $r_e = 1/2\{r_a - (3/2)\,al^2 + [(r_a - (3/2)\,al^2)^2 + 4l^2]^{1/2}\}$, where l denotes the S=O mean amplitude. Anharmonicities are: $a(S=O) = 2.0$, $a(O\cdots O) = 0.0$ Å^{-1}.

Holder, C.H., Fink, M.: J. Chem. Phys. **75** (1981) 5325.
ED: Mawhorter, R.J., Fink, M.: J. Chem. Phys. **79** (1983) 3292.

UV, IR, MW
Rotational and vibrational analysis

State	$\tilde{X}\,^1A_1$ [a])	$\tilde{A}\,^1A_2$ [a])	$\tilde{B}\,^1B_1$ [a])	$\tilde{C}\,^1B_2$	$\tilde{a}\,^3B_1$
Symmetry	C_{2v}	C_{2v}	(C_{2v}?)	C_{2v} (C_s) [b])	C_{2v}
Energy [eV]	0	3.463	3.94	5.279	3.195
Reference	[1] [2]	[3]	[4]	[5] [6]	[7]
r_0 [Å] S=O	1.432171 [c])	1.53	? [a])	1.560	1.4926(2)
θ_0 [deg] O=S=O	119.535 [c])	99	?	104	126.22(3)
r_e [Å] S=O(1)	1.4308			1.491 [b])	
S=O(2)	1.4308			1.639	
θ_e [deg] O=S=O	119.33				
Ab initio [8]					
Energy [eV]	0	3.3	3.4	5.3	2.1
r [Å]	1.504	1.585	1.564	1.59	1.547
θ [deg]	121	100	125	108	129.2

[a]) The strong absorption system extending from 3400 Å to 2600 Å, with a peak around 2900 Å, was previously thought to involve an electronically allowed transition to a single excited state $^1B_1$ labelled $\tilde{A}$ by Herzberg (1966). It is now clear that it involves the two closely adjacent states relabelled $\tilde{A}$ and $\tilde{B}$ strongly mixed through vibronic coupling. The intensity of the bands derives from $\tilde{B}\,^1B_1$ component but the only bands so far analyzable involve vibronic levels largely composed of the $\tilde{A}\,^1A_2$ component. The $\tilde{B}$ state is additionally strongly coupled to the $\tilde{X}\,^1A_1$ state so that a separate deperturbed geometry of the $\tilde{B}$ state may never be determinable.
[b]) The symmetric configuration (C_{2v}) is one of unstable equilibrium with respect to the antisymmetric S–O stretching coordinate Q_3, so that the potential surface has a shallow double-minimum section in this direction. A vibrational analysis in terms of a quadratic/quartic-cum-Gaussian model potential suggests a barrier between equivalent conformers of height 141(20) cm^{-1} (0.017 eV), some 43(20) cm^{-1} (0.0053 eV) above zero-point [6].
[c]) No estimates of uncertainty quoted.

[1] Barbe, A., Sécroun, C., Jouve, P., Duterage, B., Monnanteuil, N., Bellet, J., Steenbeckeliers, G.: J. Mol. Spectrosc. **55** (1975) 319.
[2] Saito, S.: J. Mol. Spectrosc. **30** (1969) 1; Van Riet, R., Steenbeckeliers, G.: Ann. Soc. Sci. Bruxelles, Ser. I **97** (1983) 117: The structure of previous studies was confirmed.
[3] Hamada, Y., Merer, A.J.: Can. J. Phys. **52** (1974) 1443.
[4] Hamada, Y., Merer, A.J.: Can. J. Phys. **53** (1975) 2555.
[5] Brand, J.C.D., Chiu, P.H., Hoy, A.R., Bist, H.D.: J. Mol. Spectrosc. **60** (1976) 43.
[6] Hoy, A.R., Brand, J.C.D.: Mol. Phys. **36** (1978) 1409.
[7] Brand, J.C.D., Jones, V.T., DiLauro, C.: J. Mol. Spectrosc. **40** (1971) 616.
[8] Hillier, I.H., Saunders, V.R.: Mol. Phys. **22** (1971) 193.

II/7(**2**,344), II/15(**2**,333), II/21(**2**,212)

	O$_2$S$_2$	Sulfur monoxide dimer, Disulfur dioxide	II/7(**2**,345)
	O$_2$Se	Selenium dioxide	II/7(**2**,349), II/15(**2**,334), II/21(**2**,213)
	O$_2$Te	Tellurium dioxide	II/15(**2**,335)

MW **O$_3$** **Ozone** **C$_{2v}$**
O$_3$

r_e [a])	Å		θ_e	deg
O–O	1.27156(20)		O–O–O	117.792(33)

[a]) Equivalent to Watson's r_m structure.

Depannemaecker, J.-C., Bellet, J.: J. Mol. Spectrosc. **66** (1977) 106.   II/7(**2**,337), II/15(**2**,336)

MW **O$_3$S** **Sulfur trioxide** **D$_{3h}$**

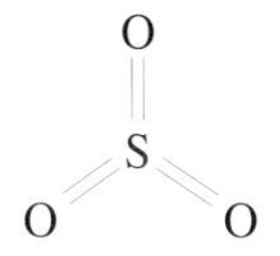

r_0	Å [a])
S=O	1.420(2)

r_e	Å [b])
S=O	1.4175(20)

[a]) Uncertainty is larger than those of the original data.
[b]) Uncertainty was not estimated in the original paper.

Meyer, V., Sutter, D.H., Dreizler, H.: Z. Naturforsch. **46a** (1991) 710.

IR

r_e	Å [a])
S=O	1.41732(50)

[a]) Uncertainties were not given by the authors.

Ortigoso, J., Escribano, R., Maki, A.G.: J. Mol. Spectrosc. **138** (1989) 602.
   II/7(**2**,346), II/15(**2**,337), II/21(**2**,214), II/23(**2**,146)

O$_3$Se	Selenium trioxide	II/7(**2**,350), II/15(**2**,338)
O$_4$Os	Osmium tetroxide	II/7(**2**,338)
O$_4$Rb$_2$W	Rubidium tungstate	II/15(**2**,339)
O$_4$ReTl	Thallium(I) perrhenate	II/7(**2**,341), II/15(**2**,340)
O$_4$Ru	Ruthenium tetroxide	II/7(**2**,342)
O$_4$STl$_2$	Thallium(I) sulfate	II/7(**2**,347), II/15(**2**,341)

MW **O$_4$S$_2$** **Sulfur dioxide dimer** **C$_s$**
(weakly bound complex) (effective symmetry class)
(SO$_2$)$_2$

r_0	Å		θ_0	deg
R_{cm}	3.822(1)		θ_1 [a])	127.0(20)
			θ_2 [a])	60.5(6)

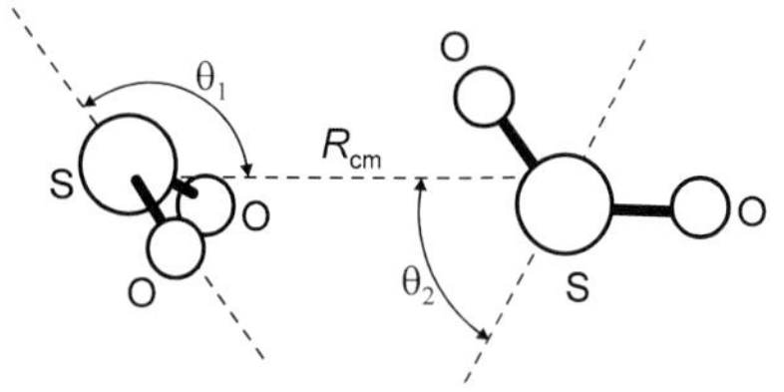

[a]) For definition see figure.

(continued)

O_4S_2 (*continued*)

Taleb-Bendiab, A., Hillig, K.W., Kuczkowski, R.L.: J. Chem. Phys. **94** (1991) 6956.
II/21(**2**,215), II/23(**2**,147)

O_4Xe	Xenon tetroxide	II/7(**2**,353)
O_6P_4	Tetraphosphorus hexaoxide	II/7(**2**,339)
	Phosphorus trioxide dimer	
O_6Sb_4	Tetraantimony hexaoxides, Antimony trioxide dimer	II/7(**2**,348)
O_7Re_2	Rhenium(VII) oxide	II/15(**2**,342)
	Dirhenium heptaoxide	
O_9W_3	Tungsten(VI) oxide trimer	II/7(**2**,352), II/15(**2**,343)
	Tritungsten nonaoxide	
$O_{10}P_4$	Phosphorus pentoxide dimer	II/7(**2**,340)
	Tetraphosphorus decaoxide	
$O_{12}Se_4$	Tetraselenium dodecaoxide	II/7(**2**,350)
P_4	Tetraphosphorus	II/15(**2**,344)
P_4S_3	Tetraphosphorus trisulfide	II/7(**2**,354)
Se_6	Hexaselenium	II/7(**2**,355)

3 Organic molecules

	CArClN	Argon – cyanogen chloride (1/1)	II/15(**3**,1)
	CArF$_2$O	Argon – carbonyl difluoride (1/1)	II/15(**3**,1a)
	CArO	Argon – carbon monoxide (1/1)	II/23(**3**,1)
	CArOS	Argon – carbonyl sulfide (1/1)	II/15(**3**,2), II/21(**3**,1), II/23(**3**,2)

	CArO$_2$		**Argon – carbon dioxide (1/1)**	**C$_{2v}$** (large-amplitude motion)
IR			(weakly bound complex)	T-shaped
				Ar$\cdots$CO$_2$

r_0	Å	θ_0	deg
C$\cdots$Ar	3.51(1) [a]	Ar$\cdots$C–O	83.1(5) [a]

r_{corr}	Å	θ_{corr}	deg
C$\cdots$Ar	3.50(1) [a]	Ar$\cdots$C–O	84.6(3) [a]

The structure of the CO$_2$ unit was supposed to be unchanged on complex formation. The structure called here r_0 is determined from the A_0 and C_0 rotational constants. Since the inertial defect Δ'' is still 2.436 u Å^2, different structures would be obtained from A_0 and B_0 or B_0 and C_0. The deviation of θ from 90° is due to averaging over vibrational motions. The structure called here r_{corr} is obtained by subtracting the harmonic part of this averaging, so that the inertial defect is reduced to –0.075 u Å^2.

[a] Uncertainties were not given in the original paper.

Sharp, S.W., Reifschneider, D., Wittig, C., Beaudet, R.A.: J. Chem. Phys. **94** (1991) 233.
See also: Randall, R.W., Walsh, M.A., Howard, B.J.: Faraday Discuss. Chem. Soc. **85** (1988) 13, and Sharp, S.W., Sheeks, R., Wittig, C., Beaudet, R.A.: Chem. Phys. Lett. **151** (1988) 267.

II/15(**3**,3), II/23(**3**,3)

CBF$_3$O	Carbonyltrifluoroboron	II/15(**3**,4)
CBrCl$_3$	Bromotrichloromethane	II/15(**3**,5)
CBrF	Bromofluoromethylene	II/23(**3**,4)
CBrF$_2$N	N–Bromodifluoromethanimine	II/23(**3**,5)
CBrF$_3$	Bromotrifluoromethane	II/7(**3**,3), II/15(**3**,6)
CBrF$_3$S	Bromo(trifluoromethyl)sulfur	II/21(**3**,2)
CBrF$_3$S$_2$	Bromo(trifluoromethyl)disulfane	II/23(**3**,6)
CBrN	Cyanogen bromide	II/7(**3**,4), II/15(**3**,7)

	CBrN$^+$		**Cyanogen bromide cation**	**C$_{\infty v}$**
LIF				Br–C≡N$^+$

State	$\tilde{\text{X}}^2\Pi_{3/2}$	$\tilde{\text{B}}^2\Pi_{3/2}$
Energy [eV]	0.0	2.326
r_s [Å] C–Br	1.745(14) [a]	[b]
C≡N	1.195(16)	[b]

(continued)

CBrN$^+$ (continued)

Rotational analysis of the $\tilde{B} - \tilde{X}$ system for the ^{79}BrCN$^+$, ^{81}BrCN$^+$, ^{79}Br^{13}CN$^+$, ^{81}Br^{13}CN$^+$, ^{79}BrC^{15}N$^+$, and ^{81}BrC^{15}N$^+$ species.

[a]) Error limits are 2σ.
[b]) No structural information has been deduced since the excited state is perturbed.

Rösslein, M., Hanratty, M.A., Maier, J.P.: Mol. Phys. **68** (1989) 823. II/21(**3**,3)

CBrNO	Bromine isocyanate	II/21(**3**,4), II/23(**3**,7)
CBrNS	Bromine thiocyanate	II/21(**3**,5)
CBrN$_3$O$_6$	Bromotrinitromethane	II/15(**3**,8)
CBr$_2$	Dibromomethylene radical	II/7(**3**,1)
CBr$_2$O	Carbonyl dibromide	II/15(**3**,9)
CBr$_2$S	Thiocarbonyl dibromide	II/15(**3**,10)
CBr$_3$	Tribromomethyl radical	II/7(**3**,2)
CBr$_3$NO$_2$	Tribromonitromethane	II/7(**3**,5)
CBr$_4$	Tetrabromomethane	II/23(**3**,8)

CCaH$_3$ **Monomethyl calcium**   **C$_{3v}$**

LIF   CaCH$_3$

State	$\tilde{X}\,^2A_1$	$\tilde{A}\,^2E$
Energy [eV]	0.0	1.828
r_0 [Å] C–H	1.100(20) [a]	1.100(20) [a]
Ca–C	2.349(13)	2.353(14)
θ_0 [deg] H–C–H	105.6(28)	109.2(30)

Rotational analysis of the 0–0 band of CaCH$_3$. An internal perturbation in the excited state permits the determination of both A′ and A″.

[a]) The error limits in parentheses are based on the assumption that r_0 (CH) = 1.10 ± 0.02 Å.

Brazier, C.R., Bernath, P.F.: J. Chem. Phys. **91** (1989) 4548. II/21(**3**,6)

CClFO	Carbonyl chloride fluoride	II/7(**3**,8), II/15(**3**,11)
CClFOS	(Fluorocarbonyl)sulfenyl chloride	II/23(**3**,9)
CClFS	Thiocarbonyl chloride fluoride	II/15(**3**,12)
CClF$_2$N	N–Chlorodifluoromethanimine	II/21(**3**,7)
	N–Chlorodifluromethyleneamine	
CClF$_3$	Trifluoromethyl chloride	II/15(**3**,13)
CClF$_3$O	Trifluoromethyl hypochlorite	II/15(**3**,14)
CClF$_3$O$_2$	(Chloroperoxy)trifluoromethane	II/15(**3**,15)
	Trifluoromethyl peroxohypochlorite	
CClF$_3$O$_2$S	Trifluoromethanesulfonyl chloride	II/15(**3**,16)
CClF$_3$S	Trifluoromethanesulfenyl chloride	II/15(**3**,17)
CClF$_3$S$_2$	Chloro(trifluoromethyl)disulfane	II/23(**3**,10)
CClF$_7$S	Chlorotetrafluoro(trifluoromethyl)sulfur	II/21(**3**,8)
	Trifluoromethylsulfur chloride tetrafluoride	

	CClN	**Chlorine cyanide**	$C_{\infty v}$
MW		Cyanogen chloride	Cl–C≡N

r_e	Å
C≡N	1.1606(28)
C–Cl	1.6290(24)

Cazzoli, G., Favero, P.G., Degli Esposti, C.: Chem. Phys. Lett. **50** (1977) 336.

UV Franck-Condon analysis

State		$\widetilde{B}$ [a]	$\widetilde{C}$ [a]
Symmetry		$C_{\infty v}$	$C_{\infty v}$
Energy [eV]		9.092	9.233
r_0 [Å]	Cl–C	1.654	1.644
	C≡N	1.210	1.213

From a Franck-Condon analysis of vibrational progressions in the stretching vibration.

[a]) Rydberg states from a $2\pi \to \sigma_x$ promotion.

King, G.W., Richardson, A.W.: J. Mol. Spectrosc. **21** (1966) 353.   II/7(**3**,10), II/15(**3**,18)

	CClN[+]	**Cyanogen chloride cation**	$C_{\infty v}$
LIF			Cl—C≡N[+]

State		$\widetilde{X}\,^2\Pi_{3/2}$	$\widetilde{B}\,^2\Pi_{3/2}$
Energy [eV]		0.0	2.792
r_s [Å]	C–Cl	1.559(12) [a]	[b]
	C≡N	1.215(12)	[b]

Rotational analysis of the $\widetilde{B}-\widetilde{X}$ system for the ^{35}ClCN[+], ^{37}ClCN[+], ^{35}Cl^{13}CN[+], ^{37}Cl^{13}CN[+], ^{35}ClC^{15}N[+], and ^{37}ClC^{15}N[+] species.

[a]) Error limits are 2σ.
[b]) No structural information has been deduced since the excited state is perturbed.

Rösslein, M., Maier, J.P.: J. Phys. Chem. **93** (1989) 7342.   II/21(**3**,9)

CClNO	Chlorine isocyanate	II/7(**3**,11), II/15(**3**,19)
CClNO$_3$S	Sulfonyl chloride isocyanate	II/15(**3**,20), II/23(**3**,11)
CClNS	Chlorine thiocyanate	II/15(**3**,21)
CClN$_3$O$_6$	Chlorotrinitromethane	II/7(**3**,12), II/15(**3**,22)
CCl$_2$	Dichloromethylene	II/21(**3**,10), II/23(**3**,12)
CCl$_2$F$_2$	Dichlorodifluoromethane	II/15(**3**,23)
CCl$_2$F$_2$S	Chlorodifluoromethanesulfenyl chloride	II/23(**3**,13)
CCl$_2$F$_6$Si$_2$	Dichlorobis(trifluorosilyl)methane	II/15(**3**,24)
CCl$_2$NO$_2$P	Dichloroisocyanatophosphine oxide	II/7(**3**,14)
CCl$_2$N$_2$O$_4$	Dichlorodinitromethane	II/15(**3**,25)

CCl$_2$O		**Carbonyl dichloride**		C_{2v}

ED, MW

r_z	Å [a])	θ_z	deg [a])
C=O	1.1789(12)	Cl–C–Cl	111.85(5)
C–Cl	1.7423(6)		

r_e	Å [a])	θ_e	deg [a])
C=O	1.1766(22)	Cl–C–Cl	111.91(12)
C–Cl	1.7365(12)		

The third-order anharmonic constants of phosgene have been determined from the rotational constants of the six fundamental vibrational states, those of eight isotopic species, and the r_z structure obtained from the ED intensity by analyzing the changes in the average structures.

[a]) Estimated limits of error.

Yamamoto, S., Nakata, M., Kuchitsu, K.: J. Mol. Spectrosc. **112** (1985) 173.
Nakata, M., Kohata, K., Fukuyama, T., Kuchitsu, K.: J. Mol. Spectrosc. **83** (1980) 105.

II/7(**3**,16), II/15(**3**,26), II/21(**3**,11)

CCl$_2$O	Carbon monoxide – dichlorine (1/1)	II/23(**3**,14)
CCl$_2$OS	Carbonylthiohypochlorite	II/21(**3**,12)
CCl$_2$OS	Thiocarbonyl dichloride S–oxide	II/23(**3**,15)
	Carbonothioic dichloride S–oxide	
CCl$_2$S	Thiocarbonyl dichloride	II/7(**3**,18), II/15(**3**,27)

CCl$_3$F		**Trichlorofluoromethane**		C_{3v}
				CFCl$_3$

ED, MW

r_g	Å [a])	θ_α^0	deg [a])
C–F	1.345(3)	Cl–C–Cl	110.5(1)
C–Cl	1.7636(10)		

The nozzle temperature was ≈ 20 °C.

[a]) Estimated limits of error.

Konaka, S., Takeuchi, M., Kimura, M.: J. Mol. Struct. **131** (1985) 317.   II/7(**3**,7), II/21(**3**,13)

CCl$_3$FS	Dichlorofluoromethanesulfenyl chloride	II/23(**3**,16)
CCl$_3$N	N,1,1–Trichloromethanimine	II/15(**3**,28)
	N–Chlorocarbonimidic dichloride	
CCl$_3$NOSi	Trichlorosilyl isocyanate	II/7(**3**,15)
CCl$_3$NO$_2$	Trichloronitromethane	II/7(**3**,13)

CCl$_4$		**Carbon tetrachloride**	T_d
			CCl$_4$

ED

r_g	Å [a])
C–Cl	1.767(3)
Cl⋯Cl	2.888(3)

(*continued*)

[a]) Estimated limits of error.

Morino, Y., Nakamura, Y., Iijima, T.: J. Chem. Phys. 32 (1960) 643.   II/7(**3**,6)

CCl$_4$F$_3$P	(Trifluoromethyl)tetrachlorophosphorane	II/21(**3**,14)
CCl$_4$O$_2$S	Trichloromethanesulfonyl chloride	II/7(**3**,17), II/15(**3**,29)
CCl$_4$S	Trichloromethanesulfenyl chloride	II/7(**3**,19)
CCl$_6$Ge	Trichloro(trichloromethyl)germane	II/7(**3**,9)
CCl$_8$Si$_2$	Dichlorobis(trichlorosilyl)methane	II/15(**3**,30)
CFN	Fluorine cyanide	II/7(**3**,22), II/15(**3**,31)
	Cyanogen fluoride	
CFNO$_2$S$_2$	S–Fluorocarbonyl–N–sulfinylthiohydroxylamine	II/23(**3**,17)
CFN$_3$O	Carbonazidic fluoride	II/23(**3**,18)
	Carbonyl azide fluoride	

CFO   **Fluoroformyl radical**   C$_s$

Fluorocarbonyl

IR

F – Ċ = O

r_0	Å		θ_0	deg
C=O	1.169(60)		F–C=O	127.3(6)
C–F	1.334(60)			
(C–F) + (C=O)	2.5026(20)			

Nagai, K., Yamada, C., Endo, Y., Hirota, E.: J. Mol. Spectrosc. 90 (1981) 249.   II/15(**3**,32)

CFP   **(Fluoromethylidyne)phosphine**   C$_{\infty v}$

MW

F – C ≡ P

r_0	Å
F–C	1.285(5)
C≡P	1.541(5)

Kroto, H.W., Nixon, J.F., Simmons, N.P.C.: J. Mol. Spectrosc. 82 (1980) 185.   II/15(**3**,33)

CF$_2$	Difluoromethylene	II/7(**3**,20), II/15(**3**,34)
CF$_2$I$_2$	Difluorodiiodomethane	II/23(**3**,19)

CF$_2$N   **Difluoromethylimino radical**   C$_{2v}$

UV   Rotational and vibrational analysis

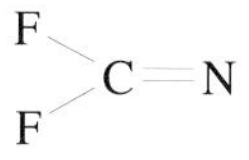

State	$\tilde{X}\,^2B_2$	$\tilde{A}\,^2A_1$
Symmetry	C$_{2v}$	C$_{2v}$
Energy [eV]	0	3.427

r_0 [Å]	C–F	1.310 [a])	1.286(20)
	C=N	1.265(20)	1.308(20)
θ_0 [deg]	F–C–F	113.5(10)	114.2(10)

[a]) Assumed.

Dixon, R.N., Duxbury, G., Mitchell, R.C., Symons, J.P.: Proc. Roy. Soc. (London) Ser. **A 300** (1967) 405.   II/7(**3**,23), II/15(**3**,35)

CF$_2$NOP	Difluoroisocyanatophosphine	II/7(**3**,28)
CF$_2$NP	Cyanodifluorophosphine	II/7(**3**,30)
CF$_2$NPS	Difluoroisothiocyanatophosphine	II/7(**3**,32)
CF$_2$NPSe	Difluoro(isoselenocyanato)phosphine	II/15(**3**,36)
CF$_2$N$_2$	Difluorocyanamide	II/7(**3**,24)
CF$_2$N$_2$	Difluorodiazirine	II/7(**3**,25), II/23(**3**,20)
CF$_2$O	Carbonyl difluoride	II/7(**3**,33), II/15(**3**,37)
CF$_2$S	Thiocarbonyl difluoride	II/7(**3**,35), II/15(**3**,38)
CF$_2$Se	Selenocarbonyl difluoride	II/15(**3**,39)

CF$_3$ **Trifluoromethyl radical** **C$_{3v}$**

MW, IR $\dot{\mathrm{CF}}_3$

r_0	Å		θ_0	deg
C–F	1.318(2)		F–C–F	110.76(40)

Yamada, C., Hirota, E.: J. Chem. Phys. **78** (1983) 1703.   II/15(**3**,40)

CF$_3$I	Trifluoromethyl iodide	II/7(**3**,21), II/15(**3**,41)
CF$_3$N	N,1,1–Trifluoromethanimine	II/15(**3**,42)
CF$_3$NO	Trifluoronitrosomethane	II/7(**3**,26), II/15(**3**,43)
CF$_3$NOS	N–(Fluorocarbonyl)imidosulfurous difluoride	II/15(**3**,44)
CF$_3$NOSi	Trifluorosilyl isocyanate	II/7(**3**,29)
CF$_3$NO$_2$	Trifluoronitromethane	II/7(**3**,27)
CF$_3$NSi	Cyanotrifluorosilane Trifluorosilyl cyanide	II/23(**3**,21)
CF$_3$NSi	Trifluoroisocyanosilane Trifluorosilyl isocyanide	II/23(**3**,22)
CF$_3$N$_3$	Azidotrifluoromethane	II/15(**3**,45)

CF$_4$ **Carbon tetrafluoride** **T$_d$**

IR, MW, Raman CF$_4$

r_e	Å
C–F	1.3151(17) [a]

A third-degree polynomial expansion of the potential function has been fitted directly to a very large set of experimental frequencies. The distance r_e is one of the fitted parameters.

[a] Estimated standard error.

Brodersen, S.: J. Mol. Spectrosc. **145** (1991) 331.

IR, MW Double resonance

r_0	Å
C–F	1.319253(3)

Takami, M.: J. Chem. Phys. **74** (1981) 4276.
ED: Bartell, L.S., Stanton, J.F.: J. Chem. Phys. **81** (1984) 3792.   II/15(**3**,46), II/23(**3**,23)

CF$_4^+$	Tetrafluoromethane(1+) ion	II/21(**3**,15)
CF$_4$N$_2$P$_2$	Bis(difluorophosphino)carbodiimide	II/7(**3**,31)

CF$_4$O	Trifluoromethyl hypofluorite		II/7(**3**,34)
CF$_4$OS	Trifluoromethanesulfinyl fluoride		II/23(**3**,24)
CF$_4$O$_2$	Trifluoro(fluoroperoxy)methane		II/15(**3**,47)
	Trifluoromethyl peroxohypofluorite		
CF$_4$S	Trifluoromethanesulfenyl fluoride		II/15(**3**,48)
CF$_4$S$_2$	Fluoro(trifluoromethyl)disulfane		II/23(**3**,25)
CF$_5$N	Pentafluoromethaneamine		II/15(**3**,49)
CF$_5$NOS	Cyanatopentafluorosulfur		II/23(**3**,26)
CF$_5$NOS	Pentafluoro(isocyanato)sulfur		II/15(**3**,50)
CF$_5$NOSe	Cyanatopentafluoroselenium		II/21(**3**,16)
CF$_5$NOTe	Pentafluoro(isocyanato)tellurium		II/15(**3**,52)
CF$_5$NS	N–(Trifluoromethyl)imidosulfurous difluoride		II/15(**3**,53)
CF$_5$NS	Cyanopentafluorosulfur		II/23(**3**,27)
CF$_6$S	Trifluoro(trifluoromethyl)sulfur	II/21(**3**,17),	II/23(**3**,28)
CF$_6$Si	Trifluoro(trifluoromethyl)silane		II/21(**3**,18)
CF$_7$P	(Trifluoromethyl)tetrafluorophosphorane		II/15(**3**,54)
CF$_8$S	Pentafluoro(trifluoromethyl)sulfur		II/21(**3**,19)
CF$_8$S$_2$	μ–(Difluoromethylene)bis(trifluorosulfur)		II/23(**3**,29)
CF$_8$S$_2$	μ–Carbido–pentafluorosulfur(VI)trifluorosulfur(VI)		II/23(**3**,30)
CF$_{12}$S$_2$	μ–(Difluoromethylene)bis[pentafluorosulfur]		II/21(**3**,20)
CHArN	Argon – hydrogen cyanide (1/1)		
		II/15(**3**,54a), II/21(**3**,21),	II/23(**3**,31)
CHAr$_2$N	Argon – hydrogen cyanide (2/1)		II/23(**3**,32)
CHBrClF	Bromochlorofluoromethane		II/15(**3**,55)
CHBrCl$_2$	Bromodichloromethane		II/15(**3**,56)
CHBrO	Carbon monoxide – hydrogen bromide (1/1)		II/15(**3**,57)
CHBrO$_2$	Carbon dioxide – hydrogen bromide (1/1)		II/23(**3**,33)
CHBr$_2$Cl	Dibromochloromethane		II/15(**3**,58)
CHBr$_3$	Bromoform, Tribromomethane	II/7(**3**,50),	II/15(**3**,59)

CHCl   **Chloromethylene** C_s
Chlorocarbene   HCCl

LIF   Dye laser induced fluorescence

r_0	Å	θ_0	deg
C–H	1.130(36)	H–C–Cl	105.1(47)
C–Cl	1.687(11)		

r_s	Å	θ_s	deg
C–H	1.1188(71)	H–C–Cl	101.4(12)
C–Cl	1.6961(25)		

Kakimoto, M., Saito, S., Hirota, E.: J. Mol. Spectrosc. **97** (1983) 194.

(*continued*)

CHCl (*continued*)

UV   Rotational and vibrational analysis

State	$\tilde{X}^1A'$	$\tilde{A}\,^1A''$
Symmetry	C_s	C_s
Energy [eV]	0	1.524

r_0 [Å]	H–C	1.12	
	C–Cl	1.689	
θ_0 [deg]	H–C–Cl	103.4	134(5) [a]

[a] From an assumed form of quasi-linear double minimum potential. Height of barrier is about 2250 cm^{-1} above the zero point.

Merer, A.J., Travis, D.N.: Can. J. Phys. **44** (1966) 525.   II/7(**3**,51), II/15(**3**,60)

CHClF$_2$	Chlorodifluoromethane	II/7(**3**,53), II/21(**3**,22)
CHClO	Carbon monoxide – hydrogen chloride (1/1)	
		II/15(**3**,61), II/23(**3**,35)
CHClO	Formyl chloride	II/15(**3**,62)
CHClOS	Carbonyl sulfide – hydrogen chloride (1/1)	II/21(**3**,23)

CHClO$_2$ **Carbon dioxide – hydrogen chloride (1/1)** $C_{\infty v}$

(weakly bound complex)   (effective symmetry class)

IR   CO$_2$···HCl

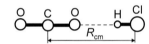

Isotopic species	CO$_2$···H^{35}Cl	CO$_2$···H^{37}Cl	CO$_2$···D^{35}Cl
r_0	Å	Å	Å
C–O	1.162 [a]	1.162 [a]	1.162 [a]
H–Cl	1.275 [a]	1.275 [a]	1.275 [a]
R_{cm}	4.555(30) [b]	4.456(30) [b]	4.516(30) [b]
O–H	2.154(5) [b]	2.154(5) [b]	2.149(5) [b]

Ab initio calculations have been performed in order to reproduce the B_0 rotational constants and the blue shifts between the ν_3 band of CO$_2$ and the corresponding ν_2 band of the complex.

[a] Assumed.
[b] Uncertainties were not estimated in the original paper.

Sharpe, S.W., Zeng, Y.P., Wittig, C., Beaudet, R.A.: J. Chem. Phys. **92** (1990) 943.

II/15(**3**,62a), II/23(**3**,36)

CHCl$_3$	Chloroform, Trichloromethane	II/7(**3**,52)
CHCl$_5$Si	(Dichloromethyl)trichlorosilane	II/15(**3**,63)

CHF   **Fluoromethylene** C_s

Fluorocarbene   HCF

LIF   Dye laser induced fluorescence

$\tilde{X}^1A'$ (000)

r_0	Å	θ_0	deg
C–H	1.138(10)	H–C–F	104.1(13)
C–F	1.305(6)		

(*continued*)

$\tilde{A}^1A''$ (000)

r_0	Å	θ_0	deg
C–H	1.063(13)	H–C–F	123.8(8)
C–F	1.308(6)		

Suzuki, T., Saito, S., Hirota, E.: J. Mol. Spectrosc. **90** (1981) 447.

UV Rotational and vibrational analysis

State		$\tilde{X}^1A'$	$\tilde{A}^1A''$
Symmetry		C_s	C_s
Energy [eV]		0	2.142
r_0 [Å]	H–C	(1.121)[a]	(1.121)[a]
	C–F	1.314	1.297
θ_0 [deg]	H–C–F	101.6	127.2

[a] Assumed.

Merer, A.J., Travis, D.N.: Can. J. Phys. **44** (1966) 1541.
Kakimoto, M., Saito, S., Hirota, E.: J. Mol. Spectrosc. **88** (1981) 300.   II/7(**3**,54), II/15(**3**,64)

CHFO		Formyl fluoride	II/7(**3**,56), II/15(**3**,65)
CHFO		Carbon monoxide – hydrogen fluoride (1/1)	
			II/15(**3**,66), II/23(**3**,37)
CHFOS		Carbonyl sulfide – hydrogen fluoride (1/1)	II/15(**3**,67)
CHFO$_2$		Carbon dioxide – hydrogen fluoride (1/1)	
			II/15(**3**,68), II/23(**3**,38)
CHF$_2$P		(Difluoromethylene)phosphine	II/15(**3**,69)

CHF$_3$ **Fluoroform** C_{3v}
MW   Trifluoromethane   CHF$_3$

r_0	Å	θ_0	deg
C–H	1.102(6)	F–C–F	108.66(14)
C–F	1.3326(12)		

r_e	Å	θ_e	deg
C–H	1.091(14)	F–C–F	108.58(34)
C–F	1.3284(31)		

r_z	Å	θ_z	deg
C–H	1.099(7)	F–C–F	108.49(15)
C–F	1.3345(13)		

Kawashima, Y., Cox, A.P.: J. Mol. Spectrosc. **72** (1978) 423.   II/7(**3**,55), II/15(**3**,70)

CHF$_3$O$_2$	Trifluoromethyl hydroperoxide	II/15(**3**,71)
CHF$_3$O$_3$S	Trifluoromethanesulfonic acid	II/15(**3**,72)
	Triflic acid	
CHF$_3$O$_6$S$_3$	Tris(fluorosulfonyl)methane	II/15(**3**,73)

CHF$_3$S	Trifluoromethanethiol	II/7(**3**,57), II/23(**3**,39)
CHF$_3$S$_2$	(Trifluoromethyl)disulfane	II/23(**3**,40)
CHIO	Carbon monoxide – hydrogen iodide (1/1)	II/23(**3**,41)
CHKrN	Krypton – hydrogen cyanide (1/1)	II/15(**3**,73a), II/23(**3**,42)

CHN **Hydrogen cyanide** $C_{\infty v}$

IR, MW H–C≡N

r_e	Å
H–C	1.06501(8) [a]
C≡N	1.15324(2) [a]

The equilibrium rotational constants B_e for eight isotopomers are derived from the corresponding B_0 values by using corrections given by variational calculations of the rovibrational energy levels.

[a] Uncertainties quoted here are 10 times the standard deviation from the least-squares calculation.

Carter, S., Mills, I.M., Handy, N.C.: J. Chem. Phys. **97** (1992) 1606.

MW

r_m^ρ [a]	Å
C–H	1.0668(2)
C≡N	1.1531(1)

[a] Multiple isotope substitution structure.

Berry, R.J., Harmony, M.D.: Struct. Chem. **1** (1990) 49.

UV Rotational and vibrational analysis

State	$\tilde{A}\,^1A''$ [d]	$\tilde{C}\,^1A'$
Symmetry	C_s	C_s
Energy [eV]	6.568	8.139
Reference	[1] [3] [d]	[2]
r_0 [Å] H–C	1.140(5) [a]	(1.14) [b]
C≡N	1.297(2) [a]	
θ_0 [deg] H–C≡N	125.0(2) [a]	141
θ_e [deg] H–C≡N	123.9 [c]	

[a] Mean value for HCN and DCN, the uncertainties quoted spanning small systematic differences arising from zero-point motion.
[b] DCN only: HCN spectra diffuse. r_0 (C≡N) assumed to be as in the $\tilde{A}$ state.
[c] From a double minimum potential containing quadratic, quartic, and Lorentzian terms defined by observed vibrational structure. Potential maximum 6809 cm^{-1}. Ref.: Johns, J.W.C.: Can. J. Phys. **45** (1967) 2639. For alternative potentials see also: Hougen, J.T., Bunker, P.R., Johns, J.W.C.: J. Mol. Spectrosc. **34** (1970) 136.
[d] The vibrational manifold of this state includes all the levels previously assigned to a separate electronic state $\tilde{B}\,^1A''$ [3].

[1] Herzberg, G., Innes, K.K.: Can. J. Phys. **35** (1957) 842.
[2] Herzberg, G., Innes, K.K., quoted in Herzberg, Molecular Spectra and Molecular Structure, Vol. III, 1966.

(*continued*)

[3] Bickel, G.A., Innes, K.K.: Can. J. Phys. **62** (1984) 1763.

II/7(**3**,58), II/15(**3**,74), II/23(**3**,43)

	CHN	**Hydrogen isocyanide**	$C_{\infty v}$
MW			HNC

r_m^ρ [a]	Å
N–H	0.9923(3)
N=C	1.1701(1)

[a]) Multiple isotope substitution structure.

Berry, R.J., Harmony, M.D.: Struct. Chem. **1** (1990) 49.

IR, MW

r_e	Å
N–H	0.9940(8)
N=C	1.16892(2)

Structure obtained from a force field obtained fron IR data of HNC and DNC.

Creswell, R.A., Robiette, A.G.: Mol. Phys. **36** (1978) 869.

II/15(**3**,75), II/23(**3**,44)

	CHN⁺	**Hydrogen cyanide cation**	$C_{\infty v}$
PES			HCN⁺

State	$\tilde{X}^2\Pi$ [a]
r [Å] C–H	1.09
C–N	1.21

[a]) Symmetry assignment based on theoretical calculations of [1].

Hollas, J.M., Sutherley, T.A.: Mol. Phys. **24** (1972) 1123.
[1] So, S.R., Richards, W.G.: J. Chem. Soc., Faraday Trans. II **71** (1975) 62.

II/7(**3**,59)

	CHNO	**Isocyanic acid**	C_s
MW, IR			HN=C=O

r_s	Å	θ_s	deg
N–H	0.9946(64)	H–N=C	123.9(17)
N=C	1.2140(24)	N=C=O	172.6(27) [a]
C=O	1.1664(8)		

[a]) O is *trans* to H.

Yamada, K.: J. Mol. Spectrosc. **79** (1980) 323.

II/7(**3**,61), II/15(**3**,76)

CHNO	Fulminic acid	II/7(**3**,62)
CHNO$_2$S	Hydrogen cyanide – sulfur dioxide (1/1)	II/21(**3**,24)

	CHNS	**Isothiocyanic acid**		C_s
MW				HN=C=S

r_s	Å	θ_s	deg
N–H	0.9928(64)	H–N=C	131.7(19)
N=C	1.2068(24)	N=C=S	173.8(23) [a]
C=S	1.5665(6)		

[a]) S is *trans* to H.

Yamada, K., Winnewisser, M., Winnewisser, G.: Szalanski, L.B., Gerry, M.C.L.: J. Mol. Spectrosc. **79** (1980) 295.   II/7(**3**,64), II/15(**3**,77)

CHNSe	**Isoselenocyanic acid**	II/15(**3**,78)

CHN$_2$	**Carbonimidoylamidogen**	C_s
UV	Rotational and vibrational analysis	

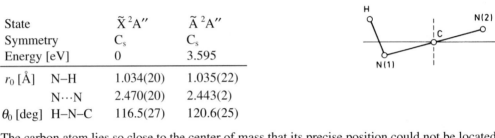

State	$\tilde{X}\,^2A''$	$\tilde{A}\,^2A''$
Symmetry	C_s	C_s
Energy [eV]	0	3.595
r_0 [Å] N–H	1.034(20)	1.035(22)
N⋯N	2.470(20)	2.443(2)
θ_0 [deg] H–N–C	116.5(27)	120.6(25)

The carbon atom lies so close to the center of mass that its precise position could not be located, and was not revealed by ^{13}C substitution. The uncertainties quoted correspond to uncertainties in the position of the carbon atom of ±0.10 Å from the center of mass. N–C–N is assumed to be linear.

Herzberg, G., Warsop, P.A.: Can. J. Phys. **41** (1963) 286.   II/7(**3**,60), II/15(**3**,79)

CHN$_3$	Hydrogen cyanide – dinitrogen (1/1)	II/23(**3**,45)
CHN$_3$O	Hydrogen cyanide – dinitrogen monoxide (1/1)	II/23(**3**,46)
CHN$_3$O$_6$	Trinitromethane	II/7(**3**,63), II/15(**3**,80)
CHO$^-$	Formyl anion	II/21(**3**,25)

	CHO	**Formyl radical**		C_s
UV, MW				

State	$\tilde{X}\,^2A'(\Pi)$	$\tilde{A}\,^2A''(\Pi)$	$\tilde{B}\,^2A'$
Symmetry	C_s	C_v	C_s
Energy [eV]	0	1.153	4.797
Reference	[1] [2]	[2]	[3]
r_0 [Å] C–H	1.1102(10) [a]	1.064(7)	(1.16) [c]
	1.125(5) [b]		
C=O	1.17115(20) [a]	1.186(2)	1.36(2)
	1.175(1) [b]		
θ_0 [deg] H–C=O	127.426(70) [a]		
	124.95(25) [b]	180	111(4)

(*continued*)

a) Microwave determination [1]. Errors are reestimated by Hirota.
b) Rotational analysis of electronic spectrum [2]. The geometric zero-point parameters are derived from the rotational constants in slightly different ways in [1] and [2].
c) Assumed.

[1] Austin, J.A., Levy, D.H., Gottlieb, C.A., Radford, H.E.: J. Chem. Phys. **60** (1974) 207.
[2] Brown, J.M., Ramsay, D.A.: Can. J. Phys. **53** (1975) 2232; reviews earlier estimates of the structural parameters, with references, and reanalyzes the electron spectrum.
[3] Dixon, R.N.: Trans. Faraday Soc. **65** (1969) 3141.

MW

r_e	Å	θ_e	deg
H–C	1.1191(50)	H–C=O	124.43(25)
C=O	1.1754(15)		

Hirota, E.: J. Mol. Struct. **146** (1986) 237. II/7(**3**,65), II/15(**3**,81), II/21(**3**,26)

CHO⁺ **Formyl cation** $C_{\infty v}$

MW H—C⁺≡O

r_e	Å
C=O	1.104738(23)
H–C	1.097247(38)

r_s	Å
C=O	1.107211(15)
H–C	1.092881(35)

Woods, R.C.: Philos. Trans. Roy. Soc. London **A 324** (1988) 141.
II/15(**3**,82), II/21(**3**,27), II/23(**3**,47)

CHO⁺ **Hydroxocarbon(1+) ion** $C_{\infty v}$
Hydroxylcarbenium ion
MW (Carbon monohydroxide ion) H—O—C⁺

r_s	Å a)
O–C	1.1595(20)
H–O	0.9641(20)

a) Uncertainties were not estimated in the original paper.

Bogey, M., Demuynck, C., Destombes, J.L.: J. Mol. Spectrosc. **115** (1986) 229.
II/15(**3**,83), II/21(**3**,28)

CHO₂⁺ **Hydroxo(oxo)carbon(1+) ion** C_s

MW

r_0 [Å] a)	Structure I	Structure II
H–O(1)	0.9715(20)	0.9766(20)
O(1)–C	1.2057(20)	1.2085(20)
C–O(2)	1.1409(20)	1.1400(20)

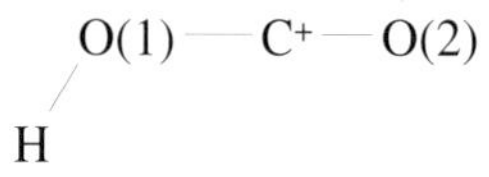

O(1)—C⁺—O(2)
  /
 H

(*continued*)

CHO$_2^+$ (*continued*)

θ_0 [deg] [a]	Structure I	Structure II
O(1)–C–O(2)	178.83(2)	174.39(2)
H–O(1)–C	122.25(2)	119.38(2)

Two possible structures are reported.

[a]) Uncertainties were not estimated in the original paper.

Bogey, M., Demuynck, C., Destombes, J.L., Krupnov, A.: J. Mol. Struct. **190** (1988) 465.

II/21(**3**,29)

CHP   **Methylidynephosphine** $C_{\infty v}$
Methinophosphide

IR, MW   H–C≡P

r_e	Å
C–H	1.0662(2)
C≡P	1.54020(3)

MW rotational constants were combined with IR α constants for HCP and DCP to get the structure.

Lavigne, J., Cabana, A.: Can. J. Phys. **60** (1982) 304.

UV, MW   Rotational and vibrational analysis

Singlet states:

State		$\tilde{X}\,^1\Sigma^-$	$\tilde{A}\,^1A''$	$\tilde{B}\,^1\Pi$	$\tilde{C}\,^1A'$
Symmetry		$C_{\infty v}$	C_s	$C_{\infty v}$	C_s
Energy [eV]		0	4.308	4.454	4.990
Reference		[1]	[2]	[2]	[2]
r_0 [Å]	H–C	1.0667(30) [a]	(1.07) [b]	(1.07) [b]	(1.07) [b]
	C≡P	1.5421(30) [a]	1.69 [c]	1.635	1.69
r_e [Å]	H–C	1.0692(8) [a] [3]			
	C≡P	1.5398(2) [a] [3]			
θ_0 [deg]	H–C≡P	180	130 [c]	180	113

Triplet states:

State		$\tilde{a}\,^3\Sigma^-$	$\tilde{b}\,^3\Pi$	$\tilde{c}\,^3\Sigma^-$	$\tilde{d}\,^3\Pi$
Symmetry		$C_{\infty v}$	$C_{\infty v}$	$C_{\infty v}$	$C_{\infty v}$
Energy [eV]		3.030	3.773	3.8471	4.460
Reference		[2]	[2]	[2]	[2]
r_0 [Å]	H–C	(1.07) [b]	(1.07) [b]	(1.07) [b]	(1.07) [b]
	C≡P	1.67	1.67	1.668	1.67
θ_0 [deg]	H–C≡P	180	180	180	180

[a]) From microwave spectroscopy; average structure of several isotopes [1]. Errors are reestimted by Hirota.
[b]) r_0 (C–H) assumed unchanged from ist ground state value.

(*continued*)

c) Rather uncertain because of large zero-point amplitudes in a quasi-linear potential. Values from DCP; r_0 (C≡P) = 1.70 Å, θ_0 = 128°.

[1] Tyler, J.K.: J. Chem. Phys. **40** (1964) 1170.
[2] Johns, J.W.C., Tyler, J.K., Shurvell, H.F.: Can. J. Phys. **47** (1969) 893.
[3] Strey, G., Mills, I.M.: Mol. Phys. **26** (1973) 129.   II/7(**3**,66), II/15(**3**,84)

	CHP+	**Methylidynephosphine(1+) ion**	$C_{\infty v}$
UV	Rotational analysis	Methinophosphide cation	H–C≡P+

State	$\tilde{X}\,^2\Pi$	$\tilde{A}\,^2\Sigma$
Symmetry	$C_{\infty v}$	$C_{\infty v}$
Energy [eV]	0	2.079
r_0 [Å] C–H a)	0.996(21)	1.072(21)
C≡P a)	1.559(20)	1.540(21)

a) Values calculated here from rotational constants B_0 given for HCP+ and DCP+.

King, M.A., Kroto, H.W., Nixon, J.F., Klapstein, D., Maier, J.P., Marthaler, O.: Chem. Phys. Lett. **82** (1981) 543.   II/15(**3**,85)

	CH₂⁻	**Methylene anion**	C_{2v}
PES			CH₂⁻

State	$\tilde{X}\,^2B_1$
θ_e [deg] H–C–H	103

The equilibrium angle was calculated by fitting a theoretical model [1] to the photoelectron spectrum of CH₂⁻ [2, 3].

[1] Bunker, P.R., Sears, T.J.: J. Chem. Phys. **83** (1985) 4866.
[2] Leopold, D.G., Murray, K.K., Lineberger, W.C.: J. Chem. Phys. **81** (1984) 1048.
[3] Leopold, D.G., Murray, K.K., Miller, A.E.S., Lineberger, W.C.: J. Chem. Phys. **83** (1985) 4849.   II/21(**3**,30)

	CH₂	**Methylene**	C_{2v}
		Carbene	H–C–H
UV, IR, Rotational, Zeeman and hfs analyses			or
LMR			$\overset{C}{H\;\;\;H}$

State		$\tilde{X}\,^3B_1$	$\tilde{a}\,^1A_1$	$\tilde{b}\,^1B_1$	$\tilde{B}\,^3\Sigma_u^-$
Symmetry		C_{2v}	C_{2v}	$C_{2v}/D_{\infty h}$ a)	$D_{\infty h}$
Energy [eV]		0	0.392(3) b)	1.273(3)	8.758
Reference		[2], [3] c)	[1], [4]	[1]	[1]
r_0 [Å]	H–C	1.085 d)	1.111	1.053 e)	1.079 f)
r_e [Å]	H–C	1.0748(4) g) h)	1.107(5) j)		
r_{lin} [Å]	H–C	1.060(5) i)			
θ_0 [deg]	H–C–H	135.5 d)	102.4	≅140	180
θ_e [deg]	H–C–H	133.84(5) g) h)	102.4(8) j)		
h(barrier) [eV]		0.241(10) g)			

(*continued*)

CH$_2$ (*continued*)

[a]) Quasi-linear. Levels of the bending vibration observed only in the range $4 \leq v \leq 18$.
[b]) From $\tilde{X} - \tilde{a}$ perturbations [3].
[c]) Contains a comprehensive review of the structure of CH$_2$, including references to a voluminous literature on *ab initio* calculations.
[d]) Calculated by averaging appropriate wave-functions over vibrational amplitudes.
[e]) Extrapolated value, see footnote [a]).
[f]) Refers to CD$_2$. Bands of CH$_2$ are too diffuse to give rotational constants accurately enough to test the linearity of the molecule in the $\tilde{B}$-state.
[g]) From a fit of rotational-vibrational energy levels of CH$_2$, CD$_2$ and ^{13}CH$_2$ to a non-rigid-bender Hamiltonian ([2], with references to the sources of data).
[h]) Parameters of a non-rigid-bender potential given in [2].
[i]) The bond-length at the potential maximum in the bending mode, at the linear configuration.
[j]) From a study of the symmetric and antisymmetric stretching bands [5].

[1] Herzberg, G., Johns, J.W.C.: Proc. Roy. Soc. (London) Ser. **A 295** (1966) 107.
[2] Bunker, P.R., Jensen, P.: J. Chem. Phys. **79** (1983) 1224.
[3] McKellar, A.R.W., Bunker, P.R., Sears, T.J., Evenson, K.M., Saykally, R.J., Langhoff, S.R.: J. Chem. Phys. **79** (1983) 5251.
[4] Sears, T.J., Bunker, P.R.: J. Chem. Phys. **79** (1983) 5265.
[5] Petek, H., Nesbitt, D.J., Darvin, D.C., Ogilby, P.R., Moore, C.B., Ramsay, D.A.: J. Chem. Phys. **91** (1989) 6566.   II/7(**3**,67), II/15(**3**,86), II/21(**3**,31)

CH$_2$ArO	Argon – formaldehyde (1/1)	II/21(**3**,32)
CH$_2$ArO$_2$	Formic acid – argon (1/1)	II/23(**3**,48)
CH$_2$BrCl	Bromochloromethane	II/15(**3**,87)
CH$_2$BrF	Bromofluoromethane	II/7(**3**,69)
CH$_2$BrN	Hydrogen bromide – hydrogen cyanide (1/1)	
		II/15(**3**,87a), II/21(**3**,33), II/23(**3**,49)

CH$_2$Br$_2$ **Dibromomethane** C$_{2v}$
MW   CH$_2$Br$_2$

r_s	Å [b])	θ_s	deg [b])
C–H	1.097(5) [a])	Br–C–Br	112.9(2)
C–Br	1.925(2)	H–C–H	110.9(8)

[a]) Assumed.
[b]) Uncertainties are slightly larger than those of the original data.

Chadwick, D., Millen, D.J.: J. Mol. Struct. **25** (1975) 216.   II/7(**3**,68), II/15(**3**,88)

CH$_2$Cl **Chloromethyl radical** C$_{2v}$
MW   CH$_2$Cl

r_0	Å	θ_0	deg
C–H	1.09($\pm$1) [a])[b])	H–C–H	122.6($\mp$20) [b])
C–Cl	1.691($\mp$4) [b])		

[a]) Assumed.
[b]) The double signs should be taken in the same order.

Endo, Y., Saito, S., Hirota, E.: Can. J. Phys. **62** (1984) 1347.   II/15(**3**,88a)

CH₂ClF	Chlorofluoromethane	II/7(**3**,71), II/15(**3**,89)
CH₂ClF₂OP	(Chloromethyl)phosphonic difluoride	II/21(**3**,34)
CH₂ClN	Hydrogen chloride – hydrogen cyanide (1/1)	
		II/15(**3**,90), II/23(**3**,50)
CH₂ClNO₂	Chloronitromethane	II/7(**3**,72)
CH₂ClP	Methylenephosphinous chloride	II/15(**3**,91)

CH₂Cl₂   **Dichloromethane** C_{2v}
Methylene dichloride
                                                CH₂Cl₂

MW, IR

r_s	Å	θ_s	deg
C–H	1.085(2)	H–C–H	112.1(2)
C–Cl	1.767(2)	Cl–C–Cl	112.2(1)

r_z	Å	θ_z	deg
C–H	1.0885(14)	H–C–H	112.10(18)
C–Cl	1.77155(27)	Cl–C–Cl	111.96(3)
δ(CH – CD)	0.0030(8)	δ(HCH – HCD)	−0.035(22)

r_e	Å	θ_e	deg
C–H	1.080(3)	H–C–H	112.10(20) [b]
C–Cl	1.766(2)	Cl–C–Cl	111.96(10) [b]

[a]) Fixed.
[b]) Assumed.

Duncan, J.L.: J. Mol. Struct. **158** (1987) 169.
See also: Davis, R.W., Robiette, A.G., Gerry, M.C.L.: J. Mol. Spectrosc. **85** (1981) 399.

MW

r_m^ρ [c])	Å	θ_m^ρ [c])	deg
C–H	1.0851(11)	H–C–H	111.90(17)
C–Cl	1.7636(3)	Cl–C–Cl	112.25(3)

[a]) Multiple isotope substitution structure.

Berry, R.J., Harmony, M.D.: Struct. Chem. **1** (1990) 49.
                                                II/7(**3**,70), II/15(**3**,92), II/21(**3**,35), II/23(**3**,51)

CH₂Cl₃OP	(Chloromethyl)phosphonic dichloride	II/15(**3**,93)
CH₂Cl₃P	Dichloro(chloromethyl)phosphine	II/15(**3**,94)
CH₂Cl₃PS	(Chloromethyl)phosphonothioic dichloride	II/15(**3**,95)
CH₂Cl₄O₂P₂	Methylenebis(phosphonic dichloride)	II/21(**3**,36)
	Bis(dichlorophosphinyl)methane	
CH₂Cl₄P₂	Methylenebis(phosphonous dichloride)	II/21(**3**,37)
CH₂Cl₄Si	(Chloromethyl)trichlorosilane	II/15(**3**,96)
CH₂Cl₄Sn	(Chloromethyl)trichlorostannane	II/7(**3**,73)
CH₂Cl₆Si₂	Bis(trichlorosilyl)methane	II/15(**3**,97)
	1,1,1,3,3,3–Hexachloro–1,3–disilapropane	

	CH$_2$F	**Fluoromethyl radical**	C$_s$ [a]
MW			CH$_2$F

r_0	Å	θ_0	deg
C–H	1.09($\pm$1) [b][c]	H–C–H	126.3($\mp$21) [c]
C–F	1.3337($\mp$49) [c]		

[a]) The barrier to inversion is low, so that the effective symmetry is C$_{2v}$.
[b]) Assumed.
[c]) The double signs should be taken in the same order.

Endo, Y., Saito, S., Hirota, E.: J. Chem. Phys. **79** (1983) 1605.   II/15(**3**,97a)

CH$_2$FN	Hydrogen cyanide – hydrogen fluoride (1/1)	II/15(**3**,98), II/21(**3**,38)
CH$_2$FP	Fluoro(methylene)phosphine	II/23(**3**,52)

	CH$_2$F$_2$	**Difluoromethane**	C$_{2v}$
MW			CH$_2$F$_2$

r_e	Å	θ_e	deg
C–H	1.084(3)	H–C–H	112.8(3)
C–F	1.3508(5)	F–C–F	108.49(6)

r_z [a])	Å	θ_z	deg
C–H	1.097(5)	H–C–H	113.67 [b]
C–F	1.3601(14)	F–C–F	108.11(16)

[a]) For ^{12}CH$_2$F$_2$. The difference between C–H in ^{12}CH$_2$F$_2$ and C–D in ^{12}CD$_2$F$_2$ is 0.0025 Å.
[b]) Assumed.

Hirota, E.: J. Mol. Spectrosc. **71** (1978) 145.   II/7(**3**,74), II/15(**3**,99)

CH$_2$F$_3$P	(Trifluoromethyl)phosphine	II/7(**3**,75)
CH$_2$F$_4$P$_2$S$_2$	Bis(difluorophosphonothionyl)methane	II/15(**3**,100)
CH$_2$F$_4$S	Tetrafluoromethylenesulfur	II/15(**3**,100a)
CH$_2$F$_6$Si$_2$	Bis(trifluorosilyl)methane	II/15(**3**,101)
CH$_2$F$_{10}$S$_2$	Methylenebis(pentafluorosulfur)	II/23(**3**,53)
CH$_2$IN	Hydrogen iodide – hydrogen cyanide (1/1)	II/23(**3**,54)
CH$_2$N$^+$	Protonated hydrogen cyanide	II/21(**3**,39)
CH$_2$N$_2$	Diazomethane	II/7(**3**,76)

	CH$_2$N$_2$	**Cyanamide**	C$_{2v}$
			(large amplitude inversion)
MW			H\\N—C≡N / H

r_s	Å	θ_s	deg
N–H	1.001(15)	α [c]	56.8(10)
N–C	1.346(5)		
(N–C) + (C≡N)	2.506(2)		

r(N–H) [Å] = 0.9994 + 0.0114ρ^2; θ(H–N–H)/2 = 60.39° – 0.1134ρ^2;
r(N–C) [Å] = 1.3301 + 0.0327ρ^2 (ρ is the inversion angle in [rad]);   *(continued)*

$r(\text{C–N}) = 1.1645$ Å is assumed.
$b^{a)} = -0.116(14)$ rad^{-1}; barrier = 510.1(56) cm^{-1}; $\phi_e^{b)} = 45.03(20)°$.
N–H = 0.9994 Å and C–N = 1.1645 Å are assumed.

^a) The NCN angle is $180° + b\rho$ where b is the backbend parameter. The cyanide nitrogen is tilted away from the hydrogens by 5.2° at an inversion angle of 45°.
^b) Equilibrium value of the inversion angle.
^c) Half of the H–N–H angle.

Brown, R.D., Godfrey, P.D., Kleibömer, B.: J. Mol. Spectrosc. **114** (1985) 257.
Read, W.G., Cohen, E.A., Pickett, H.M.: J. Mol. Spectrosc. **115** (1986) 316.

II/7(**3**,77), II/21(**3**,40)

CH$_2$N$_2$ **Diazirine** C_{2v}

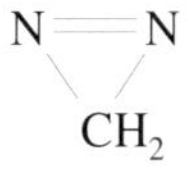

MW

r_s	Å	θ_s	deg
C–N	1.4813(24)	N–C–N	48.98(15)
C–H	1.0803(29)	H–C–H ^a)	120.54(27)
N=N	1.2280(25)		

^a) The HCH plane is perpendicular to the NCN plane.

Verma, U.P., Möller, K., Vogt, J., Winnewisser, M., Christiansen, J.J.: Can. J. Phys. **63** (1985) 1173.

II/7(**3**,78), II/21(**3**,41)

CH$_2$O   **Formaldehyde** C_{2v}
   Methanal

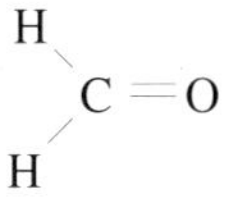

MW

$r_m^{\rho\ a)}$	Å	$\theta_m^{\rho\ a)}$	deg
C–H	1.1012(2)	H–C–H	116.25(4)
C=O	1.2031(1)		

^a) Multiple isotope substitution structure.

Berry, R.J., Harmony, M.D.: Struct. Chem. **1** (1990) 49.

UV   Rotational-vibrational analysis

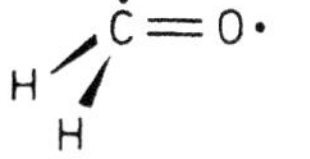

State		$\tilde{a}\ ^3A_2(A'')$ ^a)	$\tilde{A}\ ^1A_2(A'')$ ^a)
Symmetry		$C_{2v}\ (C_s)$	$C_{2v}\ (C_s)$
Energy [eV]		3.124	3.495
Reference		[1] [3] [6]	[2] [3]
r_0 [Å]	C–H	1.10	1.097(1) ^c)
	C=O	1.28	1.321(1) ^c)
θ_0 [deg]	H–C–H	116	118.6(1) ^c)
ϕ_0 [deg]	o–o–p ^b)	38	31.75(25) ^c)
r_s [Å]	C–H		1.095(1)
	C=O		1.320(1)

(*Table continued*)

CH$_2$O (*Table continued*)

State	$\tilde{a}\,^3A_2(A'')$ [a]	$\tilde{A}\,^1A_2(A'')$ [a]
Symmetry	C_{2v} (C_s)	C_{2v} (C_s)
Energy [eV]	3.124	3.495
Reference	[1] [3] [6]	[2] [3]
θ_s [deg] H–C–H		118.78(25)
ϕ_s [deg] o–o–p [b]		30.73(28)
r^0_{sri} [Å] C–H [d]	1.0835(5) [e]	1.108(12) [f]
C=O	1.307(5) [e]	1.323(3) [f]
θ^0_{sri} [deg] H–C–H	121.77(4) [e]	118.11(8) [f]
ϕ^0_{sri} [deg] o–o–p [b]	41.14 [e][g]	34.01 [f]
H_e [cm^{-1}/eV] [h]	775.6/0.096 [e][g]	350.3/0.043 [f]

[a]) The stable equilibrium configurations of the nuclei are pyramidal; only a plane C_s symmetry remains with respect to which the states are A''. Analysis of the out-of-plane potential and its energy-levels is however usually made relative to the planar (unstable) equilibrium configuration of higher symmetry C_{2v}.

[b]) o–o–p: out-of-plane.

[c]) Level $v_{1,2,3,5,6} = 0$, $v_4 = 1$. Average of values obtained from several combinations of isotopic data [2].

[d]) $(r,\theta)_{sri}$: values in a semi-rigid-inverter model in which r is allowed to vary during the course of the out-of-plane vibration v_4, i.e., in which (r,θ) are functions of the out-of-plane angle ϕ. $(r^0,\theta^0)_{sri}$: values of r, θ in the planar configuration, $\phi = 0$. Values quoted obtained by fitting the model to rotational and vibrational data for H$_2$CO, H$_2^{13}$CO and D$_2$CO [3].

[e]) Appreciably different values are obtained for the v_2(C=O stretch) = 1 vibronic level [3].

[f]) Somewhat different values were obtained by a similar calculation for H$_2$CO only. The value of r_{sri}(C–H) and θ_{sri}(H–C–H) appear to decrease by about 0.034 Å and 6° as the molecule bends from the planar to the pyramidal configuration [4].

[g]) Assuming the values for r^0_{sri}(C–H), r^0_{sri}(C=O) and θ^0_{sri} of [3] and fitting rotational constants for $\tilde{a}\,^3A_2$ in HDCO to the rigid-bender double-minimum potential of [5] gives values of $\phi_e = 30°$, $H_e = 738$ cm^{-1} [6].

[h]) Out-of-plane angle ϕ_e and barrier height H_e for the same semi-rigid-inverter potential as in footnote [d]).

[1] Raynes, W.T.: J. Chem. Phys. **44** (1966) 2755.
[2] Shah, A.K., Moule, D.C.: Spectrochim. Acta **34A** (1978) 749.
[3] Jensen, P., Bunker, P.R.: J. Mol. Spectrosc. **94** (1982) 114.
[4] Ramachandra Rao, C.V.S.: J. Mol. Spectrosc. **95** (1982) 239.
[5] Coon, J.N. Naugle M.W., McKenzie, F.D.: J. Mol. Spectrosc. **20** (1966) 107.
[6] Clouthier, D.O., Craig, A.M., Ramsay, D.A.: Can. J. Phys. **62** (1984) 973.

II/7(**3**, 79), II/15(**3**,102), II/23(**3**,55)

CH$_2$OS	Thioformic acid	II/15(**3**,103)
CH$_2$OS	Thioformaldehyde S–oxide	II/15(**3**,104)

CH₂O₂ **Formic acid** C_s

MW

r_s	Å		θ_s	deg	
	cis	trans [a]		cis	trans [a]
C–H	1.1050(43)	1.0971(30)	O=C–O	122.12(37)	124.82(30)
C=O	1.1945(31)	1.2036(30)	H–C–O	114.64(56)	111.97(30)
C–O	1.3520(28)	1.3424(30)	C–O–H	109.68(44)	106.34(50)
O–H	0.9555(53)	0.9721(50)	H–C=O	123.23(58)	123.21(30)

trans:

r_z	Å [a]	θ_z	deg [a]
C–H	1.097(3)	O=C–O	124.80(30)
C=O	1.205(3)	H–C–O	111.94(30)
C–O	1.347(3)	C–O–H	106.61(50)
O–H	0.966(5)	H–C=O	123.26(30)

r_e	Å [a]	θ_e	deg [a]
C–H	1.091(5)	O=C–O	124.80(50)
C=O	1.201(5)	H–C–O	111.94(50)
C–O	1.340(5)	C–O–H	106.61(50)
O–H	0.969(5)	H–C=O	123.26(50)

[a]) Uncertainties were not estimated in the original paper.

cis: Bjarnov, E., Hocking, W.H.: Z. Naturforsch. **33a** (1978) 610.
trans: Davis, R.W., Robiette, A.G., Gerry, M.C.L., Bjarnov, E., Winnewisser, G.: J. Mol. Spectrosc. **81** (1980) 93.
See also: Bellet, J. Deldalle, A., Samson, C., Steenbeckeliers, G., Wertheimer, R.: J. Mol. Struct. **9** (1971) 65.

UV Band contour analysis

State	$\tilde{A}^1A$
Symmetry	C_1
Energy [eV]	4.779
r [Å] C–O	1.407
θ [deg] O–C=O	111.4
out-of-plane angle ϕ [deg]	32

From rotational constants A, B and C, assuming that these are the only three geometric parameters that change on excitation from the ground state.

Ng, T.L., Bell, S.: J. Mol. Spectrosc. **50** (1974) 166.

II/7(**3**,80), II/15(**3**,105), II/23(**3**,56)

CH₂O₂	Dioxirane	II/15(**3**,106)
CH₂O₂	Carbon monoxide – water (1/1)	II/23(**3**,57)
CH₂O₂S	Carbon dioxide – hydrogen sulfide (1/1)	II/23(**3**,58)

CH_2O_3		Performic acid	II/15(**3**,106a)
CH_2O_3		Water – carbon dioxide (1/1)	II/15(**3**,106b)

CH_2S^- **Thioformaldehyde(1–) ion** C_{2v}
Methanethial anion

PES H_sCS^-

State	$\tilde{X}\,^2B_1$
r_0 [Å] C–S	1.72(2)

From fitting the intensities of the peaks in the photoelectron spectrum together with the known geometrical structure for the neutral species.

Moran, S., Ellison, G.B.: J. Phys. Chem. **92** (1988) 1794.   II/21(**3**,42)

CH_2S   **Thioformaldehyde** C_{2v}

MW $H_2C=S$

		r_s	Å	θ_s		deg
CH_2S	C=S		1.61077(1)	H–C–H		116.52(3)
	C–H		1.08692(3)			
CD_2S	C=S		1.61038(1)	D–C–D		116.47(3)
	C–D		1.08716(4)			

		r_z	Å	θ_z		deg
CH_2S	C=S		1.6138(4)	H–C–H		116.27(10)
	C–H		1.0962(6)			
CD_2S	C=S		1.6136(4)	D–C–D		116.42(8)
	C–D		1.0931(4)			

Cox, A.P., Hubbard, S.D., Kato, H.: J. Mol. Spectrosc. **93** (1982) 196.

IR, MW

r_e	Å	θ_e	deg
C=S	1.6110(8)	H–C–H	115.9(11)
C–H	1.0856(2)		

The two bond distances were estimated by Turner et al. from MW gound state constants and r_z bond lengths. The H–C–H angle was obtained from these bond distances and the (H···H) distance determined by Nakata et al.

Turner, P.H., Halonen, L., Mills, I.M.: J. Mol. Spectrosc. **88** (1981) 402.
Nakata, M., Kuchitsu, K., Mills, I.M.: J. Phys. Chem. **88** (1984) 344.

(continued)

UV   Rotational-vibrational analysis

State	$\tilde{a}\ ^3A_2(A'')$ [a]	$\tilde{A}\ ^1A_2$
Symmetry	$C_{2v}\ (C_s)$	C_{2v}
Energy [eV]	1.799	2.033
Reference	[3], [4]	[1], [2], [4]
r_0 [Å] C–H	1.082(8)	1.075(10)
C=S	1.683(5)	1.707(10)
θ_0 [deg] H–C–H	119.6(17)	121.6(17)
ϕ_0 [deg] o–o–p [b]	16(3) [c]	8.9(17) [c]
r_s [Å] C–H		1.082(10)
C=S		1.701(10) [d]
θ_s [deg] H–C–H		120.0(15)
ϕ_s [deg] o–o–p [b]		9.7(15)
r^0_{sri} [Å] C–H [e]	1.0819(5)	1.0763(3)
C=S	1.683(1)	1.694(12)
θ^0_{sri} [deg] H–C–H	119.52(1)	120.54(14)
ϕ_e [deg] o–o–p [b]	12 [f]	0 [f]
	15(2) [g]	$0 \le \phi_e \le 17$ [g]
H_e [cm^{-1}] [h]	5	0
	13(3) [i]	$0 \le H_e \le 29$ [i]

[a]) Quasiplanar; equilibrium configuration possibly non-planar (cf. formaldehyde, CH$_2$O).

[b]) o–o–p: out-of-plane.

[c]) r_0–structures are calculated from mean moments of inertia. In quasiplanar molecules vibrating anharmonically over large amplitudes, inertia defects $(\langle I \rangle^c_v - \{\langle I \rangle^a_v + \langle I \rangle^b_v\})$ can be considerable and yield "average" out-of-plane angles ϕ_0 that are non-zero even if the potential minimum of the out-of-plane motion is still at the planar configuration. Conversely, rotational constants by themselves are not a good criterion of "planarity" or otherwise.

[d]) The r_s-structures were obtained from isotopic variation of moments of inertia in H$_2$CS and D$_2$CS only. These locate the hydrogen atoms. To locate the C and S atoms, overall first- and second-moment equations were used on the moments of inertia themselves in the usual way, but the solutions found were insensitive to choice of constants used.

[e]) $(r,\theta)_{sri}$: geometric parameters in a semi-rigid-inverter model potential in which they are allowed to vary during the course of the out-of-plane bending vibration, ν_4, i.e., they are function of ϕ. $(r,\theta)^0_{sri}$: values in the planar configuration, $\phi = 0$. Based on rotational-vibrational energy-levels of H$_2$CS and D$_2$CS. As they are still averaged over the zero-point amplitudes of the motions other than ν_4, they are not quite isotopically invariant. The uncertainties quoted refer therefore to the mean and range of values for the two isotopes. Precisions needed to reproduce the data are much higher [4].

[f]) Coordinates of minima in the same semi-rigid-inverter potential as footnote [e]) [4].

[g]) Coordinates of minima in two rigid-inverter potentials of quadratic-cum Gaussian or quadratic-cum-quartic form in the out-of-plane displacement coordinate Q_4 [1].

[h]) Barrier-height between minima, same potential as footnotes [e]) and [f]). For comparison, zero-point energies in ν_4 are 120···150 cm^{-1}.

[i]) Same potential as in footnote [g]).

(continued)

CH$_2$S (*continued*)

[1] Judge, R.H., King, G.W.: J. Mol. Spectrosc. **74** (1979) 175.
[2] Judge, R.H., King, G.W.: J. Mol. Spectrosc. **78** (1979) 51.
[3] Judge, R.H., Moule, D.C.: J. Mol. Spectrosc. **81** (1980) 37.
[4] Jensen, P., Bunker, P.R.: J. Mol. Spectrosc. **95** (1982) 92.   II/7(**3**,81), II/15(**3**,107)

| CH$_2$S$_2$ | Dithioformic acid | II/15(**3**,108) |

CH$_2$Se   **Methaneselenal**   C$_{2v}$
Selenoformaldehyde
H$_2$C=Se

MW

r_0	Å [a]		θ_0	deg [a]
C=Se	1.7561(10)		H–C–H	117.35(50)
C–H	1.0805(30)			

r_s	Å [a]		θ_s	deg [a]
C=Se	1.7531(10)		H–C–H	117.93(50)
C–H	1.0904(30)			

[a] Not all the uncertainties were estimated in the original paper.

Brown, R.D., Godfrey, P.D., McNaughton, D., Taylor, P.R.: J. Mol. Spectrosc. **120** (1986) 292.

LIF

State	$\tilde{A}\ ^1A_2$ [a]
Energy [eV]	1.690

r_0 [Å]	C–H	1.075 [b]
	C=Se	1.856 (4) [c]
θ_0 [deg]	H–C–H	121.6 [b]
out-of-plane angle [deg]		12.2 (5)

Rotational analysis of the 0–0 band of D$_2$C^{80}Se. Data are also available for the 4_0^1 bands of H$_2$C^{78}Se and H$_2$C^{80}Se.

[a] Values are for the zero level.
[b] Fixed at the value for D$_2$CS.
[c] Error limits are 3σ.

Clouthier, D.J., Judge, R.H., Moule, D.C.: J. Mol. Spectrosc. **141** (1990) 175.   II/21(**3**,43)

CH$_3$ **Methyl radical**   D$_{3h}$
Raman   CH$_3$

r_e	Å
C–H	1.076(1) [a]

The B_e rotational constant is obtained from the experimental B_0 constant by using experimental values for three of the needed α_i^B parameters and a calculated one only for α_4^B. The equilibrium structure thus derived seems therefore more reliable than the 1982 structure [2] utilizing four calculated α_i^B.

(*continued*)

a) Uncertainty was not given in the original paper.

[1] Triggs, N.E., Zahedi, M., Nibler, J.W., DeBarber, P., Valentini, J.J.: J. Chem. Phys. **96** (1992) 1822.
[2] Hirota, E., Yamada, C.: J. Mol. Spectrosc. **96** (1982) 175.

IR

	r_e	Å
C–H		1.0767

Hirota, E., Yamada, C.: J. Mol. Spectrosc. **96** (1982) 175.
Spirko, V., Bunker, P.R.: J. Mol. Spectrosc. **95** (1982) 381; errata: **99** (1983) 243.

UV   Absorption and multiphoton ionization spectra, rotational analysis

State	$(e')^4 a_2''$	$(e')^4 ns\, a_1'$	$(e')^4 np\, a_2''$	$(e')^4 nd\, a_1'$
	$\tilde{X}\,^2A_2''$	$\tilde{B}\,^2A_1'(R,3s)$	$^2A_2''(R,3p)$	$\tilde{D}\,^2A_1'(R,3d)$
Symmetry	D_{3h}	D_{3h}	D_{3h}	D_{3h}
Energy [eV]	0	5.729	7.425 [a]	8.282
Reference	[1] [2]	[1] [2]	[2]	[1] [2]
r_0 [Å] C–H	1.081(2) [a]	1.130(3) [a]	1.084(2) [a]	1.053(2) [a]
θ_0 [deg] H–C–H	120	120	120	120

a) From spectra of CD_3.

[1] Herzberg, G.: Proc. Roy. Soc. (London) Ser. **A 262** (1961) 291.
[2] Hudgens, J.W., DiGiuseppe, T.D., Lin, M.C.: J. Chem. Phys. **79** (1983) 571.

II/7(**3**,82), II/15(**3**,109), II/23(**3**,59)

IR   **CH_3^+**   **Methyl cation** D_{3h}
     Methylium ion CH_3^+

	r_0	Å
C–H		1.095

	r_z	Å
C–H		1.1019

The r_0 bond distance is an average of those values obtained from the B_0 and C_0 rotational constants.

Crofton, M.W., Jagod, M.-F., Rehfuss, B.D., Kreiner, W.A., Oka, T.: J. Chem. Phys. **88** (1988) 666.

II/21(**3**,44)

CH_3ArCl	Argon – methyl chloride (1/1)	II/15(**3**,110), II/21(**3**,45)
CH_3ArNO	Argon – formamide (1/1)	II/21(**3**,46)
CH_3AsF_2	Difluoromethylarsine	II/7(**3**,83)
CH_3BBr_2	Dibromomethylborane	II/21(**3**,47)
CH_3BCl_2	Dichloromethylborane	II/21(**3**,48)
CH_3BF_2	Difluoromethylborane	II/15(**3**,111)

CH$_3$BO	Carbonyltrihydroboron		II/7(**3**,84), II/15(**3**,112)
CH$_3$BS	Methylthioxoboron		II/15(**3**,113)
	Methylborylene sulfide		

CH$_3$Br **Methyl bromide** C$_{3v}$
MW, IR, Ra CH$_3$Br

r_e	Å	θ_e	deg
C–H	1.0823(5)	H–C–H	111.157(50)
C–Br	1.9340(3)		

[a]) Uncertainties were not estimated in the original paper.

Graner, G.: J. Mol. Spectrosc. **90** (1981) 394.

II/7(**3**,85), II/15(**3**,114)

CH$_3$BrHg	Methylmercury bromide	II/7(**3**,86), II/15(**3**,115)
CH$_3$Br$_2$PS	Dibromomethylthiophosphine	II/15(**3**,116)
CH$_3$Br$_3$Ge	Methylgermyl tribromide	II/15(**3**,117)
	Tribromomethylgermane	
CH$_3$Br$_3$Si	Methyltribromosilane	II/7(**3**,87)
	Tribromomethylsilane	

CH$_3$Cl **Methyl chloride** C$_{3v}$
MW CH$_3$Cl

r_e	Å	θ_e	deg
C–H	1.0872(50)	H–C–H	110.35(32)
C–Cl	1.7756(20)		

Imachi, M., Tanaka, T., Hirota, E.: J. Mol. Spectrosc. **63** (1976) 265.

IR, Ra

r_e	Å	θ_e	deg
C–H	1.0854(5)	H–C–H	110.35(5)
C–Cl	1.7760(3)		

Jensen, P., Brodersen, S., Guelachvili, G.: J. Mol. Spectrosc. **88** (1981) 378.

II/7(**3**,88), II/15(**3**,118)

CH$_3$ClHg **Methylmercury chloride** C$_{3v}$
MW CH$_3$HgCl

r_e	Å [a])	θ_e	deg [a])
C–Hg	2.052(5)	H–C–H	109.7(10)
Hg–Cl	2.285(3)		
C–H	1.092(10)		

[a]) Uncertainties were not estimated in the original paper.

Walls, C., Lister, D.G., Sheridan, J.: J. Chem. Soc., Faraday Trans. II **71** (1975) 1091.

II/7(**3**,89), II/15(**3**,119)

CH$_3$ClN$_2$O$_2$	N–Chloro–N–methylnitramine	II/15(**3**,120)
	N–Chloro–N–nitromethylamine	
CH$_3$ClO	Methyl hypochlorite	II/7(**3**,90)
	Hypochlorous acid methyl ester	
CH$_3$ClO	Formaldehyde – hydrogen chloride (1/1)	II/21(**3**,49)
CH$_3$ClO$_2$S	Methanesulfonyl chloride	II/7(**3**,93)
CH$_3$ClO$_3$S	Chlorosulfuric acid methyl ester	II/15(**3**,122)
	Methylchlorosulfate	
CH$_3$ClS	Methanesulfenyl chloride	II/7(**3**,94)
	Thiohypochlorous acid ethyl ester	
CH$_3$Cl$_2$OP	Dichloromethoxyphosphine	II/7(**3**,91), II/15(**3**,123)
	Methyl phosphinodichloridite	
CH$_3$Cl$_2$OP	Methylphosphonic dichloride	II/7(**3**,92)
CH$_3$Cl$_2$OPS	O–Methyl dichlorothiophosphate	II/15(**3**,124)
	O–Methyl phosphorodichloridothioate	
CH$_3$Cl$_2$OPS	S–Methyl dichlorothiophosphate	II/15(**3**,125)
	S–Methyl phosphorodichloridothioate	
CH$_3$Cl$_2$O$_2$P	O–Methyl phosphorodichloridate	II/15(**3**,126)
CH$_3$Cl$_2$P	Methyldichlorophosphine	II/15(**3**,127)
CH$_3$Cl$_2$PS	Methylthiodichlorophosphine	II/15(**3**,128)
CH$_3$Cl$_3$Ge	Methyltrichlorogermane, Trichloromethylgermane	II/15(**3**,129)

CH$_3$Cl$_3$Si   **Methyltrichlorosilane**   **C$_{3v}$**

Trichloromethylsilane

MW

CH$_3$SiCl$_3$

r_s	Å [a])	θ_s	deg [a])
Si–Cl	2.026(5)	Cl–Si–Cl	108.6(5)
Si–C	1.848(5)	C–Si–Cl	110.3(5)

[a]) Uncertainties were not estimated in the original paper.

Takeo, H., Matsumura, C.: Bull. Chem. Soc. Japan **50** (1977) 1633.   II/7(**3**,95), II/15(**3**,130)

CH$_3$Cl$_3$Sn	Methyltrichlorostannane, Trichloromethylstannane	II/7(**3**,96)
CH$_3$Cl$_3$Ti	Trichloromethyltitanium, Methyltrichlorotitanium	II/21(**3**,50)

CH$_3$F   **Fluoromethane**   **C$_{3v}$**

Methyl fluoride

ED, MW, IR

CH$_3$F

r_g	Å [a])	r_z	Å [a])	θ_z	deg [a])
C–F	1.391(1)	C–F	1.391(1)	F–C–H	108.7(2)
C–H	1.108(1)	C–H	1.098(1)		

r_e	Å [a])			θ_e	deg [a])
C–F	1.383(1)			F–C–H	108.8(3)
C–H	1.086(2)			H–C–H [b])	110.2(3)

The measurement was made at room temperature.

(*continued*)

CH₃F (continued)

[a]) Estimated limits of error.
[b]) Dependent parameter.

Egawa, T., Yamamoto, S., Nakata, M., Kuchitsu, K.: J. Mol. Struct. **156** (1987) 213.
II/7(**3**,97), II/15(**3**,131), II/21(**3**,51)

CH₃FO	Formaldehyde – hydrogen fluoride (1/1)	II/15(**3**,132), II/21(**3**,52)
CH₃FO₂S	Methanesulfonyl fluoride	II/7(**3**,101)
CH₃FO₃S	Fluorosulfuric acid methyl ester Methyl fluorosulfate	II/15(**3**,133)
CH₃F₂N	N,N–Difluoromethylamine, N,N–Difluoromethanamine	II/7(**3**,98)
CH₃F₂OP	Difluoromethoxyphosphine	II/7(**3**,100), II/23(**3**,60)
CH₃F₂OP	Methylphosphonic difluoride	II/15(**3**,134), II/21(**3**,53)
CH₃F₂O₂P	Difluoromethoxyphosphine oxide	II/23(**3**,61)
CH₃F₂P	Difluoro(methyl)phosphine	II/15(**3**,135)
CH₃F₂P	(Difluoromethyl)phosphine	II/23(**3**,62)
CH₃F₂PS	Methylphosphonothioic difluoride	II/15(**3**,136)
CH₃F₂PS	Difluoro(methylthio)phosphine	II/15(**3**,137)
CH₃F₃Ge	Trifluoro(methyl)germane	II/15(**3**,138)
CH₃F₃Ge	(Trifluoromethyl)germane	II/23(**3**,63)
CH₃F₃OSi	Trifluoromethoxysilane	II/7(**3**,102), II/15(**3**,139)
CH₃F₃S	Trifluoromethylsulfurane	II/21(**3**,54)
CH₃F₃Si	Trifluoromethylsilane	II/7(**3**,104), II/21(**3**,55)
CH₃F₃Si	(Trifluoromethyl)silane	II/21(**3**,56)
CH₃F₄NP₂	N,N–Bis(difluorophosphinyl)methylamine	II/7(**3**,99)
CH₃F₄NS	Sulfur tetrafluoride methylamide Tetrafluoro[methanamidato(2–)]sulfur	II/15(**3**,140)
CH₃F₄P	Methyl phosphorus tetrafluoride Tetrafluoromethylphosphorane	II/7(**3**,103)
CH₃GeN	Germyl cyanide	II/7(**3**,105)
CH₃GeNO	Germyl isocyanate	II/7(**3**,106), II/15(**3**,141), II/21(**3**,57)
CH₃GeNS	Germyl isothiocyanate	II/15(**3**,142)
CH₃HgI	Methylmercury iodide	II/7(**3**,107), II/15(**3**,143)

CH₃I   **Methyl iodide** C_{3v}
MW, IR   CH₃I

r_s	Å		θ_s	deg
C–H	1.0840(30)		H–C–I	107.47(20)
C–I	2.1358(20)			

Mallinson, P.D.: J. Mol. Spectrosc. **55** (1975) 94.   II/7(**3**,108), II/15(**3**,144)

CH₃N	Methylnitrene radical Methylaminylene	II/21(**3**,58), II/23(**3**,64)

	CH₃N	**Methanimine**		**C$_s$**
MW				CH$_2$=NH

r_s	Å	θ_s	deg
C=N	1.273(4)	C=N–H	110.4(15)
N–H	1.021(20)	H–C–H	117.0(3)
C–H(*cis*)	1.092(20)	N=C–H(*cis*)	125.1(20)
C–H(*trans*)	1.092(20)	N=C–H(*trans*)	117.9(20)

Pearson Jr., R., Lovas, F.J.: J. Chem. Phys. **66** (1977) 4149.   II/15(**3**,145)

	CH₃NO	**Nitrosomethane**		**C$_s$**
MW				

r_s	Å	θ_s	deg
CH₃NO		CH₃NO	
C–H(s)	1.0940(20)	C–N=O	113.16(30)
C–H(a)	1.0920(20)	H(s)–C–N	111.08(20)
N=O	1.2112(20)	H(a)–C–N	107.26(20)
C–N	1.4820(20)	H(a)–C–H(a)	109.27(20)
CD₃NO		CD₃NO	
C–D(s)	1.0962(20)	C–N=O	113.25(30)
C–D(a)	1.0935(20)	D(s)–C–N	110.91(20)
N=O	1.2116(20)	D(a)–C–N	107.07(20)
C–N	1.4789(20)	D(a)–C–D(a)	109.04(20)

Turner, P.H., Cox, A.P.: J. Chem. Soc., Faraday Trans. II **74** (1978) 533.   II/7(**3**,112), II/15(**3**,146)

CH₃NO	Formaldehyde oxime, Formaldoxine	II/7(**3**,110)

	CH₃NO	**Formamide**	**C$_s$**
MW	Internal coordinates in the large-amplitude-motion (LAM) rotation model		

r [Å]		α [deg]	
$r_{12} = 1.22$		α_{123} = NCO	
$r_{23} = r_{CN} + \chi_{CN}\,\tau^2$		$\alpha_{124} = 122.5$	
$r_{24} = 1.098$		α_{235} = CNH(*syn*) – 0.15 τ^2	
$r_{35} = 1.0 + 0.010\,\tau^2$		α_{236} = CNH(*anti*) + $\chi_{CNH(anti)}\,\tau^2$	
$r_{36} = 1.0 + 0.013\,\tau^2$		$d_5 = t_5\,\tau$	

	LAM-rotation model	r_s–structure
r(C–N)	1.3558 (20)	1.352(12)
χ(C–N)	0.0315 (20)	
θ(N–C=O)	124.45 (7)	124.7(3)
θ(C–N–H(*syn*))	118.8 (14)	118.5(5)
θ(C–N–H(*anti*))	121.35 (45)	120.0(5)
θ(H(*syn*)–N–H(*anti*))	119.85 [a]	121.5 [a]
τ_5	–0.201 (29)	
t_5	0.283 (31)	0.5 [a]
b	699.97 (1)	

(*continued*)

CH₃NO (*continued*)

r and r_s are in Å, θ and α in deg, χ_{CN} is in Å rad^{-2}, $\chi_{CNH(anti)}$ in rad^{-2}, b in cm^{-1} rad^{-1}, τ in rad and t_5 is dimensionless. Internal coordinates are 1: O, 2: C, 3: N, 4: H, 5: H(syn) and 6: H(anti). Formamide has a very shallow single-minimum inversion potential. During inversion the amino group rotates around the CN bond with the *syn*-H staying closer to the NCO plane than the *anti*-H. The formyl-H moves in the opposite direction to the amino-H atoms while the CN bond lengthens as the amino-H atoms move out of plane.

[a]) Assumed.

Brown, R.D., Godfrey, P.D., Kleibömer, B.: J. Mol. Spectrosc. **124** (1987) 34.

CH₃NO	**Carbon monoxide – ammonia (1/1)**	**C$_s$**
	(weakly bound complex)	(effective symmetry class)
MW		CO...NH₃

	NH₃···CO	NH₃···¹³CO	ND₃···CO
R_{cm} [a]) [Å]	3.59(3)	3.58(3)	3.63(3)
r(N···CO) [a]) [Å]	3.54(3)	3.53(3)	3.54(3)
γ [deg]	69 [b])	69 [b])	
θ [deg]	35.3(10)	35.5(10)	33.9(10)

[a]) These constants are calculated assuming no knowledge of the CO orientation. The numbers in parentheses reflect the uncertainty due to this lack of information.
[b]) Assumed.

Fraser, G.T., Nelson, D.D., Peterson, K.I., Klemperer, W.: J. Chem. Phys. **84** (1986) 2472.

CH₃NO	**Water – hydrogen cyanide (1/1)**	**C$_{2v}$**
	(weakly bound complex)	(effective symmetry class)
MW		H₂O···HCN

Atom	a [Å]	b [Å]
O	2.1937 [a]), 2.1936 [b])	
C	0.9585 [a]), 0.9582 [b])	
N	2.0979 [a]), 2.0980 [b])	
H	2.7403 [a])	0.7859 [b])

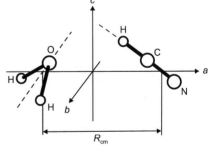

The out-of-plane H₂O bend is 20(2)° [c])[d]) and the in-plane is perhaps 10(2)° [c])[e]). The bending of the HCN is isotropic, with an amplitude of 9.4(5)° [c])[f]) in both directions.

[a]) Value from I_b.
[b]) Value from I_c.
[c]) Average angle.
[d]) Angle between the H₂O symmetry axis and the molecular plane of the complex at the equilibrium.
[e]) Angle between the H₂O symmetry axis and the plane, which is perpendicular to the complex equilibrium plane and includes R_{cm}.
[f]) Average angle between the H–C≡N and R_{cm}.

(*continued*)

Gutowsky, H.S., Germann, T.C., Augspurger, J.D., Dykstra, C.E.: J. Chem. Phys. **96** (1992) 5808.
II/15(**3**,146a), II/23(**3**,65)

| CH$_3$NOS | N–Sulfinylmethanamine | II/7(**3**,115) |
| CH$_3$NOSi | Silyl isocyanate | II/7(**3**,116), II/15(**3**,149) |

CH$_3$NO$_2$ Methyl nitrite C$_s$ (*trans, cis*)

MW

r_s	Å	
	cis	*trans*
O=N	1.182(5)	1.164(5)
N–O	1.398(5)	1.415(5)
O–C	1.437(5)	1.436(5)
C–H(s)	1.09 [a]	
C–H(a)	1.102(10)	

θ_s	deg	
	cis	*trans*
O=N–O	114.8(5)	111.8(5)
N–O–C	114.7(5)	109.9(5)
O–C–H(s)	101.8(15)	
O–C–H(a)	109.95	
H(a)–C–H(a)	108.1(15)	

[a]) Assumed.

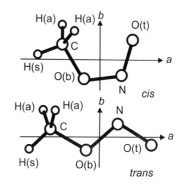

Turner, P.H., Corkhill, M.J., Cox, A.P.: J. Phys. Chem. **83** (1979) 1473.
See also: Endo, K., Kamura, Y.: Nippon Kagaku Kaishi (J. Chem. Soc. Jpn.) **1977**, 729.
II/15(**3**,147)

CH$_3$NO$_2$	Nitromethane	II/7(**3**,113), II/15(**3**,148)
CH$_3$NO$_2$	Ammonia – carbon dioxide (1/1)	II/15(**3**,148a)
CH$_3$NO$_3$	Methyl nitrate	II/7(**3**,114)
CH$_3$NS	Thioformamide	II/7(**3**,117)

CH$_3$NS Hydrogen sulfide – hydrogen cyanide (1/1) C$_s$
MW   (weakly bound complex)   (effective symmetry class)

Isotopic species	r_0 (S···C) [Å] [a]	ϕ [b] [deg] [a]
H$_2^{32}$S···HC^{14}N	3.809(5)	84.5(3)

[a]) Uncertainties were not estimated in the original paper.
[b]) See figure for definition.

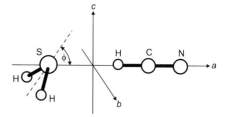

Goodwin, E.J., Legon, A.C.: J. Chem. Soc., Faraday Trans. II **80** (1984) 1669.
II/15(**3**,149a)

| CH$_3$NSSi | Silyl isothiocyanate | II/7(**3**,118), II/15(**3**,150) |
| CH$_3$NSi | Silanecarbonitrile, Silyl cyanide | II/7(**3**,119), II/21(**3**,61) |

CH₃N₃ MW

r_0	Å [a]	θ_0	deg [a]
C–H	1.079(3)	H(o)–C–H(o) [b]	106.8(3)
C–N(1)	1.483(2)	H(o)–C–N(1) [b]	112(1)
N(1)–N(2)	1.231(2)	H(i)–C–N(1) [b]	109(1)
N(2)–N(3)	1.137(2)	C–N(1)–N(2)	113.85(10)
		N(1)–N(2)–N(3)	173.1(2)

Methyl azide C_s

H₃C\
   N(1)–N(2)–N(3)

[a]) Uncertainties were not estimated in the original paper.
[b]) H(i) and H(o) are in and out of the CNNN plane, respectively.

Heineking, N., Gerry, M.C.L.: Z. Naturforsch. **44a** (1989) 669.   II/7(**3**,109), II/21(**3**,62)

CH₃N₃O₄   *N,N*–Dinitromethylamine
              Methyldinitramine   II/15(**3**,151)

CH₃O   **Methoxyl radical** C_{3v}

MW   CH₃O

r_s	Å	θ_s	deg
C–H	1.0958(12)	H–C–O	111.27(30)
C–O	1.3637(2)	H–C–H	107.61(32)

r_0	Å [a]	θ_0	deg [a]
C–H	1.0983(48)	H–C–O	111.599(6)
C–O	1.3614(12)	H–C–H	107.262(6)
δ(C–H) – (C–D)	0.003 [b]		

[a]) Uncertainties are about three times those of the original data.
[b]) Assumed.

Momose, T., Endo, Y., Hirota, E., Shida, T.: J. Chem. Phys. **88** (1988) 5338; errata **90** (1989) 4637.

LIF

State	$\tilde{X}\,^2E$	$\tilde{A}\,^2A_1$	
Energy [eV]	0.0	3.920	
Reference	[1]	[1]	[3]
r_0 [Å] C–H	1.106 [a]	1.100 [b]	1.118 [b]
C–O	1.363	1.578	1.573
θ_0 [deg] H–C–H	108.3	112.8	110

Rotational analysis of laser-excited fluorescence spectra. The ground state structure was deduced from a combination of optical and microwave data.

[a]) Error limits given in [2] are ± 0.02 Å for the bond lengths and ± 3···4° for the bond angles.
[b]) Assumed.

[1] Hsu, Y.-C., Liu, X., Miller, T.A.: J. Chem. Phys. **90** (1989) 6852.
[2] Liu, X., Damo, C.P., Lin, T.-Y.D., Foster, S.C., Misra, P., Yu, L., Miller, T.A.: J. Phys. Chem. **93** (1989) 2266.
[3] Kappert, J., Temps, F.: Chem. Phys. **132** (1989) 197.   II/15(**3**,152), II/21(**3**,63)

	CH₃OSr	Strontium monomethoxide	II/21(**3**,64)
ED	**CH₃O₃Re**	**Methyltrioxorhenium(VII)**	**C**$_{3v}$ **(staggered)** CH₃ReO₃

r_a	Å [a]		θ_a	deg [a]
Re=O	1.709(3)		Re–C–H	112(3)
Re–C	2.060(9)		C–Re=O	106.0(2)
C–H	1.105(12)		O=Re=O	113.0(3) [b]

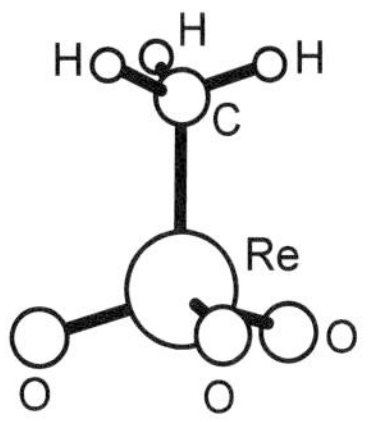

The nozzle temperature was 66(5) °C.

[a]) Twice the estimated standard errors including 0.1 % scale error.
[b]) Dependent value.

Herrmann, W.A., Kiprof, P., Rypdal, K., Tremmel, J., Blom, R., Alberto, R., Behm, J., Albach, R.W., Bock, H., Solouki, B., Mink, J., Lichtenberger, D., Gruhn, N.E.: J. Am. Chem. Soc. **113** (1991) 6527.

II/23(**3**,66)

	CH₃P	Methylenephosphine	II/15(**3**,153)
	CH₃S⁻	Methyl sulfide(1–) ion	II/21(**3**,65)
	CH₃S	Methylthio radical	II/21(**3**,66)
		Methylthiyl radical, Methanesulfenyl radical	

	CH₄	**Methane**	**T**$_d$
MW, IR			CH₄

r_e	Å
C–H	1.0870(7)

Hirota, E.: J. Mol. Spectrosc. **77** (1979) 213.

IR

r_e	Å
C–H	1.0858(10)

The bond distance was computed from a force field fitting of the IR spectrum of ^{12}CH₄, ^{13}CH₄, and all the deuterated species.

Gray, D.L., Robiette, A.G.: Mol. Phys. **37** (1979) 1901.

ED   CH₄, CD₄

r_g	Å [a]
C–H	1.1068(10)
H...H	1.811(7)
C–D	1.1027(10)
D...D	1.805(8)

The effect of anharmonic vibrations on the structure is examined.

[a]) Estimated standard errors.

Bartell, L.S., Kuchitsu, K., deNeui, R.J.: J. Chem. Phys. **35** (1961) 1211.

II/7(**3**,120), II/15(**3**,154)

CH₄⁺	Methane cation	II/23(**3**,67)
	Methaniumyl ion	
CH₄ArO	Argon – methanol (1/1)	II/21(**3**,67)
CH₄ClN	*N*–Chloromethylamine	II/7(**3**,121)
	N–Chloromethanamine	

CH_4Cl_2Si	Dichloromethylsilane		II/15(**3**,155)
CH_4F_2NP	Methylaminodifluorophosphine		II/15(**3**,156)
CH_4F_2Si	Difluoromethylsilane		II/7(**3**,122)
CH_4F_3N	Fluoroform – ammonia (1/1)		II/21(**3**, 68)
CH_4NP	Phosphine – hydrogen cyanide (1/1)		II/15(**3**,156a)
CH_4N_2	Methyldiazene		II/7(**3**,123)
CH_4N_2	Ammonia – hydrogen cyanide (1/1)		II/15(**3**,156b)
$CH_4N_2O_2$	N–Nitromethylamine, N–Methylnitramine		II/15(**3**,157)

CH_4O   **Methanol** C_s

ED, MW CH_3OH

r_z	Å [a])	θ_z	deg [a])
C–O	1.429(2)	H–C–H	109.1(2)
C–H	1.098(1)	C–O–H	107.6(9)
O–H	0.975(10)	ε [b])	3.40(5)

The nozzle was at room temperature.

[a]) Estimated limits of error.
[b]) Tilt angle of the methyl top away from the OH hydrogen.

Iijima, T.: J. Mol. Struct. **212** (1989) 137.   II/7(**3**,124), II/15(**3**,158), II/21(**3**,69)

CH_4OS	Methanesulfenic acid	II/15(**3**,159)
CH_4O_2Si	Silyl formate	II/15(**3**,160)

CH_4O_4   **Carbon dioxide – water (1/2)** C_s

MW   (weakly bound complex)   (effective symmetry class)

$CO_2 \cdots 2H_2O$

Structural parameter derived for six cases: A$\cdots$D use the six nondeuterated isotopic species and E$\cdots$F use all isotopic species. The constraints are as follows: the H_2O geometry is fixed for all cases as is CO_2 except for case C. (A) Planar; (B) planar except H_2O(456) rotated about C_{2v} axis by an amount τ_1; (C) the same as B but CO_2 geometry varied (bent); (D) planar except H(5) and H(8) are symmetrically out of plane by γ rotation about two OH bonds; (E) the same as D with all isotopes (scheme I, see original paper); and (F) the same as E (scheme II used).

Parameter	A planar	B τ_1 rotation	C bend CO_2	D γ rotation	E scheme I	F scheme II
R_1 [Å]	2.861(5)	2.859(4)	2.836(4)	2.861(4)	2.887(18)	2.903(12)
R_2 [Å]	2.901(6)	2.903(5)	2.916(2)	2.909(4)	2.887(29)	2.853(21)
θ_1 [deg]	12(5) [a])	12(5) [a])	12(8) [a])	8(5) [a])	26.5(30)	21.9(23)
θ_2 [deg]	37.4(36)	39.8(26)	29.1(16)	34.1(24)	57.2(23)	60.3(21)
α [deg]	86.9(2)	86.8(1)	85.8(2)	87.0(1)	86.9(9)	86.9(8)
β [deg]	82.6(1)	82.6(1)	82.5(1)	86.7(6)	68(3)	72.6(21)
τ_1 [deg]	0	30.1(4)	30.1(2)	0	0	0
γ [deg]	0	0	0	28.7(3)	32.9(24)	32.7(21)
CO_2 bend angle [deg]	0	0	–3.0(4)	0	0	0
r(C–O) [Å]	1.162 [b])	1.162 [b])	1.148(2)	1.162 [b])	1.162 [b])	1.162 [b])
r(H(6)$\cdots$O(7)) [Å]	2.069(6)	2.207(6)	2.212(3)	2.124(4)	1.94(4)	1.94(3)
r(H(8)$\cdots$O(3)) [Å]	2.25(4)	2.23(3)	2.37(2)	2.29(3)	2.13(19)	2.13(14)

(*continued*)

3 Organic molecules

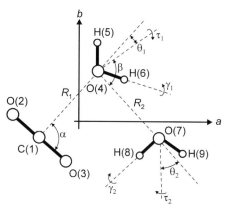

a) Value fixed within the uncertainty indicated.
b) Fixed value for both C–O bonds, and the fit value in case C.

Peterson, K.I., Suenram, R.D., Lovas, F.J.: J. Chem. Phys. **94** (1991) 106.

I/21(**3**,70), II/23(**3**,68)

CH$_4$S	Methanethiol	II/7(**3**,125)
CH$_4$S$_2$	Methyldisulfane	II/21(**3**,71)
	Methanesulfenothioic acid	
CH$_4$Se	Methaneselenol	II/7(**3**,126)
CH$_5$BO	Methoxyborane	II/21(**3**,72)
CH$_5$BrGe	Methylgermyl bromide	II/21(**3**,73)
CH$_5$BrSi	(Bromomethyl)silane	II/15(**3**,161)
CH$_5$Cl	Methane – hydrogen chloride (1/1)	II/23(**3**,69)
CH$_5$ClGe	Chloromethylgermane	II/15(**3**,162), II/23(**3**,70)
CH$_5$ClGe	(Chloromethyl)germane	II/15(**3**,163)
CH$_5$ClO	Methanol – hydrogen chloride (1/1)	II/15(**3**,163a)
CH$_5$ClSi	(Chloromethyl)silane	II/7(**3**,127)
CH$_5$ClSi	Chloro(methyl)silane	II/15(**3**,164), II/23(**3**,71)
CH$_5$F	Methane – hydrogen fluoride (1/1)	II/23(**3**,72)
CH$_5$FGe	Fluoromethylgermane	II/15(**3**,165), II/23(**3**,73)
	Methylgermyl fluoride	
CH$_5$FN$_2$	Hydrogen cyanide – hydrogen fluoride – ammonia (1/1/1)	
		II/23(**3**,74)
CH$_5$FSi	Fluoromethylsilane	II/7(**3**,128), II/15(**3**,166)
CH$_5$ISi	Methyl iodosilane	II/23(**3**,75)

CH$_5$N   **Methylamine**   C$_s$

MW   CH$_3$NH$_2$

r_0	Å	θ_0	deg
N–H	1.0181(8)	H–N–H	105.8(1)
C–N	1.471(3)	H–C–H	108.4(1)
C–H	1.0929(4)	C–N–H	111.0(3)
		tilt a)	3.4(12)

Fully deuterated species was investigated.

a) Tilt angle of the methyl group away from the H atoms of the amino group.

(*continued*)

CH$_5$N (*continued*)

Kreglewski, M., Stryjewski, D., Dreizler, H.: J. Mol. Spectrosc. **139** (1990) 182.
Kreglewski, M., Jäger, W., Dreizler, H.: J. Mol. Spectrosc. **144** (1990) 334.

II/7(**3**,129), II/21(**3**,74), II/23(**3**,76)

CH$_5$NO	*N*–Methylhydroxylamine	II/15(**3**,167)
CH$_5$NO	*O*–Methylhydroxylamine	II/15(**3**,168)

CH$_5$NO$_2$ **Formamide – water (1/1)**   **C$_s$**

MW   (weakly bound complex)   (effective symmetry class)

HC(=O)NH$_2$ · H$_2$O

r_0 [Å]	Tilt water	Tilt formamide	planar
O(1)···H(2)	2.025(10)	2.015(10)	2.020(20)
O(2)···H(1)	2.001(15)	2.023(30)	2.006(30)

θ_0 [deg]	Tilt water	Tilt formamide	planar
θ_1 [b])	107.5(10)	107.6(10)	107.5(30)
θ_2 [b])	143.3 [a])	143.3 [a])	143.3 [a])
ϕ [c])	15.3	5.0	0

[a]) Fixed.
[b]) See figure for definition.
[c]) Assumed tilt angle.

Lovas, F.J., Suenram, R.D., Fraser, G.T., Gillies, C.W., Zozom, J.: J. Chem. Phys. **88** (1988) 722.

II/21(**3**,75)

CH$_5$NSi$_2$	Disilanyl cyanide	II/15(**3**,169)
CH$_5$P	Methylphosphine	II/7(**3**,130)
CH$_6$BF$_2$P	Difluoromethylphosphine – borane (1/1)	II/15(**3**,170)
CH$_6$Ge	Methylgermane	II/7(**3**,131)
CH$_6$Ge$_2$N$_2$	Digermylcarbodiimide	II/7(**3**,132), II/15(**3**,171)

CH$_6$N$_2$ **Methylhydrazine**   **C$_1$** (inner form / outer form)

ED, MW and *ab initio*
calculations 4–31G*

NH$_2$NHCH$_3$

r_g	Å [a])	θ_z	deg [a])
C–H (mean) [b])	1.115(10)	C–N–N(i)	113.47(21)
N–H (mean) [c])	1.032 [d])	C–N–N(o)	109.46(15)
N–N(i)	1.433(12)		
N–C(i)	1.463(12)		
N–N(o)	1.431 [e])		
N–C(o)	1.466(2)		

The abundances of the inner (i) and outer (o) conformers were assumed to be i : o = 0.77 : 0.23. The nozzle was at room temperature.

[a]) Estimated limits of error.

(*continued*)

[b]) The weighted average of inequivalent C–H bond lengths.
[c]) The weighted average of inequivalent N–H bond lengths. All the differences among the inequivalent C–H and N–H parameters were assumed at their calculated values.
[d]) Assumed.
[e]) Assumed. The difference [(N–C(o)) – (N–N(o))] was assumed at an *ab initio* value, 0.0353 Å.

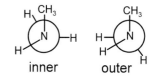

Murase, N., Yamanouchi, K., Egawa, T., Kuchitsu, K.: J. Mol. Struct. **242** (1991) 409.
Ohashi, N., Murase, N., Yamanouchi, K., Sugie, M., Takeo, H., Matsumura, C., Kuchitsu, K.: J. Mol. Spectrosc. **138** (1989) 497. II/7(**3**,133), II/23(**3**,77)

$CH_6N_2Si_2$	Disilylcarbodiimide	II/7(**3**,134)
	N,N'–Methanetetraylbissilanamine	
CH_6OSi	Methyl silyl ether	II/7(**3**,135)
CH_6P_2	Methylenebisphosphine	II/21(**3**,76)
CH_6SSi	Methylsilanethiol	II/21(**3**,77)
CH_6SSi	Methyl silyl sulfide	II/15(**3**,172)

CH_6Si **Methylsilane** C_{3v}
IR, MW H_3C-SiH_3

	r_0	Å		θ_0	deg
	C–H	1.0957(5) [a])		H–C–Si	110.88(3)
	Si–H	1.4832(4) [b])		H–Si–C	110.50(3)
	C–Si	1.8686(4)			

Improved structure results from obtaining A_0 rotational constants for the $SiHD_2CH_3$ and SiH_3CHD_2 isotopic species. All available microwave data were included in the calculations.

[a]) The assumption was made that δr_0 (CH – CD) = 0.0020 Å.
[b]) The assumption was made that δr_0 (SiH – SiD) = 0.0023 Å.

Duncan, J.L., Harvie, J.L., McKean, D.C., Cradock, S.: J. Mol. Struct. **145** (1986) 225.
Wong, M., Ozier, I., Meerts, W.L.: J. Mol. Spectrosc. **102** (1983) 89.
II/7(**3**,136), II/15(**3**,172a), II/21(**3**,78)

CH_6Sn	Methylstannane	II/7(**3**,137), II/15(**3**,172b)
CH_7B_5	Carbahexaborane(7)	II/15(**3**,173)

CH_7NO **Methanol – ammonia (1/1)** C_s
(weakly bound complex) (effective symmetry class)
MW $CH_3OH \cdot NH_3$

	r_0	Å		θ_0	deg
	R_{cm}	3.289(1)		θ_1 [a])	56.9(1)
	N–H	2.015(2)		θ_{2a} [b])	22.6(1)
				O–H···N	179.93(3)

[a]) See figure for definition.
[b]) An effective value for θ_2 is obtained from $\theta_2 \approx \theta_{2a} - 4°$ where 4° is the angle between the R_{cm} line and the *a* axis. Note that θ_{2c} is estimated to be less than 5°.

Fraser, G.T., Suenram, R.D., Lovas, F.J., Stevens, W.J.: Chem. Phys. **125** (1988) 31. II/21(**3**,79)

CH₇PSi	Silylmethylphosphine		II/7(**3**,138)
CH₈BP	Methylphosphine – borane (1/1)		II/7(**3**,139)
CH₈B₂	Methyldiborane		II/15(**3**,174), II/23(**3**,78)
CH₈B₄O	Carbonyltetraborane(8)		II/23(**3**,79)
CH₈N₂S₃	Ammonium trithiocarbonate		II/15(**3**,175)
CH₈Si₂	Disilylmethane		II/7(**3**,140), II/15(**3**,176)
CH₉B₂N	(Methylamino)diborane(6)		II/21(**3**,80)
CH₉B₅	2–Carbahexaborane(9)		II/7(**3**,141)
CH₉NSi₂	N–Methyl–N–silylsilanamine		II/7(**3**,142), II/21(**3**,81)
CH₁₀Si₃	Trisilylmethane		II/23(**3**,80)
CH₁₁AlB₂	Methylaluminum bis(tetrahydroborate) Methylbis[tetrahydroborato(1–)–H,H']aluminum		II/15(**3**,177)
CH₁₁AsB₁₀	p–Arsacarborane 1–Arsa–12–carba–*closo*–dodecaborane(11)		II/15(**3**,178)
CH₁₁B₅	1–Methylpentaborane(9)		II/7(**3**,143), II/15(**3**,179)
CH₁₁B₅	2–Methylpentaborane(9)		II/15(**3**,180)
CH₁₁B₁₀P	p–Phosphacarborane 1–Phospha–12–carba–*closo*–dodecaborane(11)		II/15(**3**,181)

ED **CH₁₂Si₄** **Tetrasilylmethane** T

C(SiH₃)₄

r ᵃ)	Å ᵇ)	θ ᵃ)	deg ᵇ)
Si–C	1.875(1)	Si–C–Si	109.47 ᶜ)
Si–H	1.486(4)	C–Si–H	108.5(6)
		Si–C–Si–H	20.0(10)
		SiH₃ tilt	0.0 ᶜ)

Temperature of the measurement was not stated, probably room temperature.

ᵃ) Unidentified, possibly r_a and θ_a.
ᵇ) Uncertainties unidentified, possibly estimated standard errors.
ᶜ) Assumed.

Hager, R., Steigelmann, O., Müller, G., Schmidbaur, H., Robertson, H.E., Rankin, D.W.H.: Angew. Chem. **102** (1990) 204; Int. Ed., Engl. **29** (1990) 201.   II/23(**3**,81)

CHgOS	Carbonyl sulfide – mercury (1/1)	II/23(**3**,82)
CHgO₂	Carbon dioxide – mercury (1/1)	II/23(**3**,83)
CIN	Iodine cyanide	II/7(**3**,36), II/15(**3**,182)
CINO	Iodine isocyanate	II/21(**3**,82)
CKN	Potassium cyanide	II/15(**3**,183)
CKrOS	Carbonyl sulfide – krypton (1/1)	II/21(**3**,83)
CKrO₂	Krypton – carbon dioxide (1/1)	II/23(**3**,84)
CLiN	Lithium isocyanide	II/15(**3**,184)
CMnN₃O₄	Carbonyltrinitrosylmanganese	II/21(**3**,84)
CNNa	Sodium cyanide	II/15(**3**,185), II/21(**3**,85)
CNO⁻	Cyanate ion	II/23(**3**,85)

3 Organic molecules

CNO — **Cyanato radical** — $C_{\infty v}$
N=C=O

UV

State	$\tilde{X}\,^2\Pi$	$\tilde{A}\,^2\Sigma^+$	$\tilde{B}\,^2\Pi$ [1]
Energy [eV]	0.0	2.821	3.937
r_0 [Å] N=C	1.200(8) [a]	1.165(8)	
C=O	1.206(8)	1.202(8)	
(N=C)+(C=O)			≤ 2.369

Rotational analysis of ^{14}NCO and ^{15}NCO.

[a]) Error limits 1σ.

Misra, P., Mathews, C.W., Ramsay, D.A.: J. Mol. Spectrosc. **130** (1988) 419.
[1] Dixon, R.N.: Can. J. Phys. **38** (1960) 10.

II/7(**3**,39); II/15(**3**,186), II/21(**3**,86)

CNOSr — **Strontium monoisocyanate** — $C_{\infty v}$
SrNCO

LIF

State	$\tilde{X}\,^2\Sigma^+$	$\tilde{A}\,^2\Pi$
Energy [eV]	0.0	1.868
r_0 [Å] Sr–N	2.26	2.24
C–N	1.19 [a]	1.19 [a]
C–O	1.23 [a]	1.23 [a]

Rotational analysis of laser excitation spectrum.

[a]) Assumed value.

O'Brien, L.C., Bernath, P.F.: J. Chem. Phys. **88** (1988) 2117.

II/21(**3**,87)

CNS	Thiocyanato radical	II/7(**3**,41), II/15(**3**,187)
CN$_2$	sym–Carbodiimide radical	II/7(**3**,37), II/15(**3**,188)

CN$_2$O — **Nitrosyl cyanide** — C_s

MW, UV

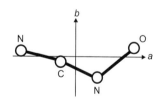

State	$\tilde{X}\,^1A'$ [b]	$\tilde{A}\,^1A''$ [c]
Energy [eV]	0.0	1.406
r_0 [Å] N≡C	1.163(5)	1.198 [d]
C–N	1.418(5)	1.316
N=O	1.217(5)	1.221 [d]
θ_s [deg] N≡C–N [a]	170(2)	162
C–N=O	113.6(10)	126.85

From microwave data for the ground state and from a rotational analysis of a Fourier transform UV spectrum for the excited state.

[a]) Terminal N is bent away from O.
[b]) From microwave data [1].
[c]) From [2]. This is cited as the most probable structure based on the rotational constants and the isotopic changes observed for NC^{15}NO.
[d]) Assumed values.

(*continued*)

CN_2O (continued)

[1] Dickinson, R., Kirby, G.W., Sweeny, J.G., Tyler, J.K.: J. Chem. Soc., Faraday Trans. II **74** (1978) 1393.
[2] Dixon, R.N., Johnson, P.: J. Mol. Spectrosc. **114** (1985) 174.

II/7(**3**,40), II/15(**3**,189), II/21(**3**,87A)

CN_4   **Cyanogen azide** C_s

MW

r_0	Å	θ_0	deg
C≡N(1)	1.164(10)	C–N(2)=N(3)	120.2(20)
C–N(2)	1.312(30)	N(1)≡C–N(2)	176(3) [a]
N(2)=N(3)	1.252(20)	N(2)=N(3)≡N(4)	180 [b]
N(3)≡N(4)	1.133(20)		

N(1)⫤C
N(2)=N(3)≡N(4)

[a] N(1) and N(4) are in a *trans* position with respect to N(2)–N(3).
[b] Assumed. This assumption is questionable, however. See the ED value.

Costain, C.C., Kroto, H.W.: Can. J. Phys. **50** (1972) 1453.

ED

r_α	Å [a]	θ_α	deg [a]
N(1)≡C	1.155(6)	N(1)≡C–N(2)	175.3(21)
C–N(2)	1.355(6)	C–N(2)=N(3)	114.5(3)
N(2)=N(3)	1.261(6)	N(2)=N(3)≡N(4)	169.2(24)
N(3)≡N(4)	1.121(6)		

The molecule is planar with a *trans* configuration with respect to the central N(2)=N(3) bond. The rotational constants calculated from the ED model are in satisfactory agreement with those from MW.

[a] Estimated standard errors. The uncertainties of the distances are three times those of the original data.

Almenningen, A., Bak, B., Jansen, P., Strand, T.G.: Acta Chem. Scand. **27** (1973) 1531.

II/7(**3**,38)

CN_4O_8	Tetranitromethane	II/15(**3**,190)
CNeOS	Carbonyl sulfide – neon (1/1)	II/21(**3**,88)
$CNeO_2$	Neon – carbon dioxide (1/1)	II/23(**3**,86)
COS	Carbonyl sulfide	II/7(**3**,44), II/15(**3**,191)
COSe	Carbonyl selenide	II/7(**3**,45), II/15(**3**,192)

CO_2   **Carbon dioxide** $D_{\infty h}$

IR

O=C=O

r_0	Å	r_m	Å	r_e	Å
C=O	1.162047(50)	C=O	1.15991(15)	C=O	1.1599588(30)

Graner, G., Rossetti, C., Bailly, D.: Mol. Phys. **58** (1986) 627.

UV C_{2v}

State	$1\,^1A_2$
Energy [eV]	6.322 [a]
r_0 [Å] C–O	1.262(10) [a]
θ_0 [deg] O–C–O	129(1) [a]

(continued)

From the analysis of a series of absorption bands in the region 1750···2000 Å. The transition from the ground state is electronically forbidden but is rendered vibronically allowed by excitation of one quantum of the asymmetric stretching vibration $v_3'(b_2)$.

a) These values pertain to the first member of a progression of 9 bands.

Cossart-Magos, C., Launay, F., Parkin, J.E.: Mol. Phys. **75** (1992) 835.

II/7(**3**,42), II/15(**3**,193), II/21(**3**,89), II/23(**3**,87)

IR | CO_2^+ | **Carbon dioxide(1+) ion** | $D_{\infty h}$
| | | O=C$^+$=O

r_0	Å
C=O	1.17682(41)

Bond distance was calculated from the B_0 rotational constant of [1].

[1] Sears, T.J.: Mol. Phys. **59** (1986) 259.
[2] Kawaguchi, K., Yamada, C., Hirota, E.: J. Chem. Phys. **82** (1984) 1174.

UV

State	$\tilde{X}\ ^2\Pi_g$	$\tilde{A}\ ^2\Pi_u$	$\tilde{B}\ ^2\Sigma_u^+$	$\tilde{C}\ ^2\Sigma_g^+$ a)
Energy [eV]	0	3.534	4.290	5.599
Reference	[1][3]	[1]	[2][3]	[4]
r_0 [Å] C=O	1.1769	1.2274	1.1805	1.1552(2) b)

a) Rotational analysis of two-photon spectrum. The $v = 0$ level is highly perturbed by another level with the same vibronic symmetry.
b) Error limits 1σ.

[1] Mrozowski, S.: Phys. Rev. **60** (1941) 730; **62** (1942) 270; **72** (1947) 682, 691.
[2] Bueso-Sanllehi, F.: Phys. Rev. **60** (1941) 556.
[3] Gauyacq, D., Horani, M., Leach, S., Rostas, J.: Can. J. Phys. **53** (1975) 2040; **57** (1979) 1634.
[4] Wyttenbach, T., Evard, D.D., Maier, J.P.: J. Chem. Phys. **90** (1989) 4645.

II/7(**3**,43), II/15(**3**,194), II/21(**3**,90)

CO_2Xe	Xenon – carbon dioxide (1/1)	II/23(**3**,88)
CO_3S	Carbon monoxide – sulfur dioxide (1/1)	II/23(**3**,89)
CSSe	Thiocarbonyl selenide	II/7(**3**,48)
CSTe	Thiocarbonyl telluride	II/7(**3**,49)

IR | CS_2 | **Carbon disulfide** | $D_{\infty h}$
| | | S=C=S

r_e	Å
C=S	1.55252(6)

Bond distance was calculated using the B_0 constants of [1] and the α constants of [2]. The bond distance given is an average obtained from both the $^{12}C^{32}S_2$ and the $^{13}C^{32}S_2$ isotopic species.

[1] Winter, F., Heyne, U., Guarnieri, A.: Z. Naturforsch. **43a** (1988) 215.
[2] Walrand, J., Humblet, V., Blanquet, G.: J. Mol. Spectrosc. **127** (1988) 304.

II/7(**3**,46), II/15(**3**,195), II/21(**3**,91)

	CS$_2^+$	Carbon disulfide(1+) ion	II/7(**3**,47), II/15(**3**,196), II/21(**3**,92)

IR **CSe$_2$** **Carbon diselenide** **D$_{\infty h}$**
Se=C=Se

r_e	Å
C=Se	1.692169(3)

Bürger, H., Willner, H.: J. Mol. Spectrosc. **128** (1988) 221. II/15(**3**,197), II/21(**3**,93)

C$_2$ArN$_2$	Argon – dicyanogen (1/1)	II/23(**3**,90)
C$_2$AsF$_6$N$_3$	Bis(trifluoromethyl)arsinous azide	II/23(**3**,91)
C$_2$BF$_3$N$_2$	Dicyan – boron trifluoride (1/1)	II/15(**3**,197a)
C$_2$BrCl	Bromochloroacetylene	II/7(**3**,145), II/15(**3**,198)
C$_2$BrF	Bromofluoroacetylene	II/15(**3**,199)

LIF **C$_2$BrH$^+$** **Bromoacetylene cation** **C$_{\infty v}$**
Br—C≡C—H$^+$

State	$\tilde{X}\,^2\Pi_{3/2}$	$\tilde{A}\,^2\Pi_{3/2}$
Energy [eV]	0.0	2.548
r_s [Å] (H···Br)	3.94(6) [a]	4.07(9)

Rotational analysis of ^{79}BrCCH$^+$, ^{81}BrCCH$^+$, ^{79}BrCCD$^+$ and ^{81}BrCCD$^+$.

[a]) The error limits represent the consistency of the data when different pairs of isotopic species are used to determine the overall length of the molecule.

King, M.A., Maier, J.P., Misev, L., Ochsner, M.: Can. J. Phys. **62** (1984) 1437. II/21(**3**,94)

C$_2$BrI	Bromoiodoacetylene	II/15(**3**,200)
C$_2$Br$_2$F$_4$	1,2–Dibromo–1,1,2,2–tetrafluoroethane	II/23(**3**,92)
C$_2$Br$_2$O$_2$	Oxalyl dibromide	II/7(**3**.146)
C$_2$Br$_4$	Tetrabromoethene	II/7(**3**,144)
C$_2$Br$_4$N$_2$	Dibromoformaldehyde azine Tetrabromoformaldazine	II/15(**3**,201), II/23(**3**,93)

LIF **C$_2$CaH** **Calcium monoacetylide** **C$_{\infty v}$**
Ca–C≡C–H

State		$\tilde{X}\,^2\Sigma^+$	$\tilde{A}\,^2\Pi$
Energy [eV]		0.0	1.924
r_0 [Å]	Ca–C	2.248 [a]	2.220 [a]
	C≡C	1.239	1.239
	C–H	1.056	1.056

Rotational analysis of laser excitation spectrum.

[a]) These values were deduced from rotational constants. All other values were fixed by comparison with related molecules.

Bopegedera, A.M.R.P., Brazier, C.R., Bernath, P.F.: J. Mol. Spectrosc. **129** (1988) 268.
II/21(**3**,95)

C$_2$ClF	Chlorofluoroacetylene	II/15(**3**,202)

	C_2ClF_3OS	Trifluoroethanethioyl chloride S–oxide Trifluorothioacetyl chloride S–oxide	II/23(**3**,94)
	C_2ClF_6NS	N–Chloro–S,S–bis(trifluoromethyl)sulfimine	II/15(**3**,203)

	C_2ClH^+	**Chloroacetylene cation**	$C_{\infty v}$
LIF			Cl–C≡C–H$^+$

State	$\tilde{X}\,^2\Pi_{3/2}$	$\tilde{A}\,^2\Pi_{3/2}$
Energy [eV]	0.0	3.350
r_s [Å] (H···Cl)	3.880(2) ^{a})	4.042(6)

Rotational analysis of $^{35}ClCCH^+$, $^{37}ClCCH^+$, $^{35}ClCCD^+$ and $^{37}ClCCD^+$.

^{a}) The error limits represent the consistency of the data when different pairs of isotopic species are used to determine the overall length of the molecule.

King, M.A., Maier, J.P., Ochsner, M.: J. Chem. Phys. **83** (1985) 3181.　　　　　II/21(**3**,96)

C_2ClI	Chloroiodoacetylene	II/7(**3**,150)
C_2ClNO_2	Chlorocarbonyl isocyanate	II/21(**3**,97)
C_2Cl_2	Dichloroacetylene Dichloroethyne	II/23(**3**,95)
$C_2Cl_2F_2$	1,1–Dichloro–2,2–difluoroethylene	II/15(**3**,204)
$C_2Cl_2F_4$	1,2–Dichloro–1,1,2,2–tetrafluoroethane	II/23(**3**,96)
$C_2Cl_2N_2O_2Si$	Dichlorosilylene diisocyanate	II/7(**3**,152)
C_2Cl_2O	Dichloroketene	II/21(**3**,98)

	$C_2Cl_2O_2$	**Oxalyl dichloride**	C_{2h}(*trans*), C_2(*gauche*)
ED		**% *trans*:** 67.6(84) ^{a}) at 0 °C 51.3(100) at 80 °C 42.4(102) at 190 °C	

r_a	Å ^{a})	θ_a	deg
C=O	1.182(2)	C–C=O	124.2(3)
C–C	1.534(5)	C–C–Cl	111.7(2)
C–Cl	1.744(2)	^{b})	125.0(58)
		^{c})	22.1(33)

$\Delta E° = E°_{gauche} - E°_{trans} = 1.38\ (\sigma = 0.35)$ kcal/mol.
$\Delta S° = S°_{gauche} - S°_{trans} = 2.3\ (\sigma = 1.0)$ kcal/mol.

^{a}) Twice estimated standard errors.
^{b}) Average torsion angle in the *gauche* form relative to an angle of 0° for the *trans* conformation.
^{c}) Root-mean-square amplitude of torsional motion in the *trans* conformer.

Hagen, K., Hedberg, K.: J. Am. Chem. Soc. **95** (1973) 1003.　　　　　II/7(**3**,153)

$C_2Cl_3F_3$	1,1,1–Trifluoro–2,2,2–trichloroethane	II/7(**3**,149)
$C_2Cl_3F_6P$	Bis(trifluoromethyl)trichlorophosphorane	II/15(**3**,205)
C_2Cl_3N	Trichloroacetonitrile Trichloromethyl cyanide	II/7(**3**,151)
C_2Cl_4	Tetrachloroethene	II/7(**3**,147)
$C_2Cl_4N_2$	1,2–(Dichloromethylidene)diazane 1,2–Bis(dichloromethylene)hydrazine	II/23(**3**,97)

C$_2$Cl$_4$O	Trichloroacetyl chloride	II/21(**3**,99)
C$_2$Cl$_6$	Hexachloroethane	II/7(**3**,148)
C$_2$Cl$_6$O$_2$S	Bis(trichloromethyl) sulfone	II/15(**3**,206)
C$_2$FNO$_2$	Carbonisocyanatidic fluoride Carbonyl fluoride isocyanate	II/23(**3**,98)
C$_2$F$_2$	Difluoroacetylene, Difluoroethyne	II/23(**3**,99)
C$_2$F$_2$OS$_2$	4,4–Difluoro–1,3–dithietan–2–one	II/23(**3**,100)
C$_2$F$_2$O$_2$S$_2$	Bis(fluorocarbonyl)disulfane	II/23(**3**,101)
C$_2$F$_2$O$_4$	Bis(fluorocarbonyl) peroxide Peroxydicarbonic difluoride	II/23(**3**,102)
C$_2$F$_2$S$_3$	4,4–Difluoro–1,3–dithietane–2–thione	II/23(**3**,103)
C$_2$F$_3$N	Trifluoroacetonitrile Trifluoromethyl cyanide	II/7(**3**,157)
C$_2$F$_3$N	Trifluoromethyl isocyanide	II/15(**3**,207)
C$_2$F$_3$NSe	Trifluoromethyl selenocyanate	II/7(**3**,160)
C$_2$F$_3$N$_2$S$_2$	4–(Trifluoromethyl)–1,2,3,5–dithiadiazole	II/21(**3**,100)
C$_2$F$_3$N$_3$S	1,3,5–Trifluoro–1λ^4,2,4,6–thiatriazine	II/23(**3**,104)
C$_2$F$_4$	Tetrafluoroethylene Tetrafluoroethene	II/7(**3**,154)
C$_2$F$_4$I$_2$	1,1,2,2–Tetrafluoro–1,2–diiodoethane	II/23(**3**,105)
C$_2$F$_4$N$_2$	1,1,4,4–Tetrafluoro–2,2–diazabuta–1,3–diene Azinobis(difluoromethane)	II/21(**3**,101)
C$_2$F$_4$O	Trifluoroacetyl fluoride	II/15(**3**,208)
C$_2$F$_4$O	Tetrafluoroethylene oxide	II/15(**3**,208a)
C$_2$F$_4$S	Tetrafluorothiirane Perfluorothiirane	II/21(**3**,102)
C$_2$F$_4$S$_2$	Tetrafluoro–1,3–dithietane	II/15(**3**,209)
C$_2$F$_4$Se$_2$	Tetrafluoro–1,3–diselenetane	II/15(**3**,210)
C$_2$F$_5$I	Pentafluoroethyl iodide, Pentafluoroiodoethane	II/7(**3**,156)
C$_2$F$_5$N$_3$S$_2$	1,3–Difluoro–5–(trifluoromethyl)–1λ^4,3λ^4,2,4,6–dithiatriazine 	II/23(**3**,106)
C$_2$F$_5$P	(Difluoromethylene)(trifluoromethyl)phosphine	II/21(**3**,103)
C$_2$F$_6$	Hexafluoroethane	II/7(**3**,155)
C$_2$F$_6$Hg	Bis(trifluoromethyl)mercury	II/15(**3**,211)
C$_2$F$_6$NO	Bis(trifluoromethyl)nitroxyl	II/7(**3**,159)
C$_2$F$_6$N$_2$	cis–Hexafluoroazomethane	II/21(**3**,104)
C$_2$F$_6$N$_2$	trans–Hexafluoroazomethane	II/7(**3**,158), II/15(**3**,212)
C$_2$F$_6$N$_2$OS	Bis(trifluoromethyl)aminyl thionitrosyl oxide	II/21(**3**,105)
C$_2$F$_6$O	Bis(trifluoromethyl) ether Perfluorodimethyl ether	II/15(**3**,213)
C$_2$F$_6$OS	Bis(trifluoromethyl) sulfoxide	II/15(**3**,214)
C$_2$F$_6$O$_2$	Bis(trifluoromethyl) peroxide	II/15(**3**,215)
C$_2$F$_6$O$_2$S	Bis(trifluoromethyl) sulfone	II/15(**3**,216)
C$_2$F$_6$S	Bis(trifluoromethyl) sulfide	II/15(**3**,217)
C$_2$F$_6$S	Trifluoroethylidynesulfur trifluoride Trifluoro(trifluoroethylidyne) sulfur	II/21(**3**,106)
C$_2$F$_6$S$_2$	Bis(trifluoromethyl) disulfide	II/15(**3**,218)
C$_2$F$_6$S$_3$	Bis(trifluoromethyl) trisulfane Bis(trifluoromethyl) trisulfide	II/23(**3**,107)
C$_2$F$_6$S$_4$	Bis(trifluoromethyl) tetrasulfane Bis(trifluoromethyl) tetrasulfide	II/23(**3**,108)

	C$_2$F$_6$Se	Bis(trifluoromethyl) selenide	II/7(**3**,161)
	C$_2$F$_6$Se$_2$	Bis(trifluoromethyl) diselenide	II/7(**3**,162)
	C$_2$F$_7$N	N–Fluorobis(trifluoromethyl)amine	II/15(**3**,219)
	C$_2$F$_8$N$_2$S	(Trifluoromethyl)(λ^6–pentafluorosulfanyl)carbodiimide	
			II/23(**3**,109)
	C$_2$F$_8$OS	Bis(trifluoromethyl)thionyl difluoride	II/15(**3**,220)
		Difluoro(oxo)bis(trifluoromethyl)sulfur	
	C$_2$F$_8$O$_2$S$_2$	2,2,3,3,3,4,4–Octafluoro–3λ^6–dithiethane 1,1–dioxide	
			II/23(**3**,110)
	C$_2$F$_8$S	Bis(trifluoromethyl)sulfur difluoride	II/15(**3**,221)
		Difluorobis(trifluoromethyl)sulfur	
	C$_2$F$_8$S	Pentafluoro(trifluorovinyl)sulfur	II/23(**3**,111)
		(Trifluorovinyl)sulfur pentafluoride	
	C$_2$F$_8$Se	Difluorobis(trifluoromethyl)selenium	II/21(**3**,107)
	C$_2$F$_9$P	Bis(trifluoromethyl)trifluorophosphorane	II/15(**3**,222)
	C$_2$F$_{10}$S	Bis(trifluoromethyl)sulfur tetrafluoride	II/21(**3**,108)
		Tetrafluorobis(trifluoromethyl)sulfur	
	C$_2$F$_{12}$S$_2$	Tetrafluoro–1,3–dithietane octafluoride	II/21(**3**,109)
	C$_2$FeN$_2$O$_4$	Dicarbonyldinitrosyliron	II/21(**3**,110)

	C$_2$H	**Ethynyl radical**	C$_{\infty v}$
MW			HC≡C

r_s	Å [a])	r_s [b])	Å [a])
C≡C	1.21652(20)	C≡C	1.21652(20)
C–H	1.04653(20)	C–H	1.04170(30)

[a]) Uncertainties are larger than those of the original data.
[b]) Using the first-moment relation to calculate the H coordinate.

Bogey, M., Demuynck, C., Destombes, J.L.: Mol. Phys. **66** (1989) 955.

IR

r_0	Å [a])	
C≡C	1.289	Ã electronic state
C–H	1.060	Ã electronic state
C≡C	1.210	X̃ ground state [a])
C–H	1.050	X̃ ground state [a])

[a]) The ground state bond distances were calculated from the microwave data of [1].

Yan, W.-B., Dane, C.B., Zeitz, D., Hall, J.L., Curl, R.F.: J. Mol. Spectrosc. **123** (1987) 486.
[1] Sastry, K.V.L.N., Helminger, P., Charo, A., Herbst, E., DeLucia, F.: Astrophys. J. **251** (1981) L119.   II/15(**3**,223), II/21(**3**,111)

C$_2$HArF$_3$	Argon – trifluoroethene (1/1)	II/23(**3**,112)
C$_2$HBF$_2$	Ethynyldifluoroborane	II/7(**3**,170)
C$_2$HBr	Bromoacetylene	II/7(**3**,171), II/15(**3**,224), II/23(**3**,113)
C$_2$HBrF$_2$	2–Bromo–1,1–difluoroethene	II/23(**3**,114)
C$_2$HBrO	Bromoketene	II/21(**3**,112)
C$_2$HCl	Chloroacetylene	II/7(**3**,172), II/15(**3**,225), II/23(**3**,115)
C$_2$HClF$_3$N	N–(Trifluoromethyl)chloromethanimine	II/21(**3**,113)
C$_2$HClO	Chloroketene	II/15(**3**,225a)

C$_2$HCl$_2$F$_3$	2,2–Dichloro–1,1,1–difluoromethane	II/23(**3**,116)
C$_2$HCl$_3$O	Dichloroacetyl chloride	II/15(**3**,226)
C$_2$HF	Fluoroacetylene	II/7(**3**,173), II/23(**3**,117)
C$_2$HFN$_2$	Cyanogen – hydrogen fluoride (1/1)	II/15(**3**,227)
C$_2$HFO	Fluoroketene	II/21(**3**,114)
C$_2$HF$_2$N	Difluoroacetonitrile	II/15(**3**,227a), II/23(**3**,118)
C$_2$HF$_3$	Trifluoroethylene, Trifluoroethene	II/7(**3**,174), II/15(**3**,228)
C$_2$HF$_3$O	Difluoroacetyl fluoride	II/15(**3**,229)
C$_2$HF$_3$O$_2$	Trifluoroacetic acid	II/15(**3**,230)
C$_2$HF$_4$N	*N*–(Trifluoromethyl)fluoromethanimine	II/21(**3**,115)
C$_2$HF$_4$P	Ethynyltetrafluorophosphorane	II/15(**3**,231)
C$_2$HF$_5$	Pentafluoroethane	II/7(**3**,175), II/15(**3**,232)
C$_2$HF$_5$S	Ethynylpentafluorosulfur	II/21(**3**,116), II/23(**3**,119)
C$_2$HF$_6$NO	*N*,*N*–Bis(trifluoromethyl)hydroxylamine	II/7(**3**,176)
C$_2$HI	Iodoacetylene	II/7(**3**,177), II/21(**3**,117)
C$_2$HI$^+$	Iodoacetylene cation	II/15(**3**,233), II/21(**3**,117A)
C$_2$HN	Cyanomethylene radical	II/15(**3**,233a), II/23(**3**,120)

MW **C$_2$HNO** **Carbon monoxide – hydrogen cyanide (1/1)** **C$_{\infty v}$**
 (weakly bound complex) (effective symmetry class)
 CO···HCN

	r_0(B···C) [Å] [a)c)]	r_s(B···C) [Å] [b)c)]	θ_{av} [deg] [c)]	ϕ_{av} [deg] [c)]
^{16}O^{12}C···H^{12}C^{15}N	3.6647(30)	3.6609(20)	11.69(30)	12.19(30)
^{16}O^{12}C···H^{12}C^{14}N	3.6641(30)	3.6601(20)	11.67(30)	12.25(30)

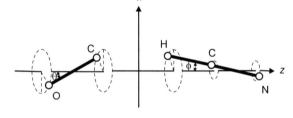

a) r_0(B···C) is the distance between the plane of the ring of the C atom of the HCN donor subunit and the plane of the ring of the C atom in CO. Uncertainties were not estimated in the original paper.
b) r_s(B···C) = |z_B − z_C| where z_B is the r_s-coordinate of the C atom in CO.
c) Uncertainties were not estimated in the original paper.

Haynes, A., Legon, A.C.: J. Mol. Struct. **189** (1988) 153. II/15(**3**,233b), II/23(**3**,121)

C$_2$HNOS	Hydrogen cyanide – carbonyl sulfide (1/1)	II/21(**3**,118)
C$_2$HNO$_2$	Hydrogen cyanide – carbon dioxide (1/1)	II/15(**3**,233c), II/21(**3**,119)
C$_2$HNO$_2$S	1,3,4–Oxathiazol–2–one	II/15(**3**,234)

MW **C$_2$HN$_3$** **Diazoacetonitrile** **C$_s$**

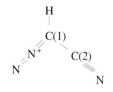

r_0	Å	θ_0	deg
C–H	1.082(4) [a)]	C(2)–C(1)=N	119.5(10)
C(2)≡N	1.165 [a)]	H–C–C	117.0(30)
C–C	1.424(10)		
C(1)=N	1.280(30)		
N=N	1.132(30)		

(*continued*)

^{a}) Assumed.

Costain, C.C., Yarwood, J.: J. Chem. Phys. **45** (1966) 1961.   II/7(**3**,178)

	C$_2$HO	**Oxoethenyl radical**	**C$_s$**
MW		Ketenyl radical	HC̈=C=O

r_0	Å		θ_0	deg
C–H	1.056(4)		H–C=C	138.7(14)
C=C	1.314(25)		C=C=O	180 ^{a})
C=O	1.150 ^{a})			

^{a}) Assumed.

Endo, Y., Hirota, E.: J. Chem. Phys. **86** (1987) 4319.   II/21(**3**,120)

C$_2$H$_2^-$	Vinylidene anion	II/21(**3**,121)

	C$_2$H$_2$	**Acetylene**	**D$_{\infty h}$**
MW, IR			HC≡CH

$r_m^{\rho\ a})$	Å
C–H	1.0631(20) ^{b})
	1.0614(20) ^{c})
C≡C	1.2026(10) ^{b})
	1.2031(10) ^{c})

^{a}) Multiple isotope substitution structure. Uncertainties are larger than those of the original paper.
^{b}) Using monodeuterated species.
^{c}) Using dideuterated species.

Berry, R.J., Harmony, M.D.: Struct. Chem. **1** (1990) 49.

UV

State	Ã^{1}A$_u$			C$_{2h}$
Energy [eV]	5.232			
r_0 [Å]	C–H	1.097		
	C=C	1.375		
θ_0 [deg]	H–C=C	122.48		

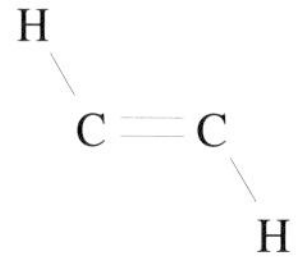

From the rotational constants for C$_2$H$_2$ and C$_2$D$_2$.
No error limits are given.

Huet, T.R., Godefroid, M., Herman, M.: J. Mol. Spectrosc. **144** (1990) 32.

LIF

State	C̃'^{1}A$_g$
Energy [eV]	7.723
r_0 [Å] C–H	1.14(3) ^{a})
C–C	1.65(2)
θ_0 [deg] H–C–C	103(4)

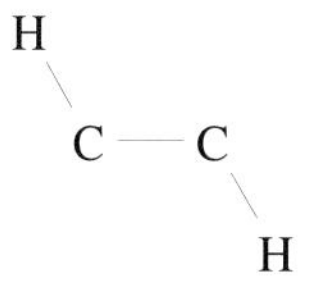

From the analysis of bands of C$_2$H$_2$ and C$_2$D$_2$.

^{a}) Error limits are 1σ.

Lundberg. J.-K., Chen, Y., Pique, J.-P., Field, R.W.: J. Mol. Spectrosc. **156** (1992) 104.
   II/7(**3**,179), II/15(**3**,235), II/23(**3**,122)

Formula	Name	Reference
$C_2H_2^+$	Acetylene ion	II/23(**3**,123)
C_2H_2Ar	Argon – acetylene (1/1)	II/21(**3**,122), II/23(**3**,124)
$C_2H_2ArF_2$	Argon – 1,1–difluoroethylene (1/1) Argon – vinylidene fluoride (1/1)	II/23(**3**,125)
$C_2H_2ArN_2$	Argon – hydrogen cyanide (1/2)	II/21(**3**,123)
C_2H_2BrClO	Bromoacetyl chloride	II/15(**3**,236)
C_2H_2BrN	Bromoacetonitrile	II/15(**3**,237)
$C_2H_2Br_2$	cis–1,2–Dibromoethene	II/7(**3**,180)
$C_2H_2Br_2O$	Bromoacetyl bromide	II/15(**3**,238)
$C_2H_2ClF_3$	1–Chloro–2,2,2–trifluoroethane	II/21(**3**,124)
C_2H_2ClN	Chloroacetonitrile	II/7(**3**,184)
$C_2H_2Cl_2$	1,1–Dichloroethylene 1,1–Dichloroethene, Vinylidene chloride	II/7(**3**,181), II/15(**3**,239)
$C_2H_2Cl_2$	cis–1,2–Dichloroethylene cis–1,2–Dichloroethene	II/7(**3**,182), II/21(**3**,125, 126)
$C_2H_2Cl_2$	trans–1,2–Dichloroethylene trans–1,2–Dichloroethene	II/15(**3**,240), II/21(**3**,126, 127)
$C_2H_2Cl_2Hg$	cis–2–Chlorovinylmercury chloride	II/7(**3**,183)
$C_2H_2Cl_2O$	Chloroacetyl chloride	II/15(**3**,241)
$C_2H_2Cl_2O_2$	Dichloroacetic acid	II/15(**3**,241b)
C_2H_2FN	Fluoroacetonitrile Fluoromethyl cyanide	II/15(**3**,241a), II/21(**3**,128)
C_2H_2FNO	Carbon monoxide – hydrogen cyanide – hydrogen fluoride (1/1/1)	II/23(**3**,126)
$C_2H_2F_2$	1,1–Difluoroethylene 1,1–Difluoroethene	II/7(**3**,185), II/15(**3**,242)
$C_2H_2F_2$	cis–1,2–Difluoroethylene cis–1,2–Difluoroethene	II/7(**3**,186), II/15(**3**,243)
$C_2H_2F_2$	trans–1,2–Difluoroethene trans–1,2–Difluoroethylene	II/7(**3**,186), II/23(**3**,127)
$C_2H_2F_2O$	Fluoroacetyl fluoride	II/7(**3**,187), II/15(**3**,244)
$C_2H_2F_2O$	cis–1,2–Difluoroethylene oxide cis–1,2–Difluorooxirane	II/15(**3**,245)
$C_2H_2F_2O$	trans–1,2–Difluoroethylene oxide trans–1,2–Difluorooxirane	II/15(**3**,245a)
$C_2H_2F_2O_2$	Difluoroacetic acid	II/15(**3**,246)
$C_2H_2F_2O_3$	cis–Difluoroethylene ozonide 3,5–Difluoro–1,2,4–trioxolane	II/15(**3**,246a)
$C_2H_2F_2O_3$	trans–Difluoroethylene ozonide trans–3,5–Difluoro–1,2,4–trioxolane	II/21(**3**,129)
$C_2H_2F_2O_3$	1,1–Difluoroethylene ozonide (3,3–Difluoro–1,2,4–trioxolane) Vinylidene fluoride ozonide	II/15(**3**,247)
$C_2H_2F_3N$	Fluoroform – hydrogen cyanide (1/1)	II/21(**3**,130)
$C_2H_2F_4$	1,1,1,2–Tetrafluoroethane	II/15(**3**,248), II/21(**3**,131)
$C_2H_2F_4$	1,1,2,2–Tetrafluoroethane	II/15(**3**,249)
$C_2H_2GeI_2$	1,1–Diiodogermacycloprop-2-ene	II/7(**3**,188)
$C_2H_2N^-$	Cyanomethanide ion	II/21(**3**,132)
C_2H_2NP	Methylenephosphinous cyanide	II/21(**3**,133)

	C₂H₂N₂	**Hydrogen cyanide dimer**			**C∞ᵥ**
MW		(weakly bound complex)			(effective symmetry class)
					HCN···HCN

	$r_0(B \cdots C)$ [Å] a)c)	$r_s(B \cdots C)$ [Å] b)c)	θ [deg] c)	ϕ_{av} [deg] c)
H¹²C¹⁵N···H¹²C¹⁵N	3.3098(30)	3.3170(20)	13.38(30)	9.15(30)

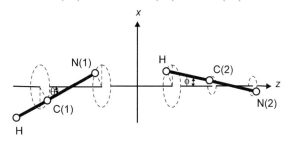

a) $r_0(B \cdots C)$ is the distance between the plane of the ring of the C atom of the HCN donor subunit and the plane of the ring of the N atom in HCN acceptor subunit.
b) $r_s(B \cdots C) = |z_B - z_C|$ where z_B is the r_s-coordinate of the N atom in HCN acceptor nearest to the HCN donor subunit.
c) Uncertainties were not estimated in the original paper.

Haynes, A., Legon, A.C.: J. Mol. Struct. **189** (1988) 153.

II/15(**3**,250), II/21(**3**,134), II/23(**3**,128)

C₂H₂N₂	N–Cyanoformimine, N–Cyanomethanimine	II/15(**3**,251)
C₂H₂N₂O	Furazan, 1,2,5–Oxadiazole	II/7(**3**,190), II/21(**3**,135)
C₂H₂N₂O	1,3,4–Oxadiazole	II/7(**3**,191)
C₂H₂N₂O	Dicyanogen –water (1/1)	II/23(**3**,129)
C₂H₂N₂O	Acetylene – nitrous oxide (1/1)	II/23(**3**,130)
C₂H₂N₂S	1,2,3–Thiadiazole	II/15(**3**,252)
C₂H₂N₂S	1,2,4–Thiadiazole	II/15(**3**,253)
C₂H₂N₂S	1,2,5–Thiadiazole	II/7(**3**,192), II/15(**3**,254)
C₂H₂N₂S	1,3,4–Thiadiazole	II/7(**3**,193), II/21(**3**,136)
C₂H₂N₂Se	1,2,5–Selenodiazole	II/7(**3**,194)
C₂H₂N₂Se	1,3,4–Selenodiazole	II/7(**3**,195)
C₂H₂N₄	Azidoacetonitrile, Azidoethanenitrile	II/21(**3**,137)
C₂H₂N₄	1,2,4,5–Tetrazine	II/7(**3**,189), II/15(**3**,255)
C₂H₂N₄	Dinitrogen – hydrogen cyanide (1/2)	II/23(**3**,131)

	C₂H₂O	**Ketene**			**C₂ᵥ**
MW					H₂C(1)=C(2)=O

r_s	Å		θ_s	deg
C(2)=O	1.1620(721)		H–C–H	122.56(1)
C(1)=C(2)	1.3137(721)			
C(1)–H	1.0825(15)			

r_m^ρ a)	Å		θ_m^ρ a)	deg
C(2)=O	1.1600(58)		H–C–H	121.58(29)
C(1)=C(2)	1.3140(62)			
C(1)–H	1.0740(19)			

a) Multiple isotope substitution structure.

Brown, R.D., Godfrey, P.D., McNaughton, D., Pierlot, A.P., Taylor, W.H.: J. Mol. Spectrosc. **140** (1990) 340.

II/7(**3**,196), II/23(**3**,132)

| C$_2$H$_2$OS | *trans*–Thioxoacetaldehyde | II/21(**3**,138) |

C$_2$H$_2$O$_2$ **s–*trans*–Glyoxal**   **C$_{2h}$**

ED, UV

r_g	Å [a)]	θ_α	deg [a)]
C–C	1.526(3)	C–C=O	121.2(2)
C=O	1.212(2)	H–C–C	112.2(17)
C–H	1.132(8)		

[a)] Estimated limits of error.

Kuchitsu, K., Fukuyama, T., Morino, Y.: J. Mol. Struct. **1** (1968) 463.
Kuchitsu, K., Fukuyama, T., Morino, Y.: J. Mol. Struct. **4** (1969) 41.

UV

State		$\tilde{X}\,^1A_g$	$\tilde{A}\,^1A_u$
Energy [eV]		0	2.726
r_0 [Å]	C–H	1.109(8)	1.115(10)
	C–C	1.527(17)	1.460(25)
	C=O	1.202(12)	1.252(16)
θ_0 [deg]	H–C–C	115.5(30)	114(4)
	C–C=O	121.15(15)	123.7(3)

Birss, F.W., Braund, D.B., Cole, A.R.H., Engleman, R., Green, A.A., Japar, S.M., Nanes, R., Orr, B.J., Ramsay, D.A., Szyszka, J.: Can. J. Phys. **55** (1977) 390.

II/7(**3**,197), II/15(**3**,256)

C$_2$H$_2$O$_2$	s–*cis*–Glyoxal	II/7(**3**,198), II/15(**3**,257)
C$_2$H$_2$O$_2$S	Acetylene – sulfur dioxide (1/1)	II/23(**3**,133)
C$_2$H$_2$O$_3$	Formic anhydride	II/7(**3**,199), II/15(**3**,258)
C$_2$H$_2$O$_3$	Formaldehyde – carbon dioxide (1/1)	II/21(**3**,140), II/23(**3**,134)
C$_2$H$_2$O$_3$	Acetylene – ozone (1/1)	II/21(**3**,139), II/23(**3**,135)
C$_2$H$_2$O$_4$	Oxalic acid	II/7(**3**.200)
C$_2$H$_2$O$_5$	Carbon dioxide – water (2/1)	II/21(**3**,141), II/23(**3**,136)
C$_2$H$_2$S	Thioketene	II/15(**3**,259), II/23(**3**,137)
C$_2$H$_2$Se	Selenoketene	II/15(**3**,260)

C$_2$H$_3$ **Vinyl radical**   **C$_s$**

IR

r_0	Å	θ_0	deg
C=C	1.3160(63)	C=C–H(3)	137.3(40)
C–H(1)	1.085(10) [a)]	C=C–H(1)	121.5(10) [a)]
C–H(3)	1.080(10) [a)]		

The molecule is of C$_{2v}$ effective symmetry, executing a double-minimum motion probably associated with the C–H(3) in-plane rocking vibration. The structure given corresponds to one of the wells and only one isotopomer, giving A_0, B_0 and C_0. The errors in r_0(C=C) and on θ_0(C=C–H(3)) are due for one half to the uncertainties in the assumed parameters and for the other one to the fact there are three choices of two rotational constants.

[a)] Assumed.

Kanamori, H., Endo, Y., Hirota, E.: J. Chem. Phys. **92** (1990) 197.

II/23(**3**,138)

C$_2$H$_3$ArF	Argon – vinyl fluoride (1/1)	II/23(**3**,139)
C$_2$H$_3$ArN	Argon – acetonitrile (1/1)	II/23(**3**,140)
C$_2$H$_3$ArN$_5$	3–Amino–s–tetrazine–argon complex	II/23(**3**,141)
C$_2$H$_3$Ar$_2$N$_5$	3–Amino–s–tetrazine–diargon complex	II/23(**3**,142)
C$_2$H$_3$BF$_2$	Difluorovinylborane	II/15(**3**,261)
C$_2$H$_3$BF$_3$N	Acetonitrile – trifluoroborane (1/1)	II/23(**3**,143)
C$_2$H$_3$Br	Vinyl bromide Bromoethylene	II/7(**3**,201), II/15(**3**,262), II/23(**3**,144)
C$_2$H$_3$BrN$_2$	3–Bromo–3–methyldiazirine	II/7(**3**,203)
C$_2$H$_3$BrO	Acetyl bromide	II/7(**3**,204)
C$_2$H$_3$Br$_3$	1,1,1–Tribromoethane	II/7(**3**,202)
C$_2$H$_3$Cl	Vinyl chloride Chloroethylene	II/7(**3**,205), II/15(**3**,263), II/23(**3**,145)
C$_2$H$_3$Cl	Acetylene – hydrogen chloride (1/1)	II/15(**3**,264)
C$_2$H$_3$ClF$_2$	1,1–Difluoro–1–chloroethane	II/7(**3**,208)
C$_2$H$_3$ClN$_2$	3–Chloro–3–methyldiazirine	II/7(**3**,209)
C$_2$H$_3$ClN$_2$	Hydrogen chloride– hydrogen cyanide (1/2)	II/21(**3**,142)

C$_2$H$_3$ClO   **Acetyl chloride** C_s

ED, MW

r_g	Å [a])		θ_z	deg [a])
C=O	1.187(3)		O=C–Cl	121.2(5)
C–Cl	1.798(2)		C–C–Cl	111.6(6)
C–C	1.506(3)		H–C–H [b])	108.6(8)
C–H [b])	1.105(5)		[c])	1.3(10)

Above data are based on experimental results of [1].
One of the C–H bonds eclipses the C=O bond.

[a]) Estimated limits of error.
[b]) For CD$_3$COCl: r_z(C–D) = 1.090(5) Å and θ_z(D–C–D) = 109.0(8)°.
[c]) Tilt angle of the methyl group.

Tsuchiya, S., Iijima, T.: J. Mol. Struct. **13** (1972) 327.
[1] Tsuchiya, S., Kimura, M.: Bull. Chem. Soc. Japan **45** (1972) 736.   II/7(**3**,210)

C$_2$H$_3$ClO	Chloroacetaldehyde	II/15(**3**,265)
C$_2$H$_3$ClOS	S–Methyl carbonochloridothioate	II/15(**3**,266)
C$_2$H$_3$ClO$_2$	Methyl chloroformate	II/15(**3**,267), II/23(**3**,146)
C$_2$H$_3$ClO$_2$	Chloroacetic acid	II/15(**3**,268)
C$_2$H$_3$ClO$_2$S	Ethenesulfonyl chloride	II/15(**3**,269)
C$_2$H$_3$ClSi	Silylchloroacetylene	II/15(**3**,270)
C$_2$H$_3$Cl$_2$OP	Vinylphosphonic dichloride	II/15(**3**,271)
C$_2$H$_3$Cl$_2$P	Dichlorovinylphosphine	II/15(**3**,272)
C$_2$H$_3$Cl$_3$	1,1,1–Trichloroethane	II/7(**3**,206), II/23(**3**,147)
C$_2$H$_3$Cl$_3$	1,1,2–Trichloroethane	II/7(**3**,207)
C$_2$H$_3$Cl$_3$Si	Vinyltrichlorosilane	II/7(**3**,211)
C$_2$H$_3$F	Vinyl fluoride Fluoroethylene	II/7(**3**,212), II/15(**3**,273), II/23(**3**,148)
C$_2$H$_3$F	Acetylene – hydrogen fluoride (1/1)	II/15(**3**,274)
C$_2$H$_3$FN$_2$	Hydrogen fluoride – hydrogen cyanide (1/2)	II/21(**3**,143)
C$_2$H$_3$FO	Acetyl fluoride	II/7(**3**,215)

C₂H₃FO₂	Hydroxyacetyl fluoride	II/21(**3**,144)
C₂H₃FO₂	Fluoroacetic acid	II/7(**3**,216), II/15(**3**,275)
C₂H₃FO₂	Methyl fluoroformate	II/23(**3**,149)
C₂H₃FO₂	Fluoromethyl formate	II/15(**3**,276)
C₂H₃FO₃	2–Fluoro–1,3,4–trioxolane	II/15(**3**,277)
	Fluoroethylene ozonide, Vinyl fluoride ozonide	
C₂H₃F₃	1,1,1–Trifluoroethane	II/7(**3**,213), II/15(**3**,278)
C₂H₃F₃	1,1,2–Trifluoroethane	II/15(**3**,279)
C₂H₃F₃Hg	Methyl(trifluoromethyl)mercury	II/15(**3**,280)
C₂H₃F₃N₂	1,1,1–Trifluoroazomethane	II/7(**3**,214)
C₂H₃F₅S	Ethenylpentafluorosulfur	II/23(**3**,150)
C₂H₃F₇S	Methyl(trifluoromethyl)sulfur tetrafluoride	II/21(**3**,145)
	Tetrafluoro(methyl)(trifluoromethyl)sulfur(VI)	
C₂H₃HgN	Methylmercury cyanide	II/21(**3**,146)
C₂H₃I	Vinyl iodide	II/21(**3**,147), II/23(**3**,151)
C₂H₃IO	Acetyl iodide	II/7(**3**,217)
C₂H₃N	Ethenimine, Vinylideneamine	II/21(**3**,148)
C₂H₃N	Acetonitrile	II/7(**3**,218), II/15(**3**,281), II/23(**3**,152)
	Methyl cyanide	
C₃H₃N	Methyl isocyanide	II/7(**3**,219), II/15(**3**,282)
C₂H₃NO	Acetonitrile oxide	II/7(**3**,220)
C₂H₃NO	Methyl cyanate	II/23(**3**,153)
C₂H₃NO	Methyl isocyanate	II/7(**3**,221), II/15(**3**,282a), II/21(**3**,149)
C₂H₃NO	Hydrogen cyanide – formaldehyde (1/1)	II/21(**3**,150)

C₂H₃NO₂   **Nitroethene** C_s

MW

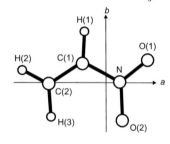

r_s	Å a)	θ_s	deg a)
C(1)=C(2)	1.3245(38)	C(2)=C(1)–N	120.93(18)
C(1)–N	1.4579(10)	C(1)–N=O(1)	115.96(16)
N=O(1)	1.2272(34)	C(1)–N=O(2)	119.12(18)
N=O(2)	1.2275(30)	C(2)=C(1)–H(1)	127.01(36)
C(1)–H(1)	1.0796(44)	C(1)=C(2)–H(2)	120.17(22)
C(2)–H(2)	1.0771(34)	C(1)=C(2)–H(3)	119.62(18)
C(2)–H(3)	1.0813(28)		

a) Uncertainties are about twice those of the original data.

Nössberger, P., Bauder, A., Günthard, H.-H.: Chem. Phys. **8** (1975) 245.

II/7(**3**,222), II/15(**3**,283)

C₂H₃NS	Methyl thiocyanate	II/7(**3**,223)
C₂H₃NS	Methyl isothiocyanate	II/7(**3**,224), II/15(**3**,283a), II/21(**3**,151)
C₂H₃NSe	Methyl selenocyanate	II/15(**3**,284), II/21(**3**,152)
C₂H₃NSe	Methyl isoselenocyanate	II/21(**3**,153), II/23(**3**,154)
C₂H₃N₃	Vinyl azide, Azidoethene	II/15(**3**,285)
C₂H₃N₃	1,2,3–Triazole	II/21(**3**,154)
C₂H₃N₃	1,2,4–Triazole	II/15(**3**,286)
C₂H₃N₃	Dicyanogen – ammonia (1/1)	II/23(**3**,155)
C₂H₃N₅	3–Amino–s–tetrazine	II/23(**3**,156)
C₂H₃O⁻	Acetaldehyde enolate anion	II/21(**3**,155)
	Ethenolate ion	

	C₂H₃O	Formylmethyl radical	II/21(**3**,156)
		2–Oxoethyl radical	
		Vinyloxyl radical	
	C₂H₃P	Ethynylphosphine	II/21(**3**,157)
	C₂H₃P	Ethylidynephosphine	II/15(**3**,287)

C₂H₄ **Ethylene** **D$_{2h}$**

MW CH$_2$=CH$_2$

r_m^ρ a)	Å		θ_m^ρ a)	deg
C–H	1.0801(5)		H–C–H	117.11(8)
C=C	1.3297(5)			

a) Multiple isotope substitution structure.

Berry, R.J., Harmony, M.D.: Struct. Chem. **1** (1990) 49.

II/7(**3**,225), II/15(**3**,288), II/23(**3**,157)

	C₂H₄⁺	Ethylene(1+) ion	II/21(**3**,158)
		Etheniumyl ion	
	C₂H₄ArO	Argon – oxirane (1/1)	II/21(**3**,159)
	C₂H₄ArS	Argon – thiirane (1/1)	II/23(**3**,158)
	C₂H₄AsBrO₂	2–Bromo–1,3,2–dioxarsolane	II/15(**3**,289)
	C₂H₄AsBrS₂	2–Bromo–1,3,2–dithiarsolane	II/15(**3**,290)
	C₂H₄AsClO₂	2–Chloro–1,3,2–dioxarsolane	II/15(**3**,291)
	C₂H₄Br	Bromoethyl radical	II/7(**3**,226)
	C₂H₄BrCl	1–Bromo–2–chloroethane	II/23(**3**,159)
	C₂H₄BrF	1–Bromo–2–fluoroethane	II/23(**3**,160)
	C₂H₄Br₂	1,2–Dibromoethane	II/7(**3**,227), II/15(**3**,292)

C₂H₄ClF **1–Chloro–2–fluoroethane** **C$_s$(anti), C$_1$(gauche)**

ED and *ab initio*
(HF/3-21G*) calculations

r_g	Å a)		θ_α	deg a)
C–H	1.107(7)		C–C–H	110.6(19)
C–C	1.504(6)		C–C–F(*anti*)	108.5(11)
C–F	1.391(3)		C–C–F(*gauche*)	111.1(4)
C–Cl	1.784(3)		C–C–Cl(*anti*)	109.3(11)
			C–C–Cl(*gauche*)	111.5(4)
			F–C–C–Cl(*gauche*)	111.9(16)
			F–C–C–Cl(*anti*)	0

The molecule exists as a mixture of *anti* and *gauche* conformers with the former the more stable (58(7) % at –25 °C). Distances and angles are averaged over three temperatures, –25, 95 and 360 °C. The analyses were aided by inclusion of experimental rotational constants.

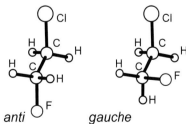

anti   *gauche*

a) Twice the estimated standard errors.

Huang, J., Hedberg, K.: J. Am. Chem. Soc. **112** (1990) 2070.

II/7(**3**,230), II/23(**3**,161)

C_2H_4ClF	Vinyl fluoride – hydrogen chloride (1/1)	II/23(**3**,162)
C_2H_4ClN	1–Chloroaziridine, *N*–Chloroethylenimine	II/7(**3**,231)
C_2H_4ClN	Acetonitrile – hydrogen chloride (1/1)	II/21(**3**,160)
C_2H_4ClNO	Chloroacetamide	II/15(**3**,293)
C_2H_4ClOPS	3–Chloro–1,3–thiaphosphetane 3–oxide	II/15(**3**,294a)
$C_2H_4ClO_2P$	2–Chloro–1,3,2–dioxaphospholane	II/7(**3**,232), II/15(**3**,294)
	Ethylene phosphonochloridite	
$C_2H_4ClO_3P$	Ethylene phosphorochloridate	II/7(**3**,233)
	2–Chloro–1,3,2–dioxaphospholane 2–oxide	
$C_2H_4ClPS_2$	2–Chloro–1,3,2–dithiaphospholane	II/7(**3**,234)
$C_2H_4ClPS_3$	Ethylene phosphorochloridotrithioate	II/7(**3**,235)
	2–Chloro–1,3,2–dithiaphospholane 2–sulfide	
$C_2H_4Cl_2$	1,1–Dichloroethane	II/7(**3**,228)

ED **$C_2H_4Cl_2$**   **1,2–Dichloroethane** C_{2h} (*anti*)
                                                   C_2 (*gauche*)

r_a	Å	θ_α	deg
C–C	1.531(6)	C–C–Cl	109.0(4)
C–Cl	1.790(4)	C–C–H	113.2(26)
C–H	1.112(10)	[b]	76.4(14)

Studies have been made at –13, 2, 40, 140, and 300 °C. The relative
amount of an *anti* (*trans*) conformer with C_{2h} symmetry compared to the amount of a *gauche* conformer with C_2 symmetry varied with temperature, but the geometrical structure is essentially unchanged with the temperature. $\Delta E = 1.05(10)$ kcal/mol; $\Delta S = 0.90(29)$ cal/K mol.

[a] Estimated standard errors, twice those of the original data.
[b] The effective torsional angle.

Kveseth, K.: Acta Chem. Scand. A **28** (1974) 482; A **29** (1975) 307.   II/7(**3**,229)

$C_2H_4Cl_2N_2O_2$	*N*–Nitrobis(chloromethyl)amine	II/23(**3**,163)
	1–Chloro–*N*–(chloromethyl)–*N*–nitromethanamine	
$C_2H_4Cl_2O$	Dichloromethyl methyl ether	II/15(**3**,295)
$C_2H_4Cl_2O$	Bis(chloromethyl) ether	II/15(**3**,296)
$C_2H_4Cl_4Si_2$	1,1,3,3–Tetrachloro–1,3–disilacyclobutane	II/7(**3**,236)
C_2H_4FN	Acetonitrile – hydrogen fluoride (1/1)	II/15(**3**,297), II/21(**3**,162)
C_2H_4FNO	Fluoroacetamine	II/15(**3**,298)
$C_2H_4F_2$	1,1–Difluoroethane	II/7(**3**,237), II/15(**3**,299)

MW **$C_2H_4F_2$**   **sc–1,2–Difluoroethane** C_2
                       gauche–1,2–Difluoroethane CH_2FCH_2F

r_0	Å	θ_0	deg
C–C	1.493(8)	C–C–F	110.6(5)
C–F	1.390(3)	C–C–H(*gauche*) [a]	108.4(6)
C–H(*gauche*) [a]	1.099(2)	C–C–H(*trans*) [a]	111.3(6)
C–H(*trans*) [a]	1.093(5)	H–C–H	109.1(5)
		F–C–H(*gauche*) [a]	109.6(3)
		F–C–H(*trans*) [a]	107.8(6)
		F–C–C–F	71.0(3)

(*continued*)

r_s	Å		θ_s	deg
C–C	1.491(2)		C–C–F	110.5(3)
C–F	1.394(2)		C–C–H(*gauche*) [a]	108.2(3)
C–H(*gauche*) [a]	1.100(1)		C–C–H(*trans*) [a]	111.4(4)
C–H(*trans*) [a]	1.093(2)		H–C–H	108.8(5)
			F–C–H(*gauche*) [a]	110.3(3)
			F–C–H(*trans*) [a]	107.7(4)
			F–C–C–F	71.3(5)

The *gauche* conformer was studied.

[a] H(*trans*) and H(*gauche*) indicate the hydrogen atoms in *trans* and *gauche* positions against the fluorine atoms attached to the other carbons, respectively.

Takeo, H., Matsumura, C., Morino, Y.: J. Chem. Phys. **84** (1986) 4205.

II/7(**3**,238), II/15(**3**,300), II/21(**3**,163)

$C_2H_4F_2O$	2,2–Difluoroethanol	II/15(**3**,301)
$C_2H_4F_3N$	1,1,1–Trifluoroethylamine	II/7(**3**,239)
C_2H_4Ge	Germylacetylene, Ethynylgermane	II/7(**3**,240)
C_2H_4INO	Iodoacetamide	II/15(**3**,302)
$C_2H_4I_2$	1,2–Diiodoethane	II/23(**3**,164)
$C_2H_4N_2$	Formaldehyde azine Formaldazine	II/15(**3**,303), II/23(**3**,165)
$C_2H_4N_2$	Aminoacetonitrile	II/7(**3**,241)
$C_2H_4N_2$	Methyldiazirine	II/7(**3**,242)
$C_2H_4N_2O$	Hydrogen cyanide – water (2/1)	II/23(**3**,166)

C_2H_4O   **Acetaldehyde** C_s
Ethanal

ED, MW

r_g	Å [a]		θ_z	deg [a]
C(1)–C(2)	1.515(4)		C(2)–C(1)=O	124.1(3)
C(1)=O	1.210(4)		O=C(1)–H	120.6(3)
C(2)–H	1.107(6)		C(2)–C(1)–H	115.3(3)
C(1)–H	1.128(4)		H–C(2)–H	109.8(9)

One of the C–H bonds of the CH_3 group is eclipsed with C=O.

[a] Estimated limits of error.

Iijima, T., Kimura, M.: Bull. Chem. Soc. Japan **42** (1969) 2159.
Kato, C., Konaka, S., Iijima, T., Kimura, M.: Bull. Chem. Soc. Japan **42** (1969) 2148.

II/7(**3**,243), II/15(**3**,304)

C_2H_4O	Vinyl alcohol, Ethenol	II/15(**3**,305)
C_2H_4O	Oxirane, Ethylene oxide	II/7(**3**,244), II/23(**3**,167)
C_2H_4O	Water – acetylene (1/1)	II/15(**3**,305a)
C_2H_4OS	*O*–Methyl thioformate	II/7(**3**,250)
C_2H_4OS	*S*–Methyl thioformate	II/15(**3**,306)
C_2H_4OS	Ethylene sulfoxide, Thiirane 1–oxide	II/7(**3**,251)

$C_2H_4OS_2$	1,3–Dithiacyclobutane 1–oxide		II/15(**3**,307)
	1,3–Dithietane 1–oxide		
$C_2H_4O_2$	Formaldehyde dimer		II/21(**3**,164), II/23(**3**,168)
$C_2H_4O_2$	Methyl formate		II/7(**3**,245), II/15(**3**,308)
$C_2H_4O_2$	Acetic acid		II/7(**3**,246), II/15(**3**,309), II/21(**3**,165)

MW **$C_2H_4O_2$** **Glycolaldehyde** C_s
Hydroxyethanal

r_s	Å	θ_s	deg
C(2)=O(2)	1.2094(30)	C(1)–C(2)=O(2)	122.7(3)
C(1)–O(1)	1.4366(50)	C(1)–C(2)–H(3)	115.3(3)
C(1)–C(2)	1.4987(30)	C(2)–C(1)–O(1)	111.5(3)
O(1)–H(1)	1.0510(50)	C(1)–O(1)–H(1)	101.6(5)
C(2)–H(3)	1.1021(50)	C(2)–C(1)–H(2)	109.2(5)
C(1)–H(2)	1.0930(50)	H(2)–C(1)–H(2′)	107.6(5)
		H(2)–C(1)–O(1)	109.7(5)

Marstokk, K.-M., Møllendal, H.: J. Mol. Struct. **16** (1973) 259.   II/7(**3**,247)

$C_2H_4O_2$	Methanol – carbon monoxide (1/1)	II/23(**3**,169)
$C_2H_4O_2S$	Ethylene sulfone, Thiirane 1,1–oxide	II/7(**3**,252)
$C_2H_4O_2S$	Ethylene – sulfur dioxide (1/1)	II/21(**3**,166), II/23(**3**,170)
$C_2H_4O_3$	Glycolic acid	II/15(**3**,310)
$C_2H_4O_3$	1,2,4–Trioxacyclopentane	II/7(**3**,248), II/15(**3**,311)
	Ethylene ozonide, 1,2,4–Trioxolane	
$C_2H_4O_3$	1,2,3–Trioxolane	II/21(**3**,167)
$C_2H_4O_3$	Ethylene – ozone (1/1)	II/21(**3**,168), II/23(**3**,171)
$C_2H_4O_3S$	Ethylene sulfite	II/7(**3**,253)
$C_2H_4O_3Se$	Ethylene selenite	II/7(**3**,254)

ED **$C_2H_4O_4$** **Formic acid dimer** C_{2h}

r_a	H(COOH)$_2$ [Å] [a]	H(COOD)$_2$ [Å] [b]	θ_a	H(COOH)$_2$ [deg] [a]	H(COOD)$_2$ [deg] [b]
C–O(1)	1.217(3)	1.217	O(1)–C–O(2)	126.2(5)	125.5
C–O(2)	1.320(3)	1.329	C–O(2)–H$\cdots$O	108.5(4)	107.8
O(2)–H	1.033(17)	1.055			
O(2)–H$\cdots$O	2.696(7)	2.715			
O$\cdots$O	2.262(4)	2.262			

The distance parameters have been reduced by about 0.25 % following the statement made in the original papers that gaseous CO_2 is used for the scale factor calibration.

[a]) Estimated standard errors.
[b]) No explicit statement in the original paper, but uncertainties similar to the corresponding hydride parameters are to be expected.

Almenningen, A., Bastiansen, O., Motzfeldt, T.: Acta Chem. Scand. **23** (1969) 2848.
Almenningen, A., Bastiansen, O., Motzfeldt, T.: Acta Chem. Scand. **24** (1970) 747.   II/7(**3**,249)

	C₂H₄O₄Si	Bis(formyloxy)silane	II/21(**3**,169)
	C₂H₄S	Ethenethiol Vinyl mercaptan	II/15(**3**,312), II/21(**3**,170)
	C₂H₄S	Thioacetaldehyde, Ethanethial	II/15(**3**,313)

C₂H₄S   **Ethylene sulfide** C_{2v}

MW   Thiirane

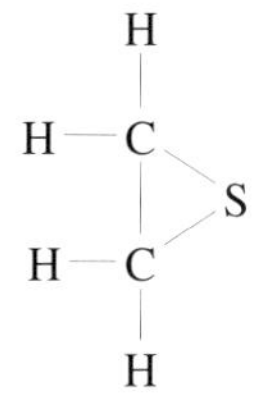

r_s	Å	θ_s	deg
C–C	1.484(5)	H–C–H	115.8(15)
C–H	1.083(5)	C–S–C	48.27(30)
C–S	1.815(5)	θ [a]	151.8(15)
		C–C–S	65.87(30)

[a] Angle between the CH₂ plane and the C–C bond.

Okije, K., Hirose, C., Lister, D.G., Sheridan, J.: Chem. Phys. Lett. **24** (1974) 111.

II/7(**3**,255)

C₂H₄S₃	1,2,4–Trithiolane	II/15(**3**,313a)
C₂H₄Se	Selenoacetaldehyde	II/15(**3**,314)

C₂H₄Si   **Ethynylsilane** C_{3v}

ED, MW, and   Silylacetylene   H–C≡C–SiH₃
liquid crystal NMR

r_{av}	Å [a]	θ_{av}	deg [a]
Si–C	1.827(1)	H–Si–C	109.2(1)
Si–H	1.480(2)		
C≡C	1.208(1)		
C–H	1.062(1)		

The Si–C≡C–H group was assumed to be linear.
The nozzle temperature was 293 K.

[a] Uncertainties were unidentified, possibly estimated standard errors.

Brookman, C.A., Cradock, S., Rankin, D.W.H., Robertson, N., Vefghi, P.: J. Mol. Struct. **216** (1990) 191.

II/7(**3**,256), II/23(**3**,172)

C₂H₅BF₂	Ethyldifluoroborane	II/15(**3**,315)
C₂H₅BO₂	1,3,2–Dioxaborolane	II/7(**3**,259)
C₂H₅B₃	1,5–Dicarba–*closo*–pentaborane(5)	II/7(**3**,257)
C₂H₅B₄Cl	2–Chloro–1,6–dicarba–*closo*–hexaborane(6)	II/7(**3**,258)
C₂H₅Br	Ethyl bromide Bromoethane	II/7(**3**,260), II/15(**3**,316), II/21(**3**,171)
C₂H₅BrO	2–Bromoethanol	II/7(**3**,261)(*gauche* form), II/23(**3**,173)
C₂H₅BrO	Oxirane – hydrogen bromide (1/1)	II/23(**3**,174)
C₂H₅BrO₂S	Bromomethyl methyl sulfone	II/15(**3**,317)
C₂H₅BrS	Thiirane – hydrogen bromide (1/1)	II/23(**3**,175)

MW	**C$_2$H$_5$Cl**	**Ethyl chloride** Chloroethane	**C$_s$** CH$_3$CH$_2$Cl

r_m^ρ	Å
C–C	1.5096(22)
C–Cl	1.7888(18)
C–H	1.0862(22)
C–H(*sym*)	1.0904(98)
C–H(*asym*)	1.0894(25)

θ_m^ρ	deg
C–C–Cl	111.02(7)
C–C–H(*asym*)	110.53(13)
H(*asym*)–C–H(*asym*)	108.26(37)
C–C–H(*sym*)	109.30(105)
C–C–H	111.81(27)
H–C–H	108.99(31)

Multiple isotope substitution structure.

Tam, H.S., Choe, J.I., Harmony, M.D.: J. Phys. Chem. **95** (1991) 9267.
See also: Hayashi, M., Inagusa, T.: J. Mol. Struct. **220** (1990) 103.

II/7(**3**,262), II/15(**3**,319), II/23(**3**,176)

C$_2$H$_5$Cl	Ethylene – hydrogen chloride (1/1)	II/15(**3**,318)
C$_2$H$_5$ClN$_2$O$_2$	N–Chloromethyl–N–nitromethylamine Methyl(chloromethyl)nitramine	II/15(**3**,320)
C$_2$H$_5$ClO	Chloromethoxymethane Chloromethyl methyl ether	II/7(**3**,263), II/15(**3**,321)
C$_2$H$_5$ClO	2–Chloroethanol Ethylene chlorohydrin	II/7(**3**,264), II/15(**3**,322)
C$_2$H$_5$ClO	Oxirane – hydrogen chloride (1/1)	II/23(**3**,177)
C$_2$H$_5$ClS	Chloromethyl methyl sulfide	II/15(**3**,323)
C$_2$H$_5$ClSi	Chlorovinylsilane	II/23(**3**,178)
C$_2$H$_5$Cl$_2$P	Ethyldichlorophosphine	II/15(**3**,324)
C$_2$H$_5$Cl$_2$PS	Ethylphosphonothioic difluoride	II/23(**3**,179)
C$_2$H$_5$F	Fluoroethane Ethyl fluoride	II/7(**3**,265), II/15(**3**,325), II/23(**3**,180)
C$_2$H$_5$F	Ethylene – hydrogen fluoride (1/1)	II/15(**3**,326)
C$_2$H$_5$FO	2–Fluoroethanol	II/7(**3**,266), II/21(**3**,172)
C$_2$H$_5$FO	Fluoromethoxymethane Fluoromethyl methyl ether	II/15(**3**,327)
C$_2$H$_5$FO	Oxirane – hydrogen fluoride (1/1) Ethylene oxide – hydrogen fluoride (1/1)	II/15(**3**,328), II/23(**3**,181)
C$_2$H$_5$FSi	Fluorovinylsilane	II/23(**3**,182)
C$_2$H$_5$F$_2$N	2,2–Difluoroethylamine	II/15(**3**,329)
C$_2$H$_5$F$_2$P	Ethyldifluorophosphine	II/15(**3**,330), II/21(**3**,173)
C$_2$H$_5$F$_2$PS	Ethylphosphonothioic difluoride	II/21(**3**,174)
C$_2$H$_5$F$_2$PS	(Ethylthio)difluorophosphine	II/23(**3**,183)
C$_2$H$_5$GeN	Methylgermyl cyanide	II/21(**3**,175)
C$_2$H$_5$I	Ethyl iodide, Iodoethane	II/7(**3**,267), II/21(**3**,176)
C$_2$H$_5$IO	2–Iodoethanol	II/23(**3**,184)
C$_2$H$_5$IO	Iodomethyl methyl ether	II/21(**3**,177)
C$_2$H$_5$N	Vinylamine	II/15(**3**,331), II/21(**3**,178), II/23(**3**,185)
C$_2$H$_5$N	Ethanimine	II/15(**3**,332)
C$_2$H$_5$N	N–Methylenemethanamine	II/7(**3**,268)

MW

C$_2$H$_5$N **Ethylenimine** C_s
Aziridine

r_s	Å		θ_s	deg
N(1)–H(1)	1.016(5)		C(2)–N(1)–C(3)	60.25(20)
N(1)–C(2)	1.475(3)		H(1)–N(1)–C(2)	109.31(50)
C(2)–C(3)	1.481(3)		H(2)–C(2)–H(2′)	115.72(50)
C(2)–H(2)	1.084(5)		H(2)–C(2)–C(3)	117.75(50)
C(2)–H(2′)	1.083(5)		H(2)–C(2)–N(1)	118.26(50)
			H(2′)–C(2)–C(3)	119.32(50)
			H(2′)–C(2)–N(1)	114.27(50)

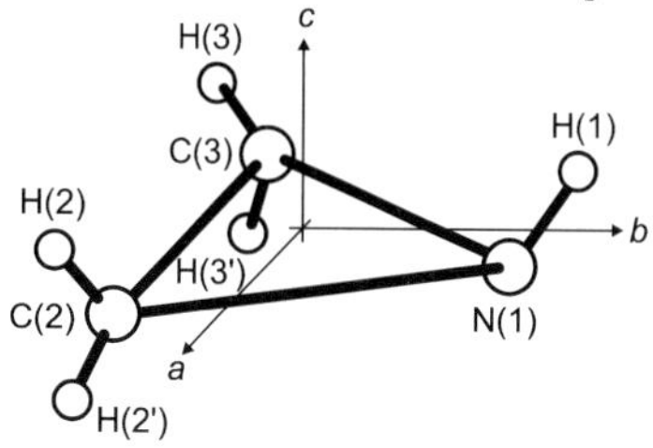

Bak, B., Skaarup, S.: J. Mol. Struct. **10** (1971) 385.

II/7(**3**,269)

| C$_2$H$_5$N | Methane – hydrogen cyanide | II/23(**3**,186) |
| C$_2$H$_5$N | Ammonia – acetylene (1/1) | II/15(**3**,332a) |

ED

C$_2$H$_5$NO **Acetamide** C_s

r_g	Å $^{a)}$		θ_α	deg $^{a)}$
C–C	1.519(6)		C–C–N	115.1(16)
C–N	1.380(4)		O=C–N	122.0(6)
C=O	1.220(3)		C–C=O	123.0
C–H $^{b)}$	1.124(10)		C–C–H $^{b)}$	109.8(20)
N–H $^{b)}$	1.022(11)			

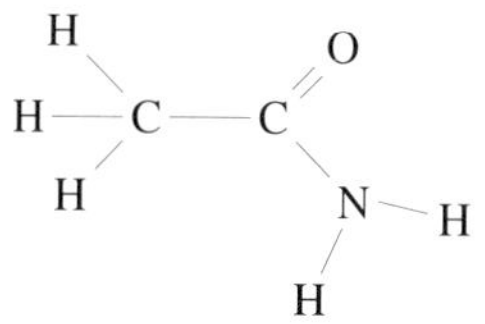

The molecule is planar except for two methyl H–atoms.

$^{a)}$ Estimated limits of error.
$^{b)}$ Average value.

Kitano, M., Kuchitsu, K.: Bull. Chem. Soc. Japan **46** (1973) 3048.

II/7(**3**,270)

C$_2$H$_5$NO	N–Methylformamide	II/7(**3**,271)
C$_2$H$_5$NOSi	Methylsilyl isocyanate	II/21(**3**,179)
C$_2$H$_5$NO$_2$	Nitroethane	II/7(**3**,272)
C$_2$H$_5$NO$_2$	Ethyl nitrite	II/15(**3**,333), II/23(**3**,187)
C$_2$H$_5$NO$_2$	Glycine	II/23(**3**,188)
C$_2$H$_5$NO$_3$	Ethyl nitrate	II/23(**3**,189)
C$_2$H$_5$NS	Thioacetamide	II/15(**3**,334)
C$_2$H$_5$NSSi	Methylsilyl isothiocyanate	II/21(**3**,180)
C$_2$H$_5$NSi	Methylsilanecarbonitrile, Methylsilyl cyanide	II/21(**3**,181)
C$_2$H$_5$N$_3$	Hydrogen cyanide – ammonia (2/1)	II/23(**3**,190)
C$_2$H$_5$O	Ethoxyl radical	II/23(**3**,191)
C$_2$H$_5$P	Phosphirane, Ethylenephosphine	II/7(**3**,273)

C₂H₆	**Ethane**	**D$_{3d}$**
ED, IR, Ra, MW		CH₃CH₃

r_g	Å [a]		r_z [c]	Å [a]		θ_z [c]	deg [a]
C–C	1.5326(20)		C–C	1.5323(20)		H–C–H	107.30(30)
C–C [b]	1.5310(20)		C–C [b]	1.5299(20)		D–C–D	107.35(30)
C–H	1.1108(20)		C–H	1.1017(20)			
C–D	1.1053(20)		C–D	1.0990(20)			
C···H	2.1964(30)						
C···D	2.1909(30)						

Above data are based on experimental work of [1] and theoretical work of [2].

[a]) Estimated standard errors in the absolute values of the parameters. Relative uncertainties are much smaller.
[b]) C–C distance of CD₃–CD₃.
[c]) Torsional motion is not considered in the averaging of the nuclear positions.

Iijima, T.: Bull. Chem. Soc. Japan **46** (1973) 2311.
[1] ED: Bartell, L.S., Higginbothan, H.K.: J. Chem. Phys. **42** (1965) 851.
[2] ED: Kuchitsu, K.: J. Chem. Phys. **49** (1968) 4456.

MW

r_z	Å		θ_z	deg
C–C	1.5351(1)		C–C–H	111.17(1)
C–H	1.0940(2)		(C–C–H)–(C–C–D)	0.040(10)
(C–H)–(C–D)	0.0012(2)			
(C–C)–(^{13}C–C)	0.00015(10)			
Δ(C–C)/D atom	−5 · 10^{-5} [a]			

[a]) Assumed. The negative sign implies a C–C bond shortening on deuterium substitution.

Hirota, E., Endo, Y., Saito, S., Duncan, J.L.: J. Mol. Spectrosc. **89** (1981) 285.

MW, IR

r_m^ρ [a])	Å		θ_m^ρ [a])	deg
C–C	1.522(2)		C–C–H	111.2(1)
C–H	1.089(1)			

[a]) Multiple isotope substitution structure.

Harmony, M.D.: J. Chem. Phys. **93** (1990) 7522.   II/7(**3**,274), II/15(**3**,335), II/23(**3**,192)

C₂H₆AsF₃Si	Dimethyl(trifluorosilyl)arsine	II/15(**3**,336)
C₂H₆BCl₂N	(Dimethylamino)dichloroborane	II/7(**3**,277)
C₂H₆BF₃O	Dimethyl ether – boron trifluoride (1/1)	II/15(**3**,337)
C₂H₆BN	Methyl isocyanide – borane (1/1)	II/15(**3**,338)
C₂H₆BN₃	Azidodimethylborane	II/23(**3**,193)
C₂H₆B₂S₃	Dimethyl–1,2,4–trithia–3,5–borolane	II/7(**3**,278)
C₂H₆B₄	1,6–Dicarba–*closo*–hexaborane(6)	II/7(**3**,275), II/15(**3**,339)
C₂H₆B₄	1,2–Dicarba–*closo*–hexaborane(6)	II/15(**3**,340)
C₂H₆B₄	2,3–Dicarba–*closo*–hexaborane(6)	II/7(**3**,276)
C₂H₆B₅F	5–Fluoro–2,4–dicarbaheptaborane(7)	II/21(**3**,182)
C₂H₆Be	Dimethylberyllium	II/7(**3**,279)

	C$_2$H$_6$Br$_2$Ge	Dimethylgermyl dibromide Dibromodimethylgermane	II/15(**3**,341)

C$_2$H$_6$Cd   **Dimethylcadmium**

Ra   Raman pure rotation

r_0	Å		θ	deg
Cd–C	2.112(4)		H–C–H	108.4 [b]
C–H	1.09 [a]			

H₃C—Cd—CH₃ (structure: H-C-Cd-C-H with three H on each C)

The molecule is a free internal rotor.
Cd–C distance determined from B_0 values of Cd(CH$_3$)$_2$ and Cd(CD$_3$)$_2$.

[a]) Assumed.
[b]) Error in this value depends on validity of above assumption.

Suryanarayana Rao, K., Stoicheff, B.P., Turner, R.: Can. J. Phys. **38** (1960) 1516.   II/7(**3**,280)

C$_2$H$_6$ClN	N–Chloro–N–methylmethanamine	II/21(**3**,183)
C$_2$H$_6$ClNO$_2$S	Dimethylsulfamoyl chloride Dimethylaminosulforyl chloride	II/7(**3**,282), II/15(**3**,342)
C$_2$H$_6$ClO$_2$PS	O,O′–Dimethyl phosphonochloridothioate	II/15(**3**,343)
C$_2$H$_6$Cl$_2$Ge	Dichlorodimethylgermane	II/15(**3**,344)
C$_2$H$_6$Cl$_2$NOP	(Dimethylamino)dichlorophosphine oxide	II/7(**3**,281)
C$_2$H$_6$Cl$_2$NP	(Dimethylamino)dichlorophosphine	II/7(**3**,283)
C$_2$H$_6$Cl$_2$Si	Dichlorodimethylsilane Dimethyldichlorosilane	II/15(**3**,345), II/21(**3**,184)
C$_2$H$_6$Cl$_2$Sn	Dimethyldichlorostannane Dimethyltin dichloride	II/7(**3**,285)
C$_2$H$_6$Cl$_3$NSi	N–(Trichlorosilyl)dimethylamine	II/7(**3**,284)
C$_2$H$_6$FN	N–Fluorodimethylamine	II/23(**3**,194)
C$_2$H$_6$FN	2–Fluoroethylamine	II/15(**3**,346)
C$_2$H$_6$FO$_2$P	Methylphosphonofluoridic acid methyl ester Fluoro(methoxy)methylphosphine oxide	II/7(**3**,289)
C$_2$H$_6$F$_2$Ge	Difluorodimethylgermane	II/15(**3**,347)
C$_2$H$_6$F$_2$NP	(Dimethylamino)difluorophosphine	II/7(**3**,286), II/23(**3**,195)
C$_2$H$_6$F$_2$Si	Dimethyldifluorosilane	II/21(**3**,185)
C$_2$H$_6$F$_3$NS	(Dimethylamino)trifluorosulfur(IV)	II/23(**3**,196)
C$_2$H$_6$F$_3$NSi	N–(Trifluorosilyl)dimethylamine	II/7(**3**,288)
C$_2$H$_6$F$_3$P	Dimethylphosphorus trifluoride	II/7(**3**,290)
C$_2$H$_6$F$_6$N$_2$P$_2$	2,2,2,4,4,4–Hexafluoro–2,2,4,4–tetrahydro–1,3– dimethyl–1,3,2,4–diazaphosphetidine	II/7(**3**,287)
C$_2$H$_6$Ge	Vinylgermane	II/15(**3**,348)
C$_2$H$_6$GeOS	Germyl thioacetate	II/15(**3**,349)
C$_2$H$_6$GeO$_2$	Germyl acetate Acetoxygermane	II/21(**3**,186)

	C₂H₆Hg	**Dimethylmercury**
Ra	Raman pure rotation	

r_0	Å		θ	deg
Hg–C	2.094(5)		H–C–H	109.3 [b]
C–H	1.09 [a]			

The molecule is a free internal rotor.
Hg–C distance determined from B_0 values of Hg(CH₃)₂ and Hg(CD₃)₂.

[a]) Assumed.
[b]) Error in this value depends on validity of above assumption.

Suryanarayana Rao, K., Stoicheff, B.P., Turner, R.: Can. J. Phys. **38** (1960) 1516.

ED

r_g	Å [a])
Hg–C	2.083(5)
C–H	1.160 [b])
Hg···H	2.712(18)

The zero-point average distance Hg–C determined by ED is significantly (0.015 Å) shorter than that derived from the spectroscopic rotational constant.

[a]) Estimated limits of error.
[b]) Assumed.

Kashiwabara, K., Konaka, S., Iijima, T., Kimura, M.: Bull. Chem. Soc. Japan **46** (1973) 407.

II/7(**3**,291)

	C₂H₆N₂	**trans–Azomethane**	**C₂ₕ** assumed
ED			

r_a	Å [a])		θ_a	deg [a])
C–H	1.105(3)		C–N=N	112.3(3)
C–N	1.482(2)		N–C–H	107.5(15)
N=N	1.247(3)		[b])	–4.1(1)

A *trans* configuration is assumed.

[a]) Estimated standard errors.
[b]) The effective tilt angle of the CH₃ group. It is defined as positive when the H atoms point away from the double bond.

Almenningen, A., Anfinsen, I.M., Haaland, A.: Acta Chem. Scand. **24** (1970) 1230.
ED: Chang, C.H., Porter, R.F., Bauer, S.H.: J. Am. Chem. Soc. **92** (1970) 5313.

II/7(**3**,292)

C₂H₆N₂	*cis*–Azomethane	II/15(**3**,350)
C₂H₆N₂	3–Methyldiaziridine	II/7(**3**,293), II/15(**3**,351)
C₂H₆N₂O	N–Nitrosodimethylamine	II/7(**3**,294), II/15(**3**,352)
C₂H₆N₂O₂	N–Nitrodimethylamine	II/7(**3**,295)
C₂H₆N₂O₂	N–Methyl–N'–methoxydiazene N–oxide	II/21(**3**,187)
C₂H₆N₂S	N,N–Dimethylsulfur diimide	II/15(**3**,353)
C₂H₆O	Ethanol	II/7(**3**,296)
C₂H₆O	Dimethyl ether	II/7(**3**,297), II/15(**3**,354)

	C_2H_6O	Ethylene – water (1/1)	II/21(**3**,188)

C_2H_6OS Dimethyl sulfoxide C_s

MW

r_s	Å	θ_s	deg
S–C	1.799(7)	C–S–C	96.6(3)
S=O	1.485(7)	C–S=O	106.5(3)
C–H	1.054(15)	S–C–H	108.3(10)
C–H'	1.097(10)	S–C–H'	108.2(7)
C–H''	1.093(10)	S–C–H''	109.6(7)
H···H'	1.800(25)	H–C–H'	113.6(10)
H···H''	1.721(20)	H–C–H''	106.6(15)
H'···H''	1.801(10)	H'–C–H''	110.6(10)

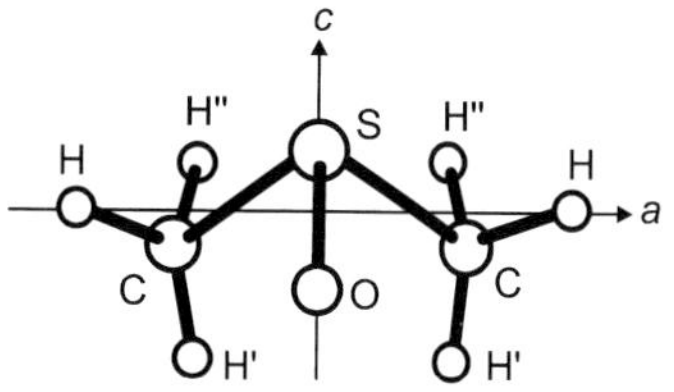

CH_3 conformation is nearly staggered with respect to the S–O bond.

Feder, W., Dreizler, H., Rudolph, H.D., Typke, V.:
  Z. Naturforsch. **24 a** (1969) 266.   II/7(**3**,298)

C_2H_6OS	2–Mercaptoethanol	II/15(**3**,355)
C_2H_6OSSi	Silyl thioacetate	II/15(**3**,356)
$C_2H_6O_2$	Dimethyl peroxide	II/15(**3**,357)
$C_2H_6O_2$	1,2–Ethanediol, Ethylene glycol	II/15(**3**,358), II/21(**3**,189)

$C_2H_6O_2S$ Dimethyl sulfone C_{2v}

MW

r_0	Å	θ_0	deg
C–S	1.777(10)	C–S–C	103.3(5)
S=O	1.431(7)	O=S=O	121.0(7)
C–H	1.091 [a]	H–C–H	109.6 [a]

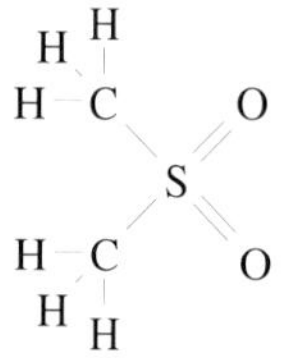

The C_2S and SO_2 planes are perpendicular to each other.

[a]) Assumed.

Saito, S., Makino, F.: Bull. Chem. Soc. Japan **45** (1972) 92.

ED

r_a	Å [a])	θ_a	deg [a])
S=O	1.435(3)	C–S–C	102.6(9)
S–C	1.771(4)	O=S=O	119.7(11)
C–H	1.114(3)	S–C–H	108.5(8)

[a]) Uncertainty estimates are total errors.

Hargittai, M., Hargittai, I.: J. Mol. Struct. **20** (1974) 283.   II/7(**3**,299)

$C_2H_6O_2Si$	Silyl acetate	II/15(**3**,359)
$C_2H_6O_3S$	Dimethyl ether – sulfur dioxide (1/1)	II/23(**3**,197)
$C_2H_6O_4S$	Dimethyl sulfate	II/15(**3**,360)

C$_2$H$_6$S	Dimethyl sulfide	II/7(**3**,300), II/15(**3**,361), II/23(**3**,198)
C$_2$H$_6$S	Ethanethiol, Ethyl mercaptan	II/15(**3**,362)
C$_2$H$_6$S$_2$	Dimethyl disulfide	II/7(**3**,301), II/15(**3**,363)
C$_2$H$_6$S$_2$	1,2–Ethanedithiol	II/7(**3**,302), II/15(**3**,363a), II/21(**3**,190)
C$_2$H$_6$Se	Ethaneselenol	II/15(**3**,364)
C$_2$H$_6$Se	Dimethyl selenide	II/7(**3**,303), II/15(**3**,365)
C$_2$H$_6$Se$_2$	Dimethyl diselenide	II/7(**3**,304)
C$_2$H$_6$Si	Vinylsilane	II/7(**3**,305), II/15(**3**,366)
C$_2$H$_6$Si$_2$	Disilylacetylene	II/23(**3**,199)
C$_2$H$_6$Te	Dimethyl telluride	II/15(**3**,367)
C$_2$H$_6$Te$_2$	Dimethylditellurane	II/23(**3**,200)
C$_2$H$_6$Zn	Dimethylzinc	II/7(**3**,306), II/15(**3**,368)
C$_2$H$_7$BN$_4$	Dimethyl–1,4–dihydro–5H–1,2,3,4,5–tetrazaborole	II/7(**3**,308)
C$_2$H$_7$B$_5$	2,4–Dicarba–*closo*–heptaborane(7)	II/7(**3**,307), II/15(**3**,369)
C$_2$H$_7$ClSi	Chlorodimethylsilane	II/15(**3**,369a)
C$_2$H$_7$ClSi	Chloroethylsilane	II/7(**3**,309), II/15(**3**,370)
C$_2$H$_7$Cl$_2$NSi	(Dimethylamino)dichlorosilane	II/21(**3**,191)
C$_2$H$_7$FSi	Ethylfluorosilane	II/15(**3**,371), II/21(**3**,192)
C$_2$H$_7$N	Dimethylamine	II/7(**3**,310)
C$_2$H$_7$N	Ethylamine	II/15(**3**,372), II/21(**3**,193)
C$_2$H$_7$NO	N,O–Dimethylhydroxylamine	II/15(**3**,373)
C$_2$H$_7$NO	2–Aminoethanol	II/7(**3**,311), II/15(**3**,374)
C$_2$H$_7$NOS	S,S–Dimethylsulfoximine	II/7(**3**,312)
C$_2$H$_7$NO$_2$	Formamide – methanol (1/1)	II/21(**3**,194)
C$_2$H$_7$NO$_2$	N–Methoxy–O–methylhydroxylamine Dimethoxylamine	II/21(**3**,195)
C$_2$H$_7$NO$_2$S	Dimethylamine – sulfur dioxide (1/1)	II/23(**3**,201)
C$_2$H$_7$NS	2–Aminoethanethiol	II/15(**3**,375), II/21(**3**,196), II/23(**3**,202)
C$_2$H$_7$P	Dimethylphosphine	II/7(**3**,313), II/15(**3**,377)
C$_2$H$_7$P	Ethylphosphine	II/15(**3**,376), II/21(**3**,197)
C$_2$H$_8$BrNSi	(Dimethylamino)bromosilane	II/21(**3**,198)
C$_2$H$_8$B$_6$	1,7–Dicarba–*closo*–octaborane(8)	II/15(**3**,378)
C$_2$H$_8$ClNSi	(Dimethylamino)chlorosilane	II/21(**3**,199)
C$_2$H$_8$Ge	Dimethylgermane	II/7(**3**,314)
C$_2$H$_8$Ge	Ethylgermane	II/15(**3**,379)
C$_2$H$_8$INSi	(Dimethylamino)iodosilane	II/21(**3**,200)
C$_2$H$_8$N$_2$	1,2–Diaminoethane Ethylenediamine, 1,2–Ethanediamine	II/7(**3**,315), II/15(**3**,380)
C$_2$H$_8$N$_2$	1,1–Dimethylhydrazine	II/21(**3**,201)
C$_2$H$_8$N$_2$	1,2–Dimethylhydrazine N,N'–Dimethylhydrazine	II/15(**3**,381), II/21(**3**,202)
C$_2$H$_8$N$_2$S	Sulfonodiimidoylbismethane	II/7(**3**,316)

	C_2H_8OSi	(Methoxymethyl)silane Methyl silylmethyl ether	II/15(**3**,382)
	C_2H_8Si	Ethylsilane	II/7(**3**,317)

	C_2H_8Si	**Dimethylsilane**	**C_{2v}**
MW			$(CH_3)_2SiH_2$

r_s	Å	θ_s	deg
Si–C	1.868(1)	C–Si–C	110.93(7)
Si–H	1.482(1)	C–Si–H	109.50(3)
C–H(s)	1.089 [a]	H–Si–H	107.83(8)
C–H(a)	1.089(3)	Si–C–H(s)	110.85(40)
		Si–C–H(a)	111.07(23)
		H(s)–C–H(a)	107.78(57)
		H(a)–C–H(a)	108.08(65)
		γ [b]	111.00(30)
		θ [c]	0.13(43)
		β [d]	–0.30(125)

[a] Assumed.

[b] Unperturbed SiCH angle defined by $(1/3)[\alpha(SiCH(s)) + 2\alpha(SiCH(a))]$, where H(s) is on the a, b plane.

[c] Tilt angle defined by $(2/3)[\alpha(SiCH(a)) - \alpha(SiCH(s))]$.

[d] $\beta = \alpha(H(s)CH(a)) - \alpha(H(a)CH(a))$

Hayashi, M., Nakata, N., Miyazaki, S.: J. Mol. Spectrosc. **135** (1989) 270.

II/7(**3**,318), II/23(**3**,203)

$C_2H_8Si_2$	1,3–Disilacyclobutane		II/21(**3**,203A)
$C_2H_8Si_2$ ($C_2H_4D_4Si_2$)	1,3–Disilacyclobutane–1,1,3,3–d_4		II/21(**3**,203B)
C_2H_8Sn	Dimethylstannane		II/7(**3**,319)
C_2H_8Sn	Ethylstannane		II/15(**3**,383)
$C_2H_9B_7$	1,6–Dicarbanonaborane(9)		II/15(**3**,384)
C_2H_9NSi	N–Silyldimethylamine	II/7(**3**,320),	II/15(**3**,385)
C_2H_9PSi	Dimethylsilylphosphine		II/7(**3**,321)
$C_2H_{10}AlB$	Dimethylaluminum tetrahydroborate		II/15(**3**,386)
$C_2H_{10}BGa$	Dimethylgallium tetrahydroborate		II/15(**3**,387)
$C_2H_{10}BP$	Ethylphosphine – borane (1/1)		II/15(**3**,388)
$C_2H_{10}BP$	Dimethylphosphine – borane (1/1)		II/15(**3**,389)
$C_2H_{10}B_2$	1,1–Dimethyldiborane(6)	II/15(**3**,390),	II/21(**3**,204)
$C_2H_{10}B_8$	1,10–Dicarba–*closo*–decaborane(10)		II/15(**3**,391)
$C_2H_{10}B_{10}I_2$	B,B'–Diiodocarborane 5,12–Diiodo–1,7–dicarba–*closo*–dodecaborane(12)		II/7(**3**,322)
$C_2H_{10}B_{10}I_2$	1,7–C,C'–Diiodoneocarborane 1,7–Diiodo–1,7–dicarba–*closo*–dodecaborane(10)	II/7(**3**,323),	II/15(**3**,392)
$C_2H_{10}B_{10}I_2$	1,12–Diiodo–1,12–dicarba–*closo*–dodecaborane(10)		II/15(**3**,393)
$C_2H_{10}OSi_2$	1,3–Dimethyldisiloxane Oxobis[methylsilane]		II/15(**3**,393a)
$C_2H_{10}SSi_2$	Bis(methylsilyl) sulfide		II/23(**3**,204)
$C_2H_{11}B_2N$	Dimethylaminodiborane		II/7(**3**,324)

	C$_2$H$_{11}$NSi$_2$	N–Ethyl–N–silylsilanamine N–Ethyldisilylamine	II/21(**3**,205)
	C$_2$H$_{12}$B$_{10}$	1,2–Dicarba–*closo*–dodecaborane(12) *ortho*–Carborane	II/7(**3**,325)
	C$_2$H$_{12}$B$_{10}$	1,7–Dicarba–*closo*–dodecaborane(12) *meta*–Carborane, Neocarborane	II/7(**3**,326)
	C$_2$H$_{12}$B$_{10}$	1,12–Dicarba–*closo*–dodecaborane(12) *para*–Carborane	II/7(**3**,327)
	C$_2$H$_{14}$AlB$_3$	Dimethylaluminum octahydrotriborate Dimethyl[octahydrotriborato(1–)]aluminum	II/15(**3**,394)
	C$_2$H$_{14}$B$_3$Ga	Dimethylgallium octahydrotriborate Dimethyl[octahydrotriborato(1–)]gallium	II/15(**3**,395)
	C$_2$I$_4$	Tetraiodoethene	II/7(**3**,163)
	C$_2$N	Carbocyanide radical Cyanomethylidyne radical	II/7(**3**,164), II/15(**3**,396)
	C$_2$N	Carboisocyanide radical Isocyanomethylidyne radical	II/7(**3**,165), II/15(**3**,397)
	C$_2$NP	Phosphinidyneacetonitrile *C*–Cyanophosphaethyne	II/15(**3**,397a)

C$_2$N$_2$ **Dicyanogen**   **D$_{\infty h}$**

IR   N≡C–C≡N

r_0	Å
C≡N	1.154(17)
C–C	1.389(30)

Structure determined from B_0 values of ^{12}C$_2^{14}$N$_2$ and ^{12}C$_2^{15}$N$_2$.

Maki, A.G.: J. Chem. Phys. **43** (1965) 3193.

ED, IR

r_g	Å [a]
C≡N	1.162(2)
C–C	1.391(2)

[a] Estimated limits of error. Original ED data have been adjusted slightly to be consistent with the rotational constant.

Morino, Y., Kuchitsu, K., Hori, Y., Tanimoto, M.: Bull. Chem. Soc. Japan **41** (1968) 2349.

                                           II/7(**3**,166), II/15(**3**,398)

C$_2$N$_2$ **Isocyanogen**   **C$_{\infty v}$**

MW   Cyanogen cyanide   C(1)≡N(1) – C(2)≡N(2)

r_0	Å [a]
C(1)≡N(1)	1.1749(18)
N(1)–C(2)	1.3160(26)
C(2)≡N(2)	1.1583(21)

Gerry, M.C.L., Stroh, F., Winnewisser, M.: J. Mol. Spectrosc. **140** (1990) 147.

                                           II/21(**3**,207), II/23(**3**,205)

| | C_2N_2O | Cyanoisocyanate | II/15(**3**,399) |
| | | Cyanogen isocyanate | |

| | C_2N_2S | **Sulfur dicyanide** | C_{2v} |

MW

r_s	Å	θ_s	deg
C–S	1.701(3)	C–S–C	98.37(30)
C≡N	1.157(3)	2ϕ [a]	108.40(30)

[a]) 2ϕ is the angle between the two C≡N bonds.

Pierce, L., Nelson, R., Thomas, C.: J. Chem. Phys. **43** (1965) 3423.   II/7(**3**,168)

C_2N_4	Azodicarbonitrile	II/7(**3**,167)
	Diazenedicarbonitrile	
C_2O	Dicarbon monoxide	II/7(**3**,169), II/15(**3**,400)
$C_2O_2S_2$	Carbonyl sulfide dimer	II/21(**3**,208)
C_2O_3	Carbon monoxide – carbon dioxide (1/1)	II/23(**3**,206)
C_2O_3S	Carbonyl sulfide – carbon dioxide (1/1)	II/21(**3**,209)

| | C_2O_4 | **Carbon dioxide dimer** | D_{2h} |

IR

r_0	Å	θ_0	deg
C(1)...C(2)	3.599(8)	O(1)=C(1)...C(2)	58.2(8)
O(1)...C(2)	3.145(26)		

Jucks, K.W., Huang, Z.S., Miller, R.E., Fraser, G.T., Pine, A.S., Lafferty, W.J.: J. Chem. Phys. **88** (1988) 2185.   II/21(**3**,210)

| | C_2Si | **Silicon dicarbide** | C_{2v} |

MW

r_s	Å [a])	θ_s	deg [a])
C≡C	1.2686(20)	C–Si–C	40.51(20)
Si–C	1.8323(30)		

[a]) Uncertainties are larger than those of the original data.

Cernicharo, J., Guelin, M., Kahane, C., Bogey, M., Demuynck, C., Destombes, J.L.: Astron. Astrophys. **246** (1991) 213.

LIF, UV

State		$\tilde{X}\,^1A_1$			$\tilde{A}\,^1B_2$	
Energy [eV]		0.0	0.0	0.0	2.490 [a])	2.490 [c])
r_0 [Å]	C–C	1.250(16) [a])	1.268(1) [b])	1.265 [c])	1.304(8)	1.329
	Si–C	1.812(12)	1.8373(1)	1.8362	1.881(6)	1.9021
θ_0 [deg]	C–Si–C	40.36(34)	40.37(2)	40.29	40.57(10)	40.91

Rotational analysis of visible spectrum and observation of astronomical lines in the source IRC + 10216.

[a]) From laser excitation in a supersonic beam [1].

(*continued*)

C₂Si (*continued*)

b) From nine astronomical lines, error limits 1σ [2]. A tenth line was subsequently observed [3].
c) From the analysis of bands observed in absorption in a flash discharge [4]. No error limits were quoted.

[1] Michalopoulos, D.L., Geusic, M.E., Langridge-Smith, P.R.R., Smalley, R.E.: J. Chem. Phys. **80** (1984) 3556.
[2] Thaddeus, P., Cunnings, S.E., Linke, R.A.: Astrophys. J. **283** (1984) L45.
[3] Snyder, L.E., Henkel, C., Hollis, J.M., Lovas, F.J.: Astrophys. J. **290** (1985) L29.
[4] Bredohl, H., Dubois, I., Leclercq, H., Melen, F.: J. Mol. Spectrosc. **128** (1988) 399.

II/15(**3**,401), II/21(**3**,210A), II/23(**3**,207)

C₃		Tricarbon			$D_{\infty h}$
IR					C=C=C

State	$\tilde{X}\,^1\Sigma_g^+$	$\tilde{a}\,^3\Pi_u$	$\tilde{b}\,^3\Pi_g$	$\tilde{A}\,^1\Pi_u$
Energy [eV]	0	2.10	2.90	3.059
Reference	[1]	[2]	[2]	[1]
r_0 [Å] C=C	1.277	1.298	1.286	1.305

No error limits are given.

[1] Gausset, L., Herzberg, G., Lagerqvist, A., Rosen, B.: Disc. Faraday Soc. **35** (1963) 113; Astrophys. J. **142** (1965) 45.
[2] Sasada, H., Amano, T., Jarman, C., Bernath, P.F.: J. Chem. Phys. **94** (1991) 2401.

II/7(**3**,328), II/15(**3**,402), II/23(**3**,208)

Formula	Name	Reference
C₃BrN	Bromocyanoacetylene Bromopropynenitrile	II/7(**3**,329)
C₃ClF₃	1–Chloro–3,3,3–trifluoropropyne	II/7(**3**,333)
C₃ClN	Chlorocyanoacetylene Chloropropynenitrile	II/7(**3**,334), II/15(**3**,403)
C₃ClN₃O₃Si	Chlorosilanetriyl triisocyanate	II/7(**3**,335)
C₃Cl₂F₆	1,3–Dichloro–1,1,2,2,3,3–hexafluoropropane	II/15(**3**,404)
C₃Cl₂F₉P	Dichlorotris(trifluoromethyl)phosphorane	II/15(**3**,405)
C₃Cl₄	Tetrachlorocyclopropene	II/7(**3**,330)
C₃Cl₆	Hexachlorocyclopropane	II/7(**3**,331)
C₃Cl₆O	Hexachloroacetone	II/7(**3**,336)
C₃Cl₈	Octachloropropane	II/7(**3**,332)
C₃CoNO₄	Tricarbonylnitrosylcobalt	II/21(**3**,211)
C₃F₂O	Difluoropropadienone	II/23(**3**,209)
C₃F₂O	Difluorocyclopropenone	II/23(**3**,210)
C₃F₃N₃	Cyanuric trifluoride	II/7(**3**,339)
C₃F₄	1,3,3,3–Tetrafluoropropyne	II/23(**3**,211)
C₃F₄O₂	Difluoromalonyl difluoride	II/23(**3**,212)
C₃F₆	Hexafluoropropene	II/15(**3**,406)
C₃F₆	Hexafluorocyclopropane	II/7(**3**,337)
C₃F₆O	Hexafluoropropanone Hexafluoroacetone	II/7(**3**,340)
C₃F₆O	2–(Trifluoromethyl)–2,3,3–trifluorooxirane	II/15(**3**,407)
C₃F₆OS	Hexafluoropropanethione *S*–oxide	II/23(**3**,213)
C₃F₆O₂	Perfluoro–1,2–dioxolane	II/23(**3**,214)

C₃F₇I		Perfluoropropyl iodide	II/7(**3**,338)
C₃F₈		Perfluoropropane	II/23(**3**,215)
C₃F₉N		Tris(trifluoromethyl)amine	II/15(**3**,408)
C₃F₉NS₃		Tris(trifluoromethylthio)amine	II/15(**3**,409)
C₃F₉P		Tris(trifluoromethyl)phosphine	II/15(**3**,410)
C₃F₁₁P		Difluorotris(trifluoromethyl)phosphorane	II/15(**3**,411)
C₃H		Cyclotricarbon monohydride	II/23(**3**,216)
C₃HClO		Propioloyl chloride	II/15(**3**,412)
C₃HCl₃F₂		1,1,2–Trichloro–3,3–difluoro–1–propene	II/15(**3**,413), II/23(**3**,217)
C₃HCl₅		1,1,2,3,3–Pentachloro–1–propene	II/23(**3**,218)
C₃HCl₇		1,1,1,2,2,3,3–Heptachloropropane	II/15(**3**,414)
C₃HFO		Propioloyl fluoride	II/15(**3**,415)
C₃HF₃		(Trifluoromethyl)acetylene 3,3,3–Trifluoro–1–propyne	II/7(**3**,344)
C₃HF₃O		3,3–Difluoroacryloyl fluoride	II/23(**3**,219)
C₃HF₆N		Hexafluoroisopropanimine	II/7(**3**,345)

C₃HN   **2–Propynenitrile** $C_{\infty v}$

MW, IR   Cyanoacetylene   H–C≡C–C≡N

r_e	Å
H–C	1.0624(5)
C≡C	1.2058(5)
C–C	1.3764(5)
C≡N	1.1605(5)

The equilibrium rotational constants B_e for six isotopomers are derived from the corresponding B_0 values by using α_i parameters obtained by *ab initio* calculations and checked against experimental values. These B_e values are then fitted to yield the equilibrium structure.

Botschwina, P., Horn, M., Seeger, S., Flügge, J.: Mol. Phys. **78** (1993) 191.

UV C_s

State		$\tilde{A}\,^1A''$
r_0 [Å]	C(2)–H	(1.08) [a]
	C(2)=C(1)	1.25
	C(1)–C(3)	1.40
	C(3)≡N	(1.159) [b]
θ_0 [deg]	C(1)=C(2)–H	164
	C(2)=C(1)–C(3)	143
	C(1)–C(3)≡N	180

[a]) Assumed.
[b]) The ground state value was assumed to hold in the excited state.

Job, V.A., King, G.W.: J. Mol. Spectrosc. **19** (1966) 155.

II/7(**3**,346), II/15(**3**,416), II/23(**3**,220)

C₃HN		Ethynyl isocyanide Isocyanoacetylene	II/23(**3**,221)

C$_3$HNO	Ethynyl isocyanate	II/23(**3**,222)
C$_3$HNO$_4$	Hydrogen cyanide – carbon dioxide (1/2)	
		II/21(**3**,213), II/23(**3**,223)
C$_3$HO	3–Oxo–1,2–Propadienyl	II/23(**3**,224)
C$_3$HP	1–Phospha–1,3–butadiyne	II/15(**3**,417)
	2–Propynylidinephosphine	

C$_3$H$_2$ **Tricarbon dihydride radical**

1,3–Didehydroallene radical   HC=C=CH

UV

State	$\tilde{X}\,^1\Sigma$?	$\tilde{A}\,^1\Sigma$?
Symmetry	D$_{\infty h}$	D$_{\infty h}$?
Energy [eV]	0	3.059

r_0 [Å]	C–H	1.087 [a]
	C=C	1.299 [b]

[a]) Assumed to be as in allene ground state.
[b]) From partially resolved rotational structure in one band.

Merer, A.J.: Can. J. Phys. **45** (1967) 4103.   II/7(**3**,347), II/15(**3**,418)

C$_3$H$_2$	Cyclopropenylidene	II/21(**3**,214), II/23(**3**,225)
C$_3$H$_2$ClF	1–Chloro–1–fluoroallene	II/23(**3**,226)
C$_3$H$_2$ClF	1–Chloro–3–fluoropropyne	II/7(**3**,348)
C$_3$H$_2$ClN	2–Chloroacrylonitrile	II/15(**3**,419)
C$_3$H$_2$Cl$_2$	1,3–Dichloropropyne	II/15(**3**,420)
C$_3$H$_2$Cl$_2$O	2–Chloropropenoyl chloride	II/21(**3**,215)
C$_3$H$_2$Cl$_4$	1,1,3,3–Tetrachloro–1–propene	II/23(**3**,227)
C$_3$H$_2$Cl$_4$	trans–1,2,3,3–Tetrachloro–1–propene	II/23(**3**,228)
C$_3$H$_2$Cl$_4$	cis–1,2,3,3–Tetrachloro–1–propene	II/23(**3**,229)
C$_3$H$_2$Cl$_6$	1,1,2,2,3,3–Hexachloropropane	II/15(**3**,421)
C$_3$H$_2$FN	Cyanoacetylene – hydrogen fluoride (1/1)	II/15(**3**,421a)
C$_3$H$_2$F$_2$	1,1–Difluoroallene	II/7(**3**,349)
C$_3$H$_2$F$_2$	3,3–Difluorocyclopropene	II/15(**3**,422)
C$_3$H$_2$F$_2$O$_2$	Malonyl difluoride	II/23(**3**,230)
C$_3$H$_2$F$_4$	1,1,2,2–Tetrafluorocyclopropane	II/21(**3**,216)
C$_3$H$_2$F$_4$	cis–1,1,2,3–Tetrafluorocyclopropane	II/23(**3**,231)
C$_3$H$_2$F$_6$	1,1,1,3,3,3–Hexafluoropropane	II/23(**3**,232)
C$_3$H$_2$N$_2$	Malononitrile	II/7(**3**,350), II/23(**3**,233)
C$_3$H$_2$N$_2$O	Carbon monoxide – hydrogen cyanide (1/2)	II/23(**3**,234)
C$_3$H$_2$N$_2$O$_2$	Carbon dioxide – hydrogen cyanide (1/2)	II/21(**3**,217)
C$_3$H$_2$N$_2$O$_2$S	2,5–Diaza–1,6–dioxa–6a–thia(6a–S^{IV})pentalene	II/21(**3**,218)
C$_3$H$_2$O	Methyleneketene	II/15(**3**,423), II/21(**3**,219)
	1,2–Propadien–1–one	

C₃H₂O   **Propiolaldehyde** C_s
Propynal

ED, MW

r_g	Å [a])		θ_z	deg [a])
C(2)–C(3)	1.4527(23)		C(2)–C(3)=O	124.2(2)
C(1)≡C(2)	1.211(6)		O=C(3)–H	122.1(8)
C(3)=O	1.214(5)		C(1)≡C(2)–C(3)	178.6(3) [b])
C(1)–H	1.085(7)			
C(3)–H	1.130(6)			

$$H-\underset{(1)}{C}\equiv\underset{(2)}{C}-\underset{(3)}{C}\diagup^{O}_{H}$$

The molecule is planar.

[a]) Estimated limits of error.
[b]) The C≡C bond is in the *anti* position to the C=O bond.

Sugié, M., Fukuyama, T., Kuchitsu, K.: J. Mol. Struct. **14** (1972) 333.

UV

State	Ã ¹A″
Symmetry	C_s
Energy [eV]	3.244
r_0 [Å] C(3)=O	1.325
C(1)–C(2)	1.238
C(1)–H	1.091

Estimates from vibrational frequencies and Clark's or Badger's rules.

Brand, J.C.D., Callomon, J.H., Watson, J.K.G.: Disc. Faraday Soc. **35** (1963) 175.
Brand, J.C.D., Chan, W.H., Liu, D.S., Callomon J.H., Watson, J.K.G.: J. Mol. Spectrosc. **50** (1974) 304.
     II/7(**3**,351), II/15(**3**,424)

C₃H₂O	Cyclopropenone	II/7(**3**,352), II/23(**3**,235)
C₃H₂O	Acetylene – carbon monoxide (1/1)	II/21(**3**,220), II/23(**3**,236)
C₃H₂O₂	Carbon dioxide – acetylene (1/1)	II/21(**3**,221), II/23(**3**,237)
C₃H₂O₃	Vinylene carbonate	II/7(**3**,353)
	Ethylene carbonate	
C₃H₂S	1,2–Propadiene–1–thione	II/21(**3**,222)
C₃H₃ArN	Argon – acrylonitrile (1/1)	II/21(**3**,223)
	Argon – vinyl cyanide (1/1)	
C₃H₃Br	Bromoallene	II/23(**3**,238)
C₃H₃Br	1–Bromopropyne	II/7(**3**,354)
C₃H₃Br	Propargyl bromide	II/7(**3**,355)
	3–Bromopropyne	
C₃H₃BrO	2–Bromopropenal	II/21(**3**,224)
C₃H₃Cl	1–Chloropropyne	II/7(**3**,356), II/23(**3**,239)
C₃H₃Cl	Propargyl chloride	II/7(**3**,357)
	3–Chloropropyne	
C₃H₃ClO	Propenoyl chloride	II/7(**3**,358), II/15(**3**,425)
	Acryloyl chloride	
C₃H₃ClO	2–Chloro–2–propenal	II/7(**3**,359), II/21(**3**,225)
	2–Chloroacrylaldehyde	
C₃H₃Cl₃	1,2,3–Trichloro–1–propene	II/21(**3**,226)

$C_3H_3Cl_3F_2Si$	Trichloro(2,2–difluorocyclopropyl)silane	II/15(**3**,426)
	1,1–Difluoro–2–(trichlorosilyl)cyclopropane	
$C_3H_3Cl_5$	1,1,2,3,3–Pentachloropropane	II/15(**3**,427)
$C_3H_3Cl_5$	1,1,3,3,3–Pentachloropropane	II/15(**3**,428)
C_3H_3F	1–Fluoro–1,2–propadiene, Fluoroallene	II/21(**3**,227)
C_3H_3F	3–Fluoro–1–propyne	II/21(**3**,228)
C_3H_3FO	Acryloyl fluoride, 2–Propenoyl fluoride	II/7(**3**,361)
$C_3H_3F_3$	3,3,3–Trifluoro–1–propene	II/7(**3**,360), II/15(**3**,429)
$C_3H_3F_3$	r–1,c–2,t–3–1,2,3–Trifluorocyclopropane	
	$cis,trans$–1,2,3–Trifluorocyclopropane	II/15(**3**,430), II/21(**3**,229)
$C_3H_3F_3N_2$	Fluoroform – hydrogen cyanide (1/2)	II/21(**3**,230)
$C_3H_3F_3O$	α,α,α–Trifluoroacetone	II/7(**3**,362)
	1,1,1–Trifluoropropanone	
$C_3H_3F_3O_4$	Formic acid – trifluoroacetic acid (1/1)	
		II/21(**3**,231), II/23(**3**,240)
$C_3H_3F_3Si$	1,1,1–Trifluoro–4–silabut–2–yne	II/15(**3**,431)
	3,3,3–Trifluoro–1–silyl–1–propyne	
$C_3H_3F_5Si$	(2,2–Difluorocyclopropyl)trifluorosilane	II/15(**3**,432)
	1,1–Difluoro–2–(trifluorosilyl)cyclopropane	
C_3H_3N	Acrylonitrile, Vinyl cyanide	II/7(**3**,363)
C_3H_3N	Vinyl isocyanide	II/21(**3**,232)
C_3H_3N	2–Propynimine	II/21(**3**,233)
	2–Propynylideneamine	
C_3H_3N	Acetylene – hydrogen cyanide (1/1)	II/15(**3**,432a), II/23(**3**,241)
C_3H_3NO	Propynamide, Propiolamide	II/15(**3**,433)
C_3H_3NO	Acetyl cyanide, Pyruvonitrile	II/7(**3**,365), II/15(**3**,434)
C_3H_3NO	Vinyl isocyanate	II/15(**3**,435)
C_3H_3NO	Oxazole	II/15(**3**,436)
C_3H_3NO	Isoxazole	II/15(**3**,437)
C_3H_3NO	Fulminic acid – acetylene (1/1)	II/23(**3**,242)
$C_3H_3NO_2$	2–Propynyl nitrite, Propargyl nitrite	II/21(**3**,234)
$C_2H_3NO_2$	Acetyl isocyanate	II/15(**3**,438), II/23(**3**,243)
$C_3H_3NO_2$	Methyl cyanoformate	II/23(**3**,244)
$C_3H_3NO_2S$	5–Methyl–1,3,4–oxathiazol–2–one	II/15(**3**,439)
C_3H_3NS	Thiazole	II/7(**3**,366)
C_3H_3NS	Isothiazole	II/21(**3**,235)
$C_3H_3N_3$	3–Azidopropyne, Propargyl azide	II/21(**3**,236)
$C_3H_3N_3$	1,3,5–Triazine	II/7(**3**,364), II/21(**3**,237)
	s–Triazine	
$C_3H_3N_3$	Hydrogen cyanide trimer	II/21(**3**,238)
$C_3H_3N_3$	Cyclic hydrogen cyanide trimer	II/21(**3**,239)
C_3H_3P	1–Phospha–3–buten–1–yne	II/15(**3**,440)
	2–Propenylidynephosphine	

C₃H₄	**Methylacetylene**	**C₃ᵥ**
MW, IR and *ab initio* calculations	Propyne	C(1)H₃–C(2)≡C(3)H

r_s	Å
C(1)–H	1.0561(3) [a]
C(3)–H	1.0895(52) [a]
C–C	1.4586(4) [a]
C≡C	1.2066(4) [a]

θ_s	deg
H–C(1)–C(2)	110.6(1) [a]

r_e	Å
C(1)–H	1.061(1)
C(3)–H	1.089(1)
C–C	1.458(2)
C≡C	1.204(1)

θ_e	deg
H–C(1)–C(2)	110.7(5)

The r_s structure was derived from the ground state molecular constants of 17 isotopomers, obtained from MW and also from IR data using normal propyne as parent molecule.
The pseudo-equilibrium structure r_e reported here is elaborated from all these experimental structures, from ED structures and from several *ab initio* ones, taking into account their known deficiencies, offsets and trends.

[a]) Uncertainty was not given in the original paper. Here the dispersion among three substitution structures is taken as uncertainty.

Le Guennec, M., Demaison, J., Wlodarczak, G., Marsden, C.J.: J. Mol. Spectrosc. **160** (1993) 471.
See also: Pekkala, K., Graner, G., Wlodarczak, G., Demaison, J., Koput, J.: J. Mol. Spectrosc. **149** (1991) 214.

II/7(**3**,367), II/15(**3**,441), II/23(**3**,245)

C₃H₄	**1,2–Propadiene**	**D₂d**
ED, IR	Allene	

r_g	Å [a])
C=C	1.3129(9)
C–H	1.102(2)

θ_z	deg [a])
H–C–H	118.3 (3)

r_e	Å [a])
C=C	1.3082(10)
C–H	1.076(3)

θ_e	deg [a])
H–C–H	118.2 (5)

The measurements were made at room temperature.

[a]) Estimated limits of error.

Ohshima, Y., Yamamoto, S., Nakata, M., Kuchitsu, K.: J. Phys. Chem. **91** (1987) 4696.

II/7(**3**,368), II/15(**3**,442), II/21(**3**,240)

C₃H₄ **Cyclopropene** C_{2v}

MW

HC=CH\
  \\ /\
  CH₂

r_m^ρ a)	Å
C=C	1.2934(3)
C–C	1.5051(3)
C–H b)	1.0719(6)
C–H c)	1.0853(8)

θ_m^ρ a)	deg
H–C=C	149.95(6)
H–C–H	114.32(12)

a) Multiple isotope substitution structure.
b) =C–H.
c) C–H in CH₂.

Berry, R.J., Harmony, M.D.: Struct. Chem. **1** (1990) 49.

II/7(**3**,369), II/15(**3**,443), II/23(**3**,246)

C₃H₄BrCl	2–Bromo–3–chloro–1–propene	II/15(**3**,444)
C₃H₄BrN	3–Bromopropionitrile	II/21(**3**,241)
C₃H₄Br₂	2,3–Dibromo–1–propene	II/15(**3**,445)
C₃H₄ClF	1–Chloro–3–fluoro–1–propene	II/15(**3**,446)
C₃H₄ClF	2–Chloro–3–fluoro–1–propene	II/15(**3**,447)
C₃H₄ClF₃	1,1,1–Trifluoro–3–chloropropane	II/21(**3**,242)
C₃H₄ClN	3–Chloropropionitrile	II/21(**3**,243)
C₃H₄ClOPS	2–Chloro–5–methyl–1,3,2–oxathiaphosphole	II/23(**3**,247)
C₃H₄Cl₂	*trans*–1,3–Dichloro–1–propene	II/21(**3**,244)
	(*E*)–1,3–Dichloro–1–propene	
C₃H₄Cl₂	*cis*–1,3–Dichloro–1–propene	II/15(**3**,448)
	(*Z*)–1,3–Dichloro–1–propene	
C₃H₄Cl₂	2,3–Dichloro–1–propene	II/15(**3**,449)
C₃H₄Cl₂	1,1–Dichlorocyclopropane	II/7(**3**,370), II/15(**3**,450)
C₃H₄Cl₃N	*N*–Methyl–2,2,2–trichloroethylidenamine	II/15(**3**,451)
C₃H₄Cl₄	1,1,3,3–Tetrachloropropane	II/15(**3**,452)
C₃H₄F₂	*cis*–1,2–Difluorocyclopropane	II/15(**3**,452a), II/21(**3**,246)
C₃H₄F₂	*trans*–1,2–Difluorocyclopropane	II/15(**3**,452a), II/21(**3**,247)
C₃H₄F₂ (C₃D₄F₂)	1,1–Fluorocyclopropane (–d_4)	II/15(**3**,453)
C₃H₄F₂NOP	3–(Difluorophosphinooxy)propionitrile	II/23(**3**,249)
C₃H₄NO₃P	2–Isocyanato–1,3,2–dioxaphospholane	II/21(**3**,248)
C₃H₄N₂	Methyl isocyanide – hydrogen cyanide (1/1)	II/23(**3**,248)
C₃H₄N₂	Pyrazole	II/7(**3**,371)
C₃H₄N₂	Imidazole	II/15(**3**,454)
C₃H₄N₂	Hydrogen cyanide – acetonitrile (1/1)	II/21(**3**,249)
C₃H₄O	Methoxyethyne	II/7(**3**,372)
	Methoxyacetylene	
C₃H₄O	2–Propyn–1–ol	II/7(**3**,373)
	Propargyl alcohol	
C₃H₄O	Methylketene	II/7(**3**,374)
C₃H₄O	Acrylaldehyde, Acrolein	II/7(**3**,375), II/15(**3**,455)
	Propenal	
C₃H₄O	Cyclopropanone	II/7(**3**,376)

C₃H₄O	Formaldehyde – acetylene (1/1)	II/21(**3**,250)
C₃H₄OS	3–Thietanone	II/15(**3**,456)
C₃H₄O₂	Vinyl formate	II/21(**3**,251)
C₃H₄O₂	Methylglyoxal	II/7(**3**,377)
C₃H₄O₂	Malonaldehyde	II/15(**3**,457)
C₃H₄O₂	2,3–Epoxypropanal Oxiranecarbaldehyde Glycidaldehyde	II/15(**3**,458)
C₃H₄O₂	3–Oxetanone	II/7(**3**,378)
C₃H₄O₃	Formic acetic anhydride	II/7(**3**,379)
C₃H₄O₃	Pyruvic acid	II/15(**3**,459)
C₃H₄O₃	Ethylene carbonate 1,3–Dioxolan–2–one	II/21(**3**,252)
C₃H₄S	Methylthioketene 1–Propene–1–thione	II/15(**3**,460)
C₃H₄S	2–Propenethial Thioacrolein	II/15(**3**,461)
C₃H₄S	Propargyl mercaptan 2–Propyne–1–thiol	II/15(**3**,462)
C₃H₄S	(Methylthio)ethyne	II/7(**3**,380)
C₃H₄S	2*H*–Thiete	II/15(**3**,462a)
C₃H₄S	Allene episulfide 2–Methylenethiirane	II/15(**3**,463)
C₃H₅	Allyl radical	II/21(**3**,253), II/23(**3**,250)
C₃H₅BF₂	Cyclopropyldifluoroborane	II/15(**3**,464)
C₃H₅Br	*trans*–1–Bromo–1–propene *trans*–1–Bromopropylene	II/7(**3**,381)
C₃H₅Br	2–Bromo–1–propene	II/15(**3**,465)
C₃H₅Br	3–Bromo–1–propene	II/15(**3**,466), II/23(**3**,251)
C₃H₅Br	Bromocyclopropane Cyclopropyl bromide	II/7(**3**,382), II/21(**3**,254)
C₃H₅BrO	Propionyl bromide	II/21(**3**,255)
C₃H₅BrO	Epibromohydrin 2–(Bromomethyl)oxirane	II/15(**3**,467)
C₃H₅BrO₂S	1–Bromovinyl methyl sulfone	II/15(**3**,468)
C₃H₅BrO₂S	3–Bromothietane 1,1–dioxide	II/21(**3**,256)
C₃H₅Br₃	1,2,3–Tribromopropane	II/7(**3**,383)
C₃H₅Cl	*cis*–1–Chloro–1–propene *trans*–1–Chloro–1–propene	II/7(**3**,384)
C₃H₅Cl	2–Chloro–1–propene	II/7(**3**,385), II/15(**3**,469)
C₃H₅Cl	3–Chloro–1–propene Allyl chloride	II/7(**3**,386), II/21(**3**,257)
C₃H₅Cl	Cyclopropyl chloride Chlorocyclopropane	II/7(**3**,387)
C₃H₅ClO	Propionyl chloride	II/15(**3**,470)
C₃H₅ClO	Chloroacetone 1–Chloro–2–propanone	II/23(**3**,252)

C₃H₅ClO	2-(Chloromethyl)oxirane Epichlorohydrin	II/15(**3**,471), II/21(**3**,258)
C₃H₅ClO₂S	*cis*-Chlorovinyl methyl sulfone	II/21(**3**,259)
C₃H₅ClO₂S	*trans*-Chlorovinyl methyl sulfone	II/21(**3**,260)
C₃H₅ClS	2-(Chloromethyl)thiirane 3-Chloropropylene sulfide	II/21(**3**,261)
C₃H₅Cl₃	1,2,3-Trichloropropane	II/7(**3**,388)
C₃H₅F	1-Fluoro-1-propene	II/7(**3**,389)
C₃H₅F	2-Fluoro-1-propene	II/7(**3**,390)
C₃H₅F	3-Fluoro-1-propene	II/7(**3**,391), II/23(**3**,253)
C₃H₅F	Propyne – hydrogen fluoride (1/1)	II/15(**3**,471a)
C₃H₅F	Propadiene – hydrogen fluoride (1/1)	II/21(**3**,262)
C₃H₅FO	Fluoroacetone	II/7(**3**,392)
C₃H₅F₃Si	Allyltrifluorosilane	II/15(**3**,472)
C₃H₅F₃Si	(Trifluorosilyl)cyclopropane	II/15(**3**,473)
C₃H₅I	3-Iodo-1-propene	II/15(**3**,474), II/23(**3**,254)
C₃H₅N	2-Propen-1-imine 3-Imino-1-propene	II/15(**3**,475)
C₃H₅N	Propargylamine 2-Propyn-1-amine	II/15(**3**,476)
C₃H₅N	Propionitrile	II/7(**3**,393)
C₃H₅N	Ethylene – hydrogen cyanide (1/1)	II/15(**3**,476a)
C₃H₅NO	Ethyl isocyanate	II/15(**3**,477), II/21(**3**,263)
C₃H₅NO	Lactonitrile	II/21(**3**,264)
C₃H₅NO	3-Hydroxypropanenitrile 3-Hydroxypropionitrile	II/21(**3**,265)
C₃H₅NO	2-Azetidinone	II/21(**3**,266)
C₃H₅NO	Hydrogen cyanide – oxirane (1/1)	II/21(**3**,267)
C₃H₅NO₂	Pyruvaldehyde oxime α-(Hydroxyimino)acetone	II/15(**3**,478)
C₃H₅NO₂	Nitrocyclopropane	II/7(**3**,394)
C₃H₅NS	Ethyl isothiocyanate	II/21(**3**,268)
C₃H₅NS	3-Mercaptopropionitrile	II/15(**3**,478a)
C₃H₅NSe	Ethyl isoselenocyanate	II/21(**3**,269)
C₃H₅N₃	Allyl azide 3-Azido-1-propene	II/21(**3**,270)
C₃H₅O₃P	2,6,7-Trioxa-1-phosphabicyclo[2.2.1]heptane	II/15(**3**,479)
C₃H₆	Propene Propylene	II/7(**3**,395), II/23(**3**,255)

C₃H₆   **Cyclopropane**   **D₃ₕ**

ED, IR

r_z	Å ᵃ)		θ_z	deg ᵃ)
C–C	1.5127(12)		H–C–H	114.5(9)
C–H	1.0840(20)			

H₂C——CH₂
    \ /
     CH₂

(*continued*)

r_e	Å [a]		θ_e	deg [a]
C–C	1.501(4)		H–C–H	114.5(9)
C–H	1.083(5)			

The measurements were made at room temperature.

[a] Estimated limits of error.

Yamamoto, S., Nakata, M., Fukuyama, T., Kuchitsu, K.: J. Phys. Chem. **89** (1985) 3298.

II/7(**3**,396), II/21(**3**,271A)

MW **C₃H₆ (C₃H₄D₂)**   **Cyclopropane–1,1–d_2**   D$_{3h}$ (C$_{2v}$)

r_0	Å		θ_0	deg
C–H	1.07739(12)		H–C–H	115.568(13)
C–C	1.515321(75)			
δ(C–H) – (C–D)	–0.001160(87)			

r_z	Å		θ_z	deg
C–H	1.080(10)		H–C–H	115.5(10)
C–C	1.5157(69)			
δ(C–H) – (C–D)	0.00029(48)			

r_e	Å		θ_e	deg
C–H	1.0742(29)		H–C–H	115.85(33)
C–C	1.5101(21)			

D$_2$C——CH$_2$
   CH$_2$

The MW spectrum of cyclopropane–1,1–d_2 has been observed.

Endo, Y., Chan, M.C., Hirota, E.: J. Mol. Spectrosc. **126** (1987) 63.   II/21(**3**,271B)

C₃H₆AsBrO₂	2–Bromo–1,3,2–dioxarsenane	II/23(**3**,256)
C₃H₆Br₂	1,2–Dibromopropane	II/21(**3**,272)
C₃H₆Br₂	1,3–Dibromopropane	II/7(**3**,397)
C₃H₆ClN₂OP	2–Chloro–3,5–dimethyl–2,3–dihydro–1,3,4,2–oxadiazaphosphole	II/7(**3**,398), II/15(**3**,480)
C₃H₆ClO₂P	2–Chloro–1,3,2–dioxaphosphorinane Trimethylene phosphorochloridite	II/7(**3**,399)
C₃H₆Cl₂	1,1–Dichloropropane	II/15(**3**,481)
C₃H₆Cl₂	1,2–Dichloropropane	II/21(**3**,273)
C₃H₆Cl₂	1,3–Dichloropropane	II/15(**3**,482)
C₃H₆Cl₂	2,2–Dichloropropane	II/15(**3**,483)
C₃H₆Cl₂Si	1,1–Dichlorosilacyclobutane	II/7(**3**,400), II/21(**3**,274)
C₃H₆F₂	1,3–Difluoropropane	II/15(**3**,484)
C₃H₆F₂	2,2–Difluoropropane	II/15(**3**,485), II/23(**3**,257)
C₃H₆F₂Si	1,1–Difluorosilacyclobutane	II/21(**3**,275)
C₃H₆F₄P₂S₂	1,3–Bis(difluorophosphinothio)propane	II/23(**3**,258)
C₃H₆NO	Isocyanatodimethylborane Dimethylboron isocyanate	II/23(**3**,259)
C₃H₆NP	3–Phosphinopropionitrile	II/15(**3**,485a)
C₃H₆N₂	(Methylamino)acetonitrile	II/21(**3**,276)
C₃H₆N₂	Dimethylcyanamide	II/7(**3**,401), II/15(**3**,486)

C₃H₆N₂	3–Aminopropionitrile	II/15(**3**,486a)
C₃H₆N₂	3,3–Dimethyldiazirine	II/7(**3**,402)
C₃H₆N₂	1–Pyrazoline	II/15(**3**,487)
C₃H₆N₂O₂	N–Methyl–N–nitrovinylamine	II/15(**3**,487a)
C₃H₆N₂O₄	2,2–Dinitropropane	II/15(**3**,488)
C₃H₆N₄O₄	1,3–Dinitroimidazolidine	II/23(**3**,260)

C₃H₆N₆O₆ **Hexahydro–1,3,5–trinitro–1,3,5–triazine** C₃

ED

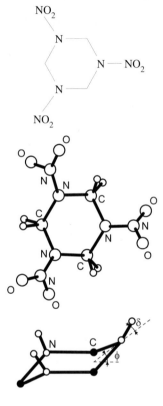

r_g	Å a)
C–N	1.464(6)
N–N	1.413(5)
N=O	1.213(2)
C–H	1.089(27)

θ_α	deg a)
N–C–N	109.4(6)
C–N–C	123.7(6)
C–N–N	116.3(5)
O=N=O	125.5(10)
H–C–H	105.1(60)
τ(NN) b)	19.1(23)
φ c)	33.9(29)
δ d)	19.9(50) f)
γ e)	356.3(50) f)

The six-membered ring has a chair conformation with an axial positions of the NO₂ groups.
The N–NO₂ fragment was assumed to be planar with C₂ᵥ local symmetry.
The nozzle temperature was 160 °C.

a) Three times the estimated standard errors.
b) Torsional angle about the N–N bond. The value 0° corresponds to a position of the NO₂ groups such that the line segments C–C and O–O for C₂N–NO₂ will lie in a single plane.
c) Dihedral angle between the CNC plane and adjacent CN···NC plane; see figure.
d) Angle between the N–N bond and the CNC plane; see figure.
e) The sum of the bond angles at the amine N atom.
f) The uncertainty was not estimated in the original paper.

Shishkov, I.F., El'fimova, T.L., Vilkov, L.V.: Zh. Strukt. Khim. **33** (1992) No.1, 41; Russ. J. Struct. Chem. (Engl. Transl.) **33** (1992) 34.
Shishkov, I.F., Vilkov, L.V., Kolonits, M., Rozsondai, B.: Struct. Chem. **2** (1991) 57.

II/23(**3**,261)

C₃H₆O **Acetone** Cₛ

ED, MW

r_g	Å a)
C–C	1.520(3)
C=O	1.214(4)
C–H	1.103(3)

θ_z	deg a)
O=C–C	122.0(3)
C–C–C	116.0(3)
H–C–H	108.4(5)
b)	2.0(5)

The effect of internal rotation is taken into account in the vibrational correction to the observed moments of inertia.

(continued)

[a]) Estimated limits of error. Those of the distances are not given in the original paper, but are estimated.

[b]) The tilt angle between the CH_3 axis and the C–C bond.

Iijima, T.: Bull Chem. Soc. Japan **45** (1972) 3526.

UV Vibrational analysis

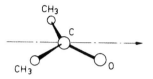

State	$\tilde{A}\,(^1A_2)(\pi^*\text{–}n)$ [a])
Energy [eV]	3.773
Reference	[1]

The vibrational intervals were fitted to effective one-dimensional polyminima potentials. The molecule is pyramidal at the formyl carbon atom, as in the analogous states of formaldehyde and acetaldehyde. The height of the barrier to inversion is 466(32) cm^{-1} (0.058 eV) from equilibrium, 246 cm^{-1} (0.031 eV) above zero-point. The barrier to concerted threefold internal rotation of the methyl groups is 740(90) cm^{-1} (0.092 eV).

[a]) Symmetry with respect to the C_2CO skeleton.

[1] Baba, H., Hanazaki, I., Nagashima, U.: J. Chem. Phys. **82** (1985) 3938.

			II/7(**3**,403), II/15(**3**,489)
C_3H_6O	Allyl alcohol		II/7(**3**,404), II/15(**3**,490), II/21(**3**,277)
	2–Propen–1–ol		
C_3H_6O	Methyl vinyl ether		II/7(**3**,405), II/15(**3**,491), II/21(**3**,278)
C_3H_6O	Propanal		II/15(**3**,492), II/21(**3**,279)
	Propionaldehyde		
C_3H_6O	Propylene oxide		II/7(**3**,406), II/15(**3**,493)
	Methyloxirane		
C_3H_6O	Trimethylene oxide		II/7(**3**,407), II/15(**3**,494)
	Oxetane		
C_3H_6OS	Trimethylene sulfoxide		II/15(**3**,495)
	Thietane 1–oxide		
$C_3H_6OS_2$	$S,S\,'$–Dimethyl dithiocarbonate		II/15(**3**,496)
$C_3H_6O_2$	Methyl acetate		II/21(**3**,280)
$C_3H_6O_2$	α–Hydroxyacetone		II/15(**3**,497)
	1–Hydroxy–2–propanone		
$C_3H_6O_2$	Propionic acid		II/7(**3**,408), II/15(**3**,498)
$C_3H_6O_2$	Oxiranemethanol		II/15(**3**,499), II/23(**3**,262)
	2,3–Epoxy–1–propanol		
	Glycidol		
$C_3H_6O_2$	1,3–Dioxolane		II/21(**3**,281)
$C_3H_6O_2S$	Methyl vinyl sulfone		II/15(**3**,500)
$C_3H_6O_2S$	Cyclopropane – sulfur dioxide (1/1)		II/23(**3**,263)
$C_3H_6O_3$	Lactic acid		II/15(**3**,500a)
$C_3H_6O_3$	Dimethyl carbonate		II/15(**3**,501)
$C_3H_6O_3$	Glycol monoformate		II/7(**3**,409)
	2–Hydroxyethyl formate		
$C_3H_6O_3$	3–Methyl–1,2,4–trioxalane		II/15(**3**,502)
	Propylene ozonide		

C$_3$H$_6$O$_3$	1,3,5–Trioxane	II/7(**3**,410), II/15(**3**,503)
C$_3$H$_6$O$_3$S	Trimethylene sulfite	II/7(**3**,411)
C$_3$H$_6$O$_3$Se	Trimethylene selenite	II/7(**3**,412)
C$_3$H$_6$O$_4$	3–Methoxy–1,2,4–trixolane	II/21(**3**,282)
C$_3$H$_6$S	Methyl vinyl sulfide	II/7(**3**,413), II/15(**3**,504)
C$_3$H$_6$S	Thietane Trimethylene sulfide	II/7(**3**,414)
C$_3$H$_6$S$_3$	Dimethyl trithiocarbonate	II/7(**3**,415)
C$_3$H$_6$S$_3$	1,3,5–Trithiane	II/15(**3**,504a), II/21(**3**,283)
C$_3$H$_6$Si	1–Silyl–1–propyne 1–Silabut–2–yne	II/15(**3**,505)
C$_3$H$_7$AsS$_2$	2–Methyl–1,3,2–dithiarsolane	II/15(**3**,506)
C$_3$H$_7$Br	1–Bromopropane	II/15(**3**,507)
C$_3$H$_7$Br	2–Bromopropane	II/7(**3**,416), II/21(**3**,284)
C$_3$H$_7$Cl	1–Chloropropane	II/15(**3**,508)
C$_3$H$_7$Cl	2–Chloropropane	II/7(**3**,417), II/15(**3**,509), II/23(**3**,264)
C$_3$H$_7$Cl	Cyclopropane – hydrogen chloride (1/1)	II/15(**3**,510)
C$_3$H$_7$ClNOP	2–Chloro–3–methyl–1,3,2–oxazaphospholane	II/15(**3**,511)
C$_3$H$_7$ClO	3–Chloro–1–propanol	II/7(**3**,418), II/15(**3**,511a)
C$_3$H$_7$ClO	Chloromethyl ethyl ether	II/15(**3**,512)
C$_3$H$_7$Cl$_2$P	Isopropyldichlorophosphine	II/15(**3**,513)
C$_3$H$_7$F	2–Fluoropropane	II/15(**3**,514), II/23(**3**,265)
C$_3$H$_7$F	1–Fluoropropane	II/21(**3**,285)
C$_3$H$_7$F	Cyclopropane – hydrogen fluoride (1/1)	II/15(**3**,515)
C$_3$H$_7$FO	1–Fluoro–2–propanol	II/15(**3**,516)
C$_3$H$_7$FO	2–Fluoro–1–propanol	II/15(**3**,517)
C$_3$H$_7$FO	3–Fluoro–1–propanol	II/15(**3**,518)
C$_3$H$_7$FO	Fluoromethyl ethyl ether	II/15(**3**,519)
C$_3$H$_7$FO	Oxetane – hydrogen fluoride (1/1)	II/15(**3**,520)
C$_3$H$_7$F$_2$P	Difluoroisopropylphosphine	II/15(**3**,521), II/21(**3**,286)
C$_3$H$_7$I	Isopropyl iodide	II/21(**3**,287)
C$_3$H$_7$N	N–Ethylidenemethylamine	II/7(**3**,419)
C$_3$H$_7$N	Allylamine	II/21(**3**,288), II/23(**3**,266)
C$_3$H$_7$N	Cyclopropylamine Aminocyclopropane	II/7(**3**,420), II/15(**3**,522), II/21(**3**,289)
C$_3$H$_7$N	2–Methylaziridine	II/7(**3**,421)
C$_3$H$_7$N	N–Methylaziridine 1–Methylaziridine	II/7(**3**,422)
C$_3$H$_7$N	Azetidine	II/7(**3**,423), II/15(**3**,523)
C$_3$H$_7$N	Ethane – hydrogen cyanide (1/1)	II/23(**3**,267)
C$_3$H$_7$NO	3–Azetidinol	II/23(**3**,268)
C$_3$H$_7$NO	N–Methylacetamide	II/7(**3**,424)
C$_3$H$_7$NO	(Z)–Propionaldehyde oxime	II/21(**3**,290), II/23(**3**,269)
C$_3$H$_7$NO	N,N–Dimethylformamide	II/7(**3**,425), II/23(**3**,270)
C$_3$H$_7$NOSi	Dimethylsilyl isocyanate	II/21(**3**,291)
C$_3$H$_7$NO$_2$	Propyl nitrite	II/21(**3**,292)

C$_3$H$_7$NO$_2$	Isopropyl nitrite	II/15(**3**,523a)
C$_3$H$_7$NO$_2$	L−Alanine	II/23(**3**,271)
C$_3$H$_7$NO$_2$	Dimethylnitromethane 2−Nitropropane	II/15(**3**,524)
C$_3$H$_7$NO$_2$	Glycine methyl ester	II/15(**3**,525)
C$_3$H$_7$NS	Thiazolidine	II/21(**3**,293)
C$_3$H$_7$NSSi	Dimethylsilyl isothiocyanate	II/21(**3**,294)
C$_3$H$_7$NSi	Dimethylsilanecarbonitrile Dimethylsilyl cyanide	II/15(**3**,526), II/21(**3**,295)
C$_3$H$_7$O$_3$P	2−Methoxy−1,3,2−dioxaphospholane	II/15(**3**,527)
C$_3$H$_7$P	2−Methylphosphirane	II/7(**3**,426)
C$_3$H$_7$P	Cyclopropylphosphine Phosphinocyclopropane	II/7(**3**,427)
C$_3$H$_8$	Propane	II/7(**3**,428), II/23(**3**,272)
C$_3$H$_8$BrClSi	(Bromomethyl)chlorodimethylsilane	II/15(**3**,528)
C$_3$H$_8$Ge	Cyclopropylgermane	II/21(**3**,296), II/23(**3**,273)
C$_3$H$_8$N$_2$	1,2−Dimethyldiaziridine	II/15(**3**,529)
C$_3$H$_8$O	1−Propanol	II/7(**3**,429)
C$_3$H$_8$O	2−Propanol	II/7(**3**,430)
C$_3$H$_8$O	Ethyl methyl ether	II/7(**3**,431), II/15(**3**,530)
C$_3$H$_8$O	Cyclopropane − water (1/1)	II/23(**3**,274)
C$_3$H$_8$OS	2−(Methylthio)ethanol	II/21(**3**,297)
C$_3$H$_8$OS	1−Mercapto−2−propanol	II/23(**3**,275)
C$_3$H$_8$O$_2$	2−Methoxyethanol	II/7(**3**,432)
C$_3$H$_8$O$_2$	Dimethoxymethane	II/7(**3**,433)
C$_3$H$_8$O$_2$	1,2−Propanediol	II/15(**3**,531)
C$_3$H$_8$O$_2$	1,3−Propanediol	II/15(**3**,532)
C$_3$H$_8$S	1−Propanethiol	II/15(**3**,533)
C$_3$H$_8$S	2−Propanethiol Isopropyl mercaptan	II/15(**3**,534)
C$_3$H$_8$S	Ethyl methyl sulfide	II/15(**3**,535)
C$_3$H$_8$S$_2$	Methyl ethyl disulfide	II/15(**3**,536)
C$_3$H$_8$Se	2−Propaneselenol	II/15(**3**,536a)
C$_3$H$_8$Si	Allylsilane	II/15(**3**,537), II/23(**3**,276)
C$_3$H$_8$Si	Methylvinylsilane	II/15(**3**,538), II/23(**3**,277)
C$_3$H$_8$Si	Dimethyl(methylene)silane	II/23(**3**,278)
C$_3$H$_8$Si	Cyclopropylsilane Silylcyclopropane	II/15(**3**,539), II/21(**3**,298)
C$_3$H$_8$Si	Silacyclobutane	II/7(**3**,434), II/15(**3**,540)
C$_3$H$_9$Al	Trimethylaluminum	II/7(**3**,435)
C$_3$H$_9$AlCl$_3$N	Aluminumtrichloride − trimethylamine (1/1)	II/7(**3**,436), II/15(**3**,541)
C$_3$H$_9$As	Trimethylarsine	II/7(**3**,437), II/15(**3**,542)
C$_3$H$_9$AsF$_2$	Trimethylarsenic difluoride	II/21(**3**,299)
C$_3$H$_9$AsO	Trimethylarsine oxide	II/15(**3**,543)
C$_3$H$_9$AsS	Trimethylarsine sulfide	II/15(**3**,544)
C$_3$H$_9$B	Trimethylborane	II/7(**3**,438)

C$_3$H$_9$BBr$_3$N	Trimethylamine – boron tribromide (1/1)	II/15(**3**,545)
C$_3$H$_9$BBr$_3$P	Trimethylphosphine – boron tribromide (1/1)	II/15(**3**,546)
C$_3$H$_9$BCl$_3$N	Trimethylamine – boron trichloride (1/1)	II/15(**3**,547)
C$_3$H$_9$BCl$_3$P	Trimethylphosphine – boron trichloride (1/1)	II/15(**3**,548)
C$_3$H$_9$BF$_3$N	Trimethylamine – boron trifluoride (1/1)	
		II/7(**3**,439), II/15(**3**,549)
C$_3$H$_9$BI$_3$N	Trimethylamine – boron triiodide (1/1)	II/15(**3**,550)
C$_3$H$_9$BI$_3$P	Trimethylphosphine – boron triiodide (1/1)	II/15(**3**,551)
C$_3$H$_9$BO	(Methoxy)dimethylborane	II/15(**3**,552)
C$_3$H$_9$BO$_2$	(Dimethoxy)methylborane	II/15(**3**,553)
C$_3$H$_9$BO$_3$	Trimethoxyborane	II/15(**3**,554)
C$_3$H$_9$BS	Dimethyl(methylthio)borane	II/7(**3**,440)
C$_3$H$_9$BS$_2$	Methylbis(methylthio)borane	II/15(**3**,555)
C$_3$H$_9$BS$_3$	Tris(methylthio)borane	II/7(**3**,441)
C$_3$H$_9$Bi	Trimethylbismuth	II/7(**3**,442), II/15(**3**,556)
C$_3$H$_9$BrGe	Bromotrimethylgermane	II/7(**3**,443)
C$_3$H$_9$BrSi	Bromotrimethylsilane	II/7(**3**,444), II/15(**3**,557)
C$_3$H$_9$Br$_2$N	Dibromine – trimethylamine (1/1)	II/21(**3**,300)
C$_3$H$_9$ClGe	Chlorotrimethylgermane	II/15(**3**,558)
C$_3$H$_9$ClOSi	Chloromethoxydimethylsilane	II/23(**3**,279)
C$_3$H$_9$ClSi	Chlorotrimethylsilane	II/7(**3**,445)
C$_3$H$_9$ClSn	Chlorotrimethylstannane	II/7(**3**,446)
C$_3$H$_9$Cl$_2$Sb	Dichlorotrimethylantimony	II/21(**3**,301)
	Trimethylantimony(V) dichloride	
C$_3$H$_9$FSi	Trimethylfluorosilane	II/21(**3**,302)
C$_3$H$_9$F$_2$P	Trimethylphosphine difluoride	II/7(**3**,447)
C$_3$H$_9$Ga	Trimethylgallium	II/7(**3**,448)
C$_3$H$_9$In	Trimethylindium	II/7(**3**,449), II/15(**3**,559)
C$_3$H$_9$N	Trimethylamine	II/7(**3**,450), II/15(**3**,560)
C$_3$H$_9$NO	*N,N,O*–Trimethylhydroxylamine	II/15(**3**,561)
C$_3$H$_9$NO	3–Amino–1–propanol	II/15(**3**,562)
C$_3$H$_9$NO	2–Methoxyethylamine	II/15(**3**,563)
C$_3$H$_9$NO	2–Amino–1–propanol	II/15(**3**,564)
C$_3$H$_9$NO	Trimethylamine *N*–oxide	II/23(**3**,280)
C$_3$H$_9$NO$_2$S	*N,N*–Dimethylmethanesulfonamide	II/15(**3**,565)
	Methanesulfonic acid dimethylamide	
C$_3$H$_9$NO$_2$S	Trimethylamine – sulfur dioxide (1/1)	II/21(**3**,303), II/23(**3**,281)
C$_3$H$_9$N$_3$Si	Azidotrimethylsilane	II/7(**3**,451)
C$_3$H$_9$OP	Trimethylphosphine oxide	II/7(**3**,452), II/15(**3**,566)
C$_3$H$_9$O$_3$P	Trimethyl phosphite	II/15(**3**,567)
C$_3$H$_9$O$_4$P	Trimethyl phosphate	II/7(**3**,453)
C$_3$H$_9$P	Isopropylphosphine	II/15(**3**,568)
C$_3$H$_9$P	Trimethylphosphine	II/7(**3**,454), II/15(**3**,569), II/23(**3**,282)
C$_3$H$_9$PS	Dimethyl(methylthio)phosphine	II/23(**3**,283)
C$_3$H$_9$PS	Trimethylphosphine sulfide	II/15(**3**,570), II/21(**3**,304)
C$_3$H$_9$PS$_3$	Trimethyl trithiophosphite	II/15(**3**,571), II/21(**3**,305)
	Trimethyl phosphorotrithioite	

C$_3$H$_9$PSe	Trimethylphosphine selenide	II/15(**3**,572)
C$_3$H$_9$Sb	Trimethylantimony	II/15(**3**,573)
C$_3$H$_9$Tl	Trimethylthallium	II/15(**3**,574)
C$_3$H$_{10}$BN	Dimethyl(methylamino)borane	II/15(**3**,575)
C$_3$H$_{10}$BrN	Trimethylamine – hydrogen bromide (1/1)	II/23(**3**,284)
C$_3$H$_{10}$ClN	Trimethylamine – hydrogen chloride (1/1)	II/21(**3**,306)
C$_3$H$_{10}$ClP	Trimethylphosphine – hydrogen chloride (1/1)	II/23(**3**,285)
C$_3$H$_{10}$FN	Trimethylamine – hydrogen fluoride (1/1)	II/21(**3**,307)
C$_3$H$_{10}$Ge	Trimethylgermane	II/7(**3**,455)
C$_3$H$_{10}$O	Propane – water (1/1)	II/23(**3**,286)
C$_3$H$_{10}$OSi	Methoxydimethylsilane	II/23(**3**,287)
C$_3$H$_{10}$Si	Ethylmethylsilane	II/7(**3**,456)
C$_3$H$_{10}$Si	Trimethylsilane	II/7(**3**,457)
C$_3$H$_{10}$Si	Propylsilane	II/15(**3**,576)
C$_3$H$_{10}$Sn	Trimethylstannane	II/7(**3**,458)
C$_3$H$_{11}$BFN	Trimethylamine – fluoroborane (1/1)	II/15(**3**,577)
C$_3$H$_{11}$BN$_2$	Methylbis(methylamino)borane	II/15(**3**,578)
C$_3$H$_{11}$NSi	*N,N*,1–Trimethylsilanamine	II/15(**3**,579)
	(Dimethylamino)methylsilane	
C$_3$H$_{12}$AlN	Trimethylamine – aluminum hydride (1/1)	II/7(**3**,459)
C$_3$H$_{12}$BN	Trimethylamine – borane (1/1)	II/7(**3**,460), II/15(**3**,580)
C$_3$H$_{12}$BN$_3$	Tris(methylamino)borane	II/15(**3**,581)
C$_3$H$_{12}$BP	Trimethylphosphine – borane (1/1)	II/7(**3**,461), II/21(**3**,308)
C$_3$H$_{12}$GaN	Trimethylamine – gallium hydride (1/1)	II/15(**3**,582)
C$_3$H$_{12}$GaN	Trimethylgallium – ammonia (1/1)	II/23(**3**,288)
C$_3$H$_{13}$NSi$_2$	*N,N*–Bis(methylsilyl)methylamine	II/21(**3**,309)
C$_3$H$_{13}$NSi$_2$	2–Isopropyldisilazane	II/21(**3**,310)
	N–Isopropyldisilylamine	
C$_3$H$_{15}$NSi$_3$	Tris(methylsilyl)amine	II/15(**3**,583)
C$_3$IN	Iodocyanoacetylene	II/7(**3**,341)
	3–Iodo–2–propynenitrile	
C$_3$N$_2$O	Carbonyl cyanide	II/7(**3**,342), II/15(**3**,584)
C$_3$N$_3$P	Tricyanophosphine	II/15(**3**,584a)
C$_3$O	Tricarbon monoxide	II/21(**3**,311)

C$_3$OS   3–Thioxo–1,2–propadien–1–one $C_{\infty v}$
Tricarbon oxide sulfide   O=C(1)=C(2)=C(3)=S
MW   Allene oxide sulfide

r_s ^{a})	Å ^{b})
O=C(1)	1.1343(5)
C(1)=C(2)	1.2696(5)
C(2)=C(3)	1.2540(5)
C(3)=S	1.5825(5)

^{a}) Contributions of the v_7 mode were corrected for.
^{b}) Uncertainties were not estimated in the original paper.

Winnewisser, M., Peau, E.W.: Acta Phys. Hung. **55** (1984) 33.   II/15(**3**,584b)

C_3O_2		**Tricarbon dioxide**	quasilinear
ED, IR		Carbon suboxide	O=C=C=C=O

r_g	Å [a]
O=C	1.1640(15)
C=C	1.286(4)

Available spectroscopic data such as the energy-level intervals and the rotational constants have also been taken into the analysis of the large-amplitude CCC bending vibration using a model which allows the C=C bond lengths and the C=C=O bond angles to vary with the bending displacement ρ (= 180° − C=C=C). The effective bending potential function determined for the vibrational ground state has a barrier of 27(16) cm^{-1} and a minimum at ρ = 20(2)°. The bond lengths corresponding to the linear configuration, averaged over all the small-amplitude vibrations, are determined to be r_{lin}(CO) = 1.1602(15) Å and r_{lin}(CC) = 1.2761(12) Å. The present analysis shows that the C=C bonds are slightly stretched by the CCC bending, and the CCO angles are slightly bent in the direction of the CCC bending.
The nozzle temperature was ≈ 23 °C.

[a]) Estimated limits of error.

Ohshima, Y., Yamamoto, S., Kuchitsu, K.: Acta Chem. Scand. Ser. A **42** (1988) 307.

II/7(**3**,343), II/21(**3**,312)

C_3O_6	Carbon dioxide trimer	II/21(**3**,313)
C_3S	Tricarbon monosulfide	II/23(**3**,289)
C_3S_2	Tricarbon disulfide	II/23(**3**,290)
C_4	Tetracarbon	II/23(**3**,291)
$C_4Br_2O_2$	3,4–Dibromo–3–cyclobutene–1,2–dione	II/21(**3**,314)
$C_4Cl_2F_4$	1,2–Dichloro–3,3,4,4–tetrafluorocyclobutene	II/23(**3**,292)
$C_4Cl_2F_6$	1,1–Dichloro–2,2,3,3,4,4–hexafluorocyclobutane	II/7(**3**,462)
$C_4Cl_2O_2$	1,2–Dichlorocyclobutene–3,4–dione	II/21(**3**,315)
$C_4Cl_2O_3$	Dichloromaleic anhydride	II/15(**3**,585)
C_4Cl_6	Hexachloro–1,3–butadiene	II/21(**3**,316)
$C_4Cu_2F_6O_4$	Dicopper bis(trifluoroacetate)	II/21(**3**,317)
	Bis–μ–trifluoroacetato–dicopper(I)	
$C_4F_4O_3$	Tetrafluorosuccinic anhydride	II/15(**3**,586)
C_4F_6	Perfluoro–2–butyne	II/7(**3**,463)
C_4F_6	Perfluoro–1,3–butadiene	II/7(**3**,464)
C_4F_6	Hexafluorocyclobutene	II/7(**3**,465), II/23(**3**,293)
$C_4F_6NS_2$	4,5–Bis(trifluoromethyl)–1,3,2–dithiazol–2–yl	II/21(**3**,318)
$C_4F_6O_3$	Perfluoroacetic anhydride	II/7(**3**,468)
C_4F_6S	Perfluoro–2,5–dihydrothiophene	II/21(**3**,319)
	Perfluoro–3–thiolene	
$C_4F_6S_2$	3,4–Bis(trifluoromethyl)–1,2–dithiete	II/15(**3**,587)
C_4F_7NO	1,2,2,3,3,4,4–Heptafluoro–1–nitrosocyclobutane	II/21(**3**,320)
	Perfluoronitrosocyclobutane	
C_4F_8	Octafluorocyclobutane	II/7(**3**,466), II/21(**3**,321)
$C_4F_8N_3P$	2,2–Difluoro–4,6–bis(trifluoromethyl)–	II/21(**3**,322)
	1,3,5,2–triazaphosphorine–2–P^V	
C_4F_8OS	Perfluorotetrahydrothiophene 1–oxide	II/21(**3**,323)

C₄F₈O₂S	Perfluorotetrahydrothiophene 1,1–dioxide	II/21(**3**,324)
C₄F₈S	Perfluorotetrahydrothiophene	II/21(**3**,325)
C₄F₉I	Perfluoro–*t*–butyl iodide	II/15(**3**,588)
C₄F₁₀P₂	2,2,4,4–Tetrafluoro–1,3–bis(trifluoromethyl)–1,3–diphosphetane	II/21(**3**,326)
	Perfluoro–2–phosphapropene dimer	
C₄F₁₂Ge	Tetrakis(trifluoromethyl)germane	II/15(**3**,589)
C₄F₁₂N₂	Tetrakis(trifluoromethyl)hydrazine	II/7(**3**,467)
C₄F₁₂P₂	Tetrakis(trifluoromethyl)diphosphane	II/15(**3**,590)
C₄F₁₂S	Octafluorotetrahydrothiophene tetrafluoride	II/21(**3**,327)
C₄F₁₂Sn	Tetrakis(trifluoromethyl)tin	II/15(**3**,591)
C₄HBr	1–Bromo–1,3–butadiyne	II/7(**3**,470)
C₄HCl	1–Chloro–1,3–butadiyne	II/7(**3**,471)
C₄HCoO₄	Tetracarbonylhydrocobalt	II/15(**3**,592)
	Hydridotetracarbonylcobalt(I)	
C₄HF₉	Tris(fluoromethyl)methane	II/7(**3**,472)
	1,1,1,3,3,3–Hexafluoro–2–(trifluoromethyl)propane	
C₄HF₉O	Perfluoro–*t*–butyl alcohol	II/15(**3**,593)
C₄HNO₆	Hydrogen cyanide – carbon dioxide (1/3)	II/21(**3**,328)
C₄H₂	1,3–Butadiyne	II/7(**3**,473), II/23(**3**,294)

C₄H₂⁺   **Diacetylene cation**   **D**$_{\infty h}$
                   1,3–Butadiyne cation   H–C≡C–C≡C–H⁺

LIF

State	$\tilde{X}\,^2\Pi_g$	$\tilde{A}\,^2\Pi_u$
Energy [eV]	0.0	2.445
r_s [Å] C–H	1.046(10) [a]	1.045(10)
C≡C	1.234(10)	1.243(10)
C–C	1.346(10)	1.410(10)

Rotational analysis of HCCCCH⁺, H¹³CCCCH⁺, H¹³C¹³CCCH⁺ and DCCCCD⁺.

[a]) Error limits are 2σ.

Maier, J.P.: Int. Rev. Phys. Chem. **9** (1990) 281.   II/21(**3**,329)

C₄H₂Cl₂N₂	3,6–Dichloropyridazine	II/15(**3**,594)
C₄H₂Cl₂O₂	Fumaroyl dichloride	II/21(**3**,330)
C₄H₂Cl₄	*trans,trans*–1,2,3,4–Tetrachloro–1,3–butadiene	II/21(**3**,331)
C₄H₂F₄	3,3,4,4–Tetrafluorocyclobutene	II/23(**3**,295)
C₄H₂F₆	*trans*– and *cis*–1,1,1,4,4,4–Hexafluoro–2–butene	II/15(**3**,595)
C₄H₂F₆	Bis(trifluoromethyl)ethene	II/7(**3**,474)
C₄H₂FeO₄	Tetracarbonyldihydridoiron(II)	II/15(**3**,596)
C₄H₂O	1,2,3–Butatrien–1–one	II/15(**3**,597)
C₄H₂O₃	Maleic anhydride	II/7(**3**,475), II/15(**3**,598)
C₄H₃BrO	2–Bromofuran	II/15(**3**,599), II/21(**3**,332)
C₄H₃BrO	3–Bromofuran	II/15(**3**,600), II/21(**3**,333)
C₄H₃BrS	2–Bromothiophene	II/7(**3**,476)

C$_4$H$_3$BrS	3–Bromothiophene	II/15(**3**,601)
C$_4$H$_3$ClO	2–Chlorofuran	II/15(**3**,602)
C$_4$H$_3$ClO$_2$S$_2$	2–Thiophenesulfonyl chloride	II/15(**3**,603)
C$_4$H$_3$ClS	2–Chlorothiophene	II/7(**3**,477)
C$_4$H$_3$ClS	3–Chlorothiophene	II/21(**3**,334)
C$_4$H$_3$Cl$_2$PS	Dichloro–2–thienylphosphine	II/15(**3**,604)
C$_4$H$_3$CoGeO$_4$	Tetracarbonylgermylcobalt	II/15(**3**,605)
C$_4$H$_3$CoO$_4$Si	Tetracarbonylsilylcobalt	II/7(**3**,478)
C$_4$H$_3$F$_3$	1,1,1–Trifluoro–2–butyne	II/7(**3**,479)
C$_4$H$_3$N	2–Propynyl isocyanide 1–Isocyano–2–propyne	II/21(**3**,335)
C$_4$H$_3$N	Methylcyanoacetylene 2–Butynenitrile	II/7(**3**,480)
C$_4$H$_3$N	3–Butynenitrile	II/21(**3**,336), II/23(**3**,296)
C$_4$H$_3$N	3–Cyanocyclopropene	II/15(**3**,605a), II/21(**3**,337)
C$_4$H$_3$NO$_2$	Maleimide	II/21(**3**,338)
C$_4$H$_3$NS	3–Thiocyanato–1–propyne Propargyl thiocyanate	II/21(**3**,339)

C$_4$H$_4$ **Acetylene dimer** **C$_s$**
MW (weakly bound complex) (effective symmetry class)
(C$_2$H$_2$)$_2$

r_0	Å	θ_0	deg
R_{cm}	4.38(2)	θ_T	27(2)
		θ_L	0 [a]

[a]) Assumed.

Prichard, D.G., Nandi, R.N., Muenter, J.S.: J. Chem. Phys. **89** (1988) 115.

IR, MW

r_0	Å
R_{cm}	4.404(20)

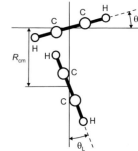

The complex has a T-shaped structure with interconversion tunneling between four isoenergetic hydrogen-bonded minima. The barrier to the tunneling is quite small (33.2 cm^{-1}). The distance R_{cm} is the separation between the two acetylene monomer units. The assumption was made that the structure of the monomer units is unchanged in the complex.

[a]) Uncertainties were not given in the original paper.

Fraser, G.T., Suenram, R.D., Lovas, F.J., Pine, A.S., Hougen, J.T., Lafferty, W.J., Muenter, J.S.: J. Chem. Phys. **89** (1988) 6028.
Ohshima, Y., Matsumoto, Y., Takami, M., Kuchitsu, K.: Chem. Phys. Lett. **147** (1988) 1.
Ohshima, Y., Matsumoto, Y., Takami, M., Kuchitsu, K.: Chem. Phys. Lett. **152** (1988) 116.

II/21(**3**,340)

C$_4$H$_4$	Butatriene	II/7(**3**,481)
C$_4$H$_4$	Vinylacetylene, 1–Buten–3–yne	II/7(**3**,482)

| **C₄H₄** | **Methylenecyclopropene** | **C₂ᵥ** |

MW

r_s	Å
C(1)=C(2)	1.323(3)
C(1)–C(3)	1.441(6)
C(3)=C(4)	1.332(6)

r_s	Å		θ_s	deg
C(1)–H	1.080 [a]		C(1)–C(3)–C(2)	54.7 [b]
C(4)–H	1.085 [a]		C(2)=C(1)–H	147.5 [a]
			H–C(4)–H	118.0 [a]

[a] Assumed.
[b] Derived.

Norden, T.D., Staley, S.W., Taylor, W.H., Harmony, M.D.: J. Am. Chem. Soc. **108** (1986) 7912.
II/21(**3**,341)

C₄H₄ArO	Furan – argon (1/1)	II/15(**3**,605b)
C₄H₄Ar₂O	Argon – furan (2/1)	II/23(**3**,297)
C₄H₄Br₄	1,2,3,4–Tetrabromocyclobutane	II/21(**3**,342)
C₄H₄Cl₂	cis,cis–1,4–Dichloro–1,3–butadiene	II/21(**3**,343)
C₄H₄Cl₂	cis,trans–1,4–Dichloro–1,3–butadiene	II/21(**3**,344)
C₄H₄Cl₂	trans,trans–1,4–Dichloro–1,3–butadiene	II/21(**3**,345)
C₄H₄Cl₂	2,3–Dichloro–1,3–butadiene	II/21(**3**,346)
C₄H₄Cl₂	3,4–Dichlorocyclobutene	II/7(**3**,483)
C₄H₄Cl₂OSi	2,2–Dichloro–2–sila–1–oxa–3,5–cyclohexadiene	II/15(**3**,606)
C₄H₄Cl₄Ge₂	1,1,4,4–Tetrachloro–1,4–digerma–2,5–cyclohexadiene	II/7(**3**,484)
C₄H₄F₂	1,4–Difluoro–2–butyne	II/21(**3**,348)
C₄H₄F₄O₂P₂	1,4–Bis(difluorophosphinooxy)–2–butyne	II/23(**3**,298)
C₄H₄F₆	1,1,1,4,4,4–Hexafluorobutane	II/23(**3**,299)
C₄H₄Ge₂I₄	1,1,4,4–Tetraiodo–1,4–digerma–2,5–cyclohexadiene	II/7(**3**,485)
C₄H₄N₂	Succinonitrile, Butanedinitrile 1,2–Dicyanoethane	II/15(**3**,607)
C₄H₄N₂	Ethylene diisocyanide 1,2–Diisocyanoethane	II/21(**3**,349)
C₄H₄N₂	Pyridazine	II/7(**3**,486), II/15(**3**,608)
C₄H₄N₂	Pyrazine	II/15(**3**,609), II/21(**3**,350)
C₄H₄N₂	Pyrimidine	II/15(**3**,610), II/21(**3**,351)
C₄H₄N₂O₂	Uracil	II/21(**3**,352)
C₄H₄N₆	2,3–Diazido–1,3–butadiene	II/21(**3**,353)
C₄H₄N₆O₆	2–Methyl–4–trinitromethyl–2H–1,2,3–triazole	II/23(**3**,301)
C₄H₄O	3,4–Epoxy–1–butyne	II/7(**3**,487)
C₄H₄O	Furan	II/7(**3**,488), II/15(**3**,611), II/21(**3**,354)
C₄H₄O	Diacetylene – water (1/1)	II/23(**3**,302)
C₄H₄O	Acetylene – ketene (1/1)	II/23(**3**,303)

C$_4$H$_4$O$_2$	*trans*–2–Butenedial Fumaraldehyde	II/7(**3**,489)
C$_4$H$_4$O$_2$	1,2–Cyclobutanedione	II/15(**3**,612)
C$_4$H$_4$O$_2$	1,4–Dioxin	II/15(**3**,613)
C$_4$H$_4$O$_3$	Succinic anhydride	II/7(**3**,490)
C$_4$H$_4$O$_3$	Cyclobutadiene ozonide 2,3,7–Trioxabicyclo[2.2.1]hept–5–ene	II/21(**3**,355)
C$_4$H$_4$O$_3$	2,4–Dioxabicyclo[3.1.0]hexan–3–one	II/15(**3**,614)
C$_4$H$_4$O$_3$S	Furan – sulfur dioxide (1/1)	II/23(**3**,304)
C$_4$H$_4$S	Thiophene	II/7(**3**,491), II/21(**3**,356)
C$_4$H$_4$Se	Selenophene	II/7(**3**,492)
C$_4$H$_5$ArN	Argon – pyrrole (1/1)	II/21(**3**,357)
C$_4$H$_5$BrCl$_2$	1,1–Dichloro–2–(bromomethyl)cyclopropane	II/21(**3**,358)
C$_4$H$_5$Cl	*cis*–1–Chloro–1,3–butadiene	II/15(**3**,615), II/23(**3**,305)
C$_4$H$_5$Cl	*trans*–1–Chloro–1,3–butadiene	II/15(**3**,615), II/23(**3**,306)

C$_4$H$_5$Cl   **2–Chloro–1,3–butadiene**   **C$_s$** (*anti*)
ED, MW and   **C$_1$** (*synclinal*)
ab initio calculations

r_a	Å [a])	θ_α	deg [a])
C(1)=C(2)	1.344(2)	C=C(2)–C	123.5(1)
C(3)=C(4)	1.344(2)	C=C(3)–C	125.6(2)
C(2)–C(3)	1.469(3)	C(3)–C(2)–Cl	117.2(1)
C–Cl	1.742(2)	θ (*anti*) [b])	180
C(1)–H(1)	1.098(2)	θ (*synclinal*) [b])	27(11)

Some MO constraints were used in defining the positions of hydrogen atoms and for a difference $r(C(1)=C(2)) - r(C(3)=C(4))$ parameters.
$E(synclinal) - E(anti) = 6.4(12)$ kJ mol^{-1},
$S(synclinal) - S(anti) = -6(3)$ J K^{-1} mol^{-1}.
Mole fractions of the *synclinal* conformer at different nozzle temperatures: 25(5) % at 655 K, 18(5) % at 565 K, 7(4) % at 298 K. Data at 298 K are listed.

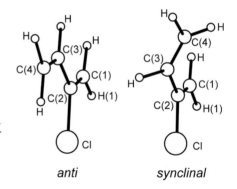

anti   synclinal

[a]) Twice the estimated standard errors including the experimental scale error.
[b]) Torsional angle around the C(2)–C(3) bond.

Gundersen, G., Thomassen, H.G., Boggs, J.E., Peng, C.: J. Mol. Struct. **243** (1991) 385.

II/15(**3**,616), II/23(**3**,307)

C$_4$H$_5$Cl	4–Chloro–1,2–butadiene	II/15(**3**,617)
C$_4$H$_5$Cl	1–Chloro–2–butyne	II/7(**3**,493)
C$_4$H$_5$Cl	Vinylacetylene – hydrogen chloride (1/1)	II/23(**3**,308)
C$_4$H$_5$ClO	2–Butenoyl chloride Crotonyl chloride	II/21(**3**,359)
C$_4$H$_5$ClO	Cyclopropanecarbonyl chloride	II/7(**3**,494)

C₄H₅ClO	Methacryloyl chloride 2–Methyl–2–propenoyl chloride	II/23(**3**,309), II/23(**3**,310)	
C₄H₅ClO	Furan – hydrogen chloride (1/1)	II/15(**3**,617a)	
C₄H₅F	Vinylacetylene – hydrogen fluoride (1/1)	II/23(**3**,311)	
C₄H₅F₃	3,3,3–Trifluoro–2–methylpropene	II/21(**3**,360)	
C₄H₅N	3–Butenenitrile Allyl cyanide	II/15(**3**,618)	
C₄H₅N	Crotononitrile 2–Butenenitrile	II/7(**3**,495)	
C₄H₅N	Methacrylonitrile 2–Cyano–1–propene	II/7(**3**,496)	
C₄H₅N	Cyclopropyl cyanide Cyclopropanecarbonitrile	II/7(**3**,497), II/15(**3**,619)	
C₄H₅N	Isocyanocyclopropane Cyclopropyl isocyanide	II/15(**3**,619a)	
C₄H₅N	Pyrrole	II/7(**3**,498)	
C₄H₅N	Acetylene – acetonitrile (1/1)	II/21(**3**,361)	
C₄H₅N	Methyl isocyanide – acetylene (1/1)	II/23(**3**,312)	
C₄H₅N	Diacetylene – ammonia (1/1)	II/23(**3**,313)	
C₄H₅NO	3–Methoxy–2–propenenitrile	II/23(**3**,314)	
C₄H₅NO	Cyclopropyl isocyanate	II/21(**3**,362)	
C₄H₅NO	2–Methyloxazole	II/23(**3**,315)	
C₄H₅NO	4–Methyloxazole	II/23(**3**,316)	
C₄H₅NO	5–Methyloxazole	II/23(**3**,317)	
C₄H₅NO	3–Methylisoxazole	II/23(**3**,318)	
C₄H₅NO	5–Methylisoxazole	II/23(**3**,319)	
C₄H₅NO₂S	3–(Methylsulfonyl)–2–propenenitrile	II/23(**3**,320)	
C₄H₅NS	Cyclopropyl isothiocyanate	II/21(**3**,363)	
C₄H₅N₃	2–Azido–1,3–butadiene	II/21(**3**,364)	
C₄H₅N₃	1–Azido–2–butyne	II/21(**3**,365)	
C₄H₅N₃	2–Aminopyrimidine	II/15(**3**,620)	
C₄H₆	1,2–Butadiene	II/21(**3**,366), II/23(**3**,321)	
C₄H₆	Dimethylacetylene 2–Butyne	II/7(**3**,499)	

ED

C₄H₆ **1,3–Butadiene** **C₂ₕ** (*anti*)

r_a	Å ᵃ)		θ_a	deg ᵃ)
C(2)–C(3)	1.468(2)		C(1)=C(2)–C(3)	124.3(1)
C(1)=C(2)	1.348(1)		C–C–H	120.7(3)
C–H	1.107(1)			

(*continued*)

C$_4$H$_6$ (*continued*)

The predominant conformer at all temperatures is *anti*.
The measurements were made at 25···900 °C.
The results at 500 °C are listed.

^{a}) Estimated standard errors including systematic error.

Kveseth, K., Seip, R., Kohl, D.A.: Acta Chem.
  Scand. Ser. **A 34** (1980) 31.

II/7(**3**,500), II/15(**3**,621)

C$_4$H$_6$ (C$_4$H$_4$D$_2$)	1,3–Butadiene–1,1–d_2	II/21(**3**,366A)
C$_4$H$_6$	Methylenecyclopropane	II/7(**3**,501)
C$_4$H$_6$	1–Methylcyclopropene	II/7(**3**,502)
C$_4$H$_6$	Cyclobutene	II/7(**3**,503)
C$_4$H$_6$	Bicyclo[1.1.0]butane	II/7(**3**,504)

MW | **C$_4$H$_6$** | **Ethylene – acetylene (1/1)** | **C$_{2v}$**
   | | (weakly bound complex) | (effective symmetry class)
   | | | H$_2$C=CH$_2$···HCCH

	$r_0(R_{cm})$ [Å]	r_0(H) [Å] ^{a})
H$_2$C=CH$_2$···HCCH	4.443	2.78
H$_2$C=CH$_2$···DCCD	4.442	2.78
H$_2$C=CH$_2$···DCCH	4.380	2.78
t–DHC=CDH···HCCH	4.441	2.78
	4.442(2) (average)	2.780(2) (average)

This complex has a C$_{2v}$ structure in which the HCCH unit hydrogen-bonds to ethylene π cloud, with the HCCH axis normal to the plane of the ethylene. Tunneling splitting due to the hindered internal rotation of the ethylene unit about its C=C bond led to a barrier of 240 cm^{-1}.

^{a}) Hydrogen-bond length.

Fraser, G.T., Lovas, F.J., Suenram, R.D., Gillies, J.Z., Gillies, C.W.: Chem. Phys. **163** (1992) 91.

II/23(**3**,322)

C$_4$H$_6$BF$_6$N	(Dimethylamino)bis(trifluoromethyl)borane	II/21(**3**,367)
C$_4$H$_6$B$_2$	2,3,4,5–Tetracarbahexaborane(6)	II/7(**3**,505)
C$_4$H$_6$BrCl	1–Bromo–3–chlorocyclobutane	II/7(**3**,507)
C$_4$H$_6$BrN	4–Bromobutanenitrile	II/21(**3**,368)
	4–Bromobutyronitrile	
C$_4$H$_6$Br$_2$	1,3–Dibromocyclobutane	II/7(**3**,506)
C$_4$H$_6$ClN	4–Chlorobutanenitrile	II/21(**3**,369)
	4–Chlorobutyronitrile	
C$_4$H$_6$ClOP	Divinylphosphinic chloride	II/15(**3**,622)
C$_4$H$_6$ClOP	1–Oxo–1–chloro–1λ^5–phosphacyclopent–2–ene	II/7(**3**,508)
	1–Chloro–2,3–dihydro–1H–phosphole 1–oxide	
C$_4$H$_6$ClOP	1–Oxo–1–chloro–1λ^5–phosphacyclopent–3–ene	II/7(**3**,509)
	1–Chloro–2,5–dihydro–1H–phosphole 1–oxide	

C$_4$H$_6$ClO$_2$P	2–Chloro–4,5–dimethyl–1,3–dioxaphosphole	
		II/21(**3**,370), II/23(**3**,323)
C$_4$H$_6$Cl$_2$	cis–1,4–Dichloro–2–butene	II/15(**3**,623)
C$_4$H$_6$Cl$_2$	3–Chloro–2–chloromethyl–1–propene	II/15(**3**,624)
C$_4$H$_6$Cl$_2$Si	1,1–Dichlorosilacyclopent–3–ene	II/15(**3**,625)
C$_4$H$_6$Cu$_2$O$_4$	Dicopper diacetate	II/21(**3**,371)
	Di–μ–acetato–dicopper(I)	
C$_4$H$_6$F$_2$Si	1,1–Difluorosilacyclopent–3–ene	II/15(**3**,626)
C$_4$H$_6$N$_4$	2–Methyl–5–vinyl–2H–tetrazole	II/21(**3**,372)
C$_4$H$_6$O	Divinyl ether	II/21(**3**,373)
C$_4$H$_6$O	trans–Crotonaldehyde	II/7(**3**,510)
C$_4$H$_6$O	2,3–Butadien–1–ol	II/15(**3**,626a)
C$_4$H$_6$O	1–Methoxypropadiene	II/7(**3**,511), II/21(**3**,374)
	Methoxyallene	
C$_4$H$_6$O	3–Buten–2–one	II/21(**3**,375)
	Methyl vinyl ketone	
C$_4$H$_6$O	Dimethylketene	II/7(**3**,512)
C$_4$H$_6$O	2–Methylpropenal	II/7(**3**,513), II/21(**3**,376)
	Methacrylaldehyde	
	Methacrolein	
C$_4$H$_6$O	Methyl propargyl ether	II/15(**3**,627)
C$_4$H$_6$O	3–Butyn–2–ol	II/21(**3**,377)
C$_4$H$_6$O	Cyclopropylmethanal	II/7(**3**,514)
	Cyclopropanecarbaldehyde	
C$_4$H$_6$O	3,4–Epoxy–1–butene	II/15(**3**,628)
	Butadiene monoxide	
C$_4$H$_6$O	3–Methyleneoxetane	II/7(**3**,515)
C$_4$H$_6$O	Cyclobutanone	II/7(**3**,516), II/15(**3**,629)
C$_4$H$_6$O	2,5–Dihydrofuran	II/15(**3**,630)
C$_4$H$_6$O	1–Oxaspiro[2.2]pentane	II/15(**3**,631)
C$_4$H$_6$O	Ketene – ethylene (1/1)	II/23(**3**,324)
C$_4$H$_6$OSi	2–Furylsilane	II/21(**3**,378)
C$_4$H$_6$O$_2$	d,l–1,2:3,4–Diepoxybutane	II/7(**3**,517)
C$_4$H$_6$O$_2$	trans–2–Butenoic acid	II/21(**3**,379)
	Crotonic acid	
C$_4$H$_6$O$_2$	2,3–Butanedione	II/7(**3**,518), II/15(**3**,632)
	Biacetyl, Dimethylglyoxal	
C$_4$H$_6$O$_2$	3,6–Dihydro–1,2–dioxin	II/15(**3**,633)
C$_4$H$_6$O$_2$	2,3–Dihydro–1,4–dioxin	II/15(**3**,634)
C$_4$H$_6$O$_2$	2,2'–Bioxirane	II/21(**3**,380)
C$_4$H$_6$O$_2$	Cyclopropanecarboxylic acid	II/23(**3**,325)
C$_4$H$_6$O$_2$	4–Methyl–2–oxetanone	II/23(**3**,326)
C$_4$H$_6$O$_2$S	Divinyl sulfone	II/15(**3**,635)
C$_4$H$_6$O$_2$S	2,5–Dihydrothiophene 1,1–dioxide	II/23(**3**,327)
C$_4$H$_6$O$_2$S	1,3–Butadiene – sulfur dioxide (1/1)	II/23(**3**,328)
C$_4$H$_6$O$_3$	Acetic anhydride	II/7(**3**,519)

C$_4$H$_6$O$_3$	1,3–Dioxan–2–one Trimethylene carbonate	II/21(**3**,381)
C$_4$H$_6$O$_3$	2,3,7–Trioxabicyclo[2.2.1]heptane	II/21(**3**,382)
C$_4$H$_6$S	Methyl 1,2–propadienyl sulfide (Methylthio)allene	II/7(**3**,520)
C$_4$H$_6$S$_2$	Bis(methylthio)ethyne Bis(methylthio)acetylene	II/15(**3**,636)
C$_4$H$_7$Br	3–Bromo–2–methyl–1–propene	II/15(**3**,637)
C$_4$H$_7$Br	4–Bromo–1–butene	II/21(**3**,383)
C$_4$H$_7$Br	(Bromomethyl)cyclopropane	II/15(**3**,638)
C$_4$H$_7$Br	Cyclobutyl bromide Bromocyclobutane	II/7(**3**,521)
C$_4$H$_7$BrO	3–Bromotetrahydrofuran	II/15(**3**,639)
C$_4$H$_7$Cl	1–Chloro–2–methyl–1–propene 2,2–Dimethyl–1–chloroethene	II/15(**3**,640)
C$_4$H$_7$Cl	3–Chloro–2–methyl–1–propene	II/15(**3**,641)
C$_4$H$_7$Cl	*trans*–1–Chloro–2–butene	II/21(**3**,384)
C$_4$H$_7$Cl	3–Chloro–1–butene	II/15(**3**,642)
C$_4$H$_7$Cl	4–Chloro–1–butene	II/21(**3**,385)
C$_4$H$_7$Cl	(Chloromethyl)cyclopropane	II/15(**3**,643)
C$_4$H$_7$Cl	Chlorocyclobutane Cyclobutyl chloride	II/7(**3**,522), II/21(**3**,386)
C$_4$H$_7$F	3–Fluoro–2–methyl–1–propene	II/23(**3**,329)
C$_4$H$_7$F	4–Fluoro–1–butene	II/23(**3**,330)
C$_4$H$_7$FO	2–Methylpropionyl fluoride Isobutyryl fluoride	II/23(**3**,331)
C$_4$H$_7$FO	2,5–Dihydrofuran – hydrogen fluoride (1/1)	II/21(**3**,387)
C$_4$H$_7$N	Isopropyl cyanide Isobutyronitrile	II/15(**3**,644)
C$_4$H$_7$N	*N*–Methyl–2–propynylamine *N*–Methylpropargyl amine	II/21(**3**,388)
C$_4$H$_7$N	3–Butynylamine 3–Butyn–1–amine	II/21(**3**,389)
C$_4$H$_7$N	3,4–Dihydro–2*H*–pyrrole 1–Pyrroline	II/21(**3**,390)
C$_4$H$_7$N	Cyclopropane – hydrogen cyanide (1/1)	II/15(**3**,644a)
C$_4$H$_7$NO	3–Methoxypropionitrile 3–Methoxypropanenitrile	II/15(**3**,645)
C$_4$H$_7$NO	Propyl isocyanate	II/15(**3**,646)
C$_4$H$_7$NO	Isopropyl isocyanate	II/21(**3**,391)
C$_4$H$_7$NO	*N*–Acetylaziridine *N*–Acetylethylenimine	II/7(**3**,523), II/15(**3**,647)
C$_4$H$_7$NO$_2$	Diacetamide	II/15(**3**,648)
C$_4$H$_7$NS	Isopropyl isothiocyanate	II/21(**3**,392)
C$_4$H$_7$NSi	*N*–Silylpyrrole	II/7(**3**,524)
C$_4$H$_7$P	2,5–Dihydro–1*H*–phosphole 3–Phospholene	II/15(**3**,649)

	C$_4$H$_8$	1–Butene	II/7(**3**,525), II/15(**3**,650)
	C$_4$H$_8$	cis–2–Butene	II/7(**3**,526), II/15(**3**,651)
	C$_4$H$_8$	trans–2–Butene	II/7(**3**,526)
	C$_4$H$_8$	Isobutylene, Isobutene	II/7(**3**,527), II/23(**3**,332)
	C$_4$H$_8$	Methylcyclopropane	II/15(**3**,652)

MW **C$_4$H$_8$** **Cyclobutane** **D$_{2d}$** (CH$_2$)$_4$

r_0	Å		θ_0	deg
C–C	1.5549(5)		θ	28.32(23)
C–H(ax)	1.0934(19)		γ(ax)	119.93(14)
C–H(eq)	1.0910(9)		γ(eq)	130.74(11)

The microwave spectra of cyclobutane–d_1 and –1,1–d_2 were measured.

Caminati, W., Vogelsanger, B., Meyer, R., Grassi, G., Bauder, A.: J. Mol. Spectrosc. **131** (1988) 172.
Vogelsanger, B., Caminati, W., Bauder, A.: Chem. Phys. Lett. **141** (1987) 245.

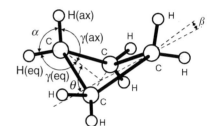

ED, IR

r_z	Å ^{a})		θ_z	deg ^{a})
C–C	1.552(1)		θ ^{b})	27.9(16)
C–H	1.093(3)		α ^{b})	106.4(13)
			β ^{b})	6.2(12)

r_g	Å ^{a})
C–C	1.554(1)
C–H	1.109(3)

Mean amplitudes were fixed at calculated values.
The measurements were made at room temperature; the temperature of the reservoir was 230 K.

^{a}) Estimated limits of error.
^{b}) See figure for the definition.

Egawa, T., Fukuyama, T., Yamamoto, S., Takabayashi, F., Kambara, H., Ueda, T., Kuchitsu, K.:
J. Chem. Phys. **86** (1987) 6018. II/7(**3**,528), II/21(**3**,393)

C$_4$H$_8$AsCl	1–Chloroarsolane	II/15(**3**,653)
C$_4$H$_8$BrCl	2–Bromo–1–chloro–2–methylpropane	II/15(**3**,654)
C$_4$H$_8$ClN	N–Chloropyrrolidine	II/21(**3**,394)
C$_4$H$_8$F$_2$Si	1,1–Difluorosilacyclopentane	II/21(**3**,395)
C$_4$H$_8$Ge	1,1–Dimethyl–1H–germirene	II/7(**3**,529)
C$_4$H$_8$N$_2$	3–(Methylamino)propanenitrile	II/21(**3**,396)
C$_4$H$_8$N$_2$	Acetaldehyde azine Acetaldazine	II/15(**3**,655)
C$_4$H$_8$N$_2$	1,1'–Biaziridine	II/7(**3**,530)
C$_4$H$_8$N$_2$O$_2$	N–Nitropyrrolidine	II/23(**3**,333)
C$_4$H$_8$O	2–Buten–1–ol Crotyl alcohol	II/21(**3**,397)

C₄H₈O	3–Buten–2–ol	II/21(**3**,398)
C₄H₈O	Methyl ethyl ketone 2–Butanone	II/7(**3**,531)
C₄H₈O	Isobutyraldehyde 2–Methylpropanal	II/7(**3**,532), II/21(**3**,399)
C₄H₈O	2–Methoxy–1–propene	II/15(**3**,656)
C₄H₈O	2–Methyl–2–propen–1–ol 2–Methylallyl alcohol	II/21(**3**,400)
C₄H₈O	3–Buten–1–ol	II/15(**3**,657)
C₄H₈O	1,2–Epoxybutane 1–Butene oxide	II/15(**3**,658)
C₄H₈O	*cis*– and *trans*–2,3–Dimethyloxirane 2,3–Epoxybutane	II/7(**3**,533), II/23(**3**,334)
C₄H₈O	2–Methyloxetane	II/15(**3**,659)
C₄H₈O	Tetrahydrofuran	II/7(**3**,535)
C₄H₈OS	Tetrahydrothiophene 1–oxide	II/21(**3**,401)
C₄H₈OS	1,4–Thioxane 1,4–Oxathiane	II/7(**3**,537), II/15(**3**,660)
C₄H₈OS	3–Methoxythietane	II/23(**3**,335)
C₄H₈O₂	2–Methyl–1,3–dioxolane	II/21(**3**,402)

C₄H₈O₂ **Ethyl acetate** C_s (*trans*), C_1 (*gauche*)

ED and *ab initio* calculations 4-21G

trans conformer:

r_g	Å [a])	θ_α	deg [a])
C(3)–H(7)	1.105(3)	O(2)=C(1)–C(3)	124.1(10)
C(1)=O(2)	1.203(2)	O(2)=C(1)–O(4)	124.0(3)
C(1)–C(3)	1.508(2)	C(3)–C(1)–O(4) [b])	111.9
C(1)–O(4)	1.345(3)	C(1)–O(4)–C(5)	115.7(5)
O(4)–C(5)	1.448(3)	O(4)–C(5)–C(6)	108.2(11)
C(5)–C(6)	1.515(2) [g])	C(1)–C(3)–H	107.7(13)
		H–C(5)–H [b])	108.1
		O(4)–C(5)–H [b])	108.3
		C(5)–C(6)–H	108.1(13)
		ϕ_1 [c])	0 [e])
		ϕ_2 [d])	0 [e])
		ϕ_3 [f])	180 (*trans*)
			79(6) [g]) (*gauche*)
		ϕ_4 [h])	180 [e])

(*continued*)

Two stable conformers, *trans* and *gauche*, with respect to ϕ_3 were observed; the fraction of the *trans* conformer was estimated to be 38(7) %. With the aid of *ab initio* calculations [1], the potential function for the torsional motion of the ethyl group, $V(\phi_3) = 1/2[V_1 (1 + \cos \phi_3) + V_2 (1 - \cos 2\phi_3) + (h - V_1) (1 + \cos 3\phi_3)]$ was estimated to be (in units of kcal mol^{-1}) $V_1 = 3.5(4)$ and $V_2 = -2.7(3)$, where h was assumed to be 5.7.

The nozzle was at room temperature.

a) Estimated limits of error.
b) Dependent parameter.
c) Dihedral angle H(7)–C(3)–C(1)=O(2).
d) Dihedral angle O(2)=C(1)–O(4)–C(5).
e) Assumed.
f) Dihedral angle C(1)–O(4)–C(5)–C(6).
g) Uncertainty was not estimated in the original paper.
h) Dihedral angle O(4)–C(5)–C(6)–H(8).

Sugino, M., Takeuchi, H., Egawa, T., Konaka, S.: J. Mol. Struct. **245** (1991) 357.
[1] Manning, J., Klimkowski, V.J., Siam, K., Ewbank, J.D., Schafer, L.: J. Mol. Struct. (Theochem) **139** (1986) 305.   II/23(**3**,336)

$C_4H_8O_2$	1,3–Dioxane	II/15(**3**,661)
$C_4H_8O_2$	1,4–Dioxane	II/7(**3**,536)
$C_4H_8O_2$	1–(2–Oxiranyl)ethanol α–Methyloxiranemethanol	II/23(**3**,337)
$C_4H_8O_2$	2,3–Epoxy–1–butanol	II/23(**3**,338)
$C_4H_8O_2S$	Sulfolane Tetrahydrothiophene 1,1–dioxide	II/7(**3**,538)
$C_4H_8O_3$	Ethyl glycolate	II/23(**3**,339)
$C_4H_8O_3S$	4,5–Dimethyl–1,3,2–dioxathiolane 2–oxide 2,3–Buthanediyl sulfite	II/15(**3**,662)
$C_4H_8O_4$	Acetic acid dimer	II/7(**3**,246)
$C_4H_8O_4$	1,3,5,7–Tetraoxacyclooctane	II/15(**3**,663)
C_4H_8S	Ethyl vinyl sulfide	II/23(**3**,340)
C_4H_8S	3–Butene–1–thiol	II/21(**3**,403)
C_4H_8S	Allyl methyl sulfide	II/23(**3**,341)
C_4H_8S	Cyclopropyl methyl sulfide (Methylthio)cyclopropane	II/15(**3**,664)
C_4H_8S	Cyclopropanemethanethiol	II/23(**3**,342)
C_4H_8S	3–Methylthietan	II/21(**3**,404)
C_4H_8S	Tetrahydrothiophene	II/7(**3**,539), II/15(**3**,665)
$C_4H_8S_2$	1,1–Bis(methylthio)ethylene	II/15(**3**,666)
$C_4H_8S_2$	1,3–Dithiane	II/15(**3**,667)
C_4H_8Se	Tetrahydroselenophene	II/7(**3**,540), II/15(**3**,668), II/23(**3**,343)
C_4H_8Si	Silacyclopent–3–ene	II/15(**3**,669)
C_4H_9Br	1–Bromobutane Butyl bromide	II/7(**3**,541)

C₄H₉Br	t–Butyl bromide	II/15(**3**,670)
	2–Bromo–2–methylpropane	
C₄H₉Cl	1–Chlorobutane	II/7(**3**,542), II/15(**3**,671)
	Butyl chloride	
C₄H₉Cl	2–Chlorobutane	II/7(**3**,543)
	sec–Butyl chloride	
C₄H₉Cl	t–Butyl chloride	II/7(**3**,544), II/15(**3**,672)
	2–Chloro–2–methylpropane	
C₄H₉Cl	Isobutyl chloride	II/7(**3**,545), II/15(**3**,673), II/21(**3**,405)
	1–Chloro–2–methylpropane	
C₄H₉ClO	4–Chloro–1–butanol	II/21(**3**,406)
C₄H₉ClSi	Chlorodimethylvinylsilane	II/15(**3**,674)
C₄H₉Cl₂P	t–Butyldichlorophosphine	II/15(**3**,675)
C₄H₉F	t–Butyl fluoride	II/7(**3**,546)
C₄H₉F₂P	t–Butyldifluorophosphine	II/15(**3**,676)
C₄H₉F₃IN	Trimethylamine – trifluoromethyl iodide (1/1)	II/15(**3**,677)
C₄H₉GeN	Trimethylcyanogermane	II/7(**3**,547)
	Trimethylgermyl cyanide	
C₄H₉N	N–Methylallylamine	II/21(**3**,407)
C₄H₉N	3–Butenylamine	II/21(**3**,408)
	3–Buten–1–amine	
C₄H₉N	2,2–Dimethylaziridine	II/7(**3**,548)
C₄H₉N	Pyrrolidine	II/15(**3**,677a), II/21(**3**,409)
C₄H₉N	Cyclopropylmethylamine	II/15(**3**,677b)
	(Aminomethyl)cyclopropane	
C₄H₉NOSi	Trimethylsilyl isocyanate	II/7(**3**,549), II/21(**3**,410)
C₄H₉NO₂	2–Methyl–2–nitropropane	II/15(**3**,678)
C₄H₉NO₂	t–Butyl nitrite	II/15(**3**,679), II/23(**3**,344)
C₄H₉NO₂	Alanine methyl ester	II/21(**3**,411)
C₄H₉NS	Tetrahydro–1,4–thiazine	II/15(**3**,680)
	Thiomorpholine	
C₄H₉NSSi	Trimethylsilyl isothiocyanate	II/21(**3**,412)
C₄H₉NSi	Trimethylsilyl cyanide	II/7(**3**,550), II/15(**3**,681)
C₄H₉OP	Acetyldimethylphosphine	II/15(**3**,682)

C₄H₁₀   **Butane** C_{2h} (*trans*)
C_2 (*gauche*)
CH₃CH₂CH₂CH₃

ED

r_g	Å [a])	θ_α	deg [a])
C–C	1.531(2)	C–C–C	113.8(4)
C–H	1.117(5)	(C–C–H)$_{average}$	111.0(5)
		τ [b])	64.9(60) [c])

Amount of *trans* conformer: 53.5(90) %.
ΔG^0 (*gauche – trans*) = 497(220) cal mol⁻¹.
The nozzle temperature was room temperature; the temperature of the reservoir was –63.5 °C.

(*continued*)

a) Three times the estimated standard errors including systematic error.
b) *Gauche* dihedral angle.
c) Uncertainty represents 2.5 times the least-squares estimate.

Bradford, W.F., Fitzwater, S., Bartell, L.S.: J. Mol. Struct. **38** (1977) 185.
See also: Heenan, R.K., Bartell, L.S.: J. Chem. Phys. **78** (1983) 1270. II/7(**3**,551), II/15(**3**,683)

C_4H_{10}	Isobutane	II/7(**3**,552)
$C_4H_{10}BClN_2$	1,3–Dimethyl–2–chloro–1,3,2–diazaboracyclopentane	II/7(**3**,553)
$C_4H_{10}ClN$	N–Chloro–N–ethylethanamine	II/23(**3**,345)
$C_4H_{10}ClN_2P$	2–Chloro–1,3–dimethyl–1,3,2–diazaphospholane	II/15(**3**,684)
$C_4H_{10}ClP$	Chlorodiethylphosphine	II/15(**3**,685)
$C_4H_{10}Cl_2Si$	Bis(chloromethyl)dimethylsilane	II/15(**3**,686)
$C_4H_{10}Ge$	Cyclobutylgermane, Germylcyclobutane	II/21(**3**,413)
$C_4H_{10}Ge$	Germanocyclopentane	II/7(**3**,554)
$C_4H_{10}NP$	Trimethylphosphine – hydrogen cyanide (1/1)	II/15(**3**,686a)
$C_4H_{10}N_2$	Acetaldehyde dimethylhydrazone	II/15(**3**,687)
$C_4H_{10}N_2$	1,2–Dimethyl–1,2–diazetidine	II/15(**3**,688)
$C_4H_{10}N_2$	Piperazine	II/7(**3**,555)
$C_4H_{10}N_2$	Trimethylamine – hydrogen cyanide (1/1)	II/23(**3**,346)
$C_4H_{10}O$	Methyl propyl ether	II/15(**3**,689)
$C_4H_{10}O$	Isopropyl methyl ether	II/15(**3**,689a)

$C_4H_{10}O$ **Diethyl ether** C_s (*trans–trans*)

MW

$C(1)H_3–C(2)H_2–O–C(2)H_2–C(1)H_3$

r_s	Å	θ_s	deg
C(1)–C(2)	1.517(5)	C(1)–C(2)–O	108.4(3)
C(2)–O	1.411(3)	C(2)–O–C(2)	112.1(3)
C(1)–H_s	1.092(4)	C(2)–C(1)–H_s	110.2(4)
C(1)–H_a	1.090(5)	C(2)–C(1)–H_a	110.2(5)
C(2)–H	1.100(3)	H_s–C(1)–H_a	108.9(4)
		H_a–C(1)–H_a	108.2(5)
		γ a)	110.2(5)
		θ b)	0.0(6)
		C(1)–C(2)–H	110.4(5)
		H–C(2)–O	110.0(3)
		H–C(2)–H	107.6(4)

a) The corrected C–C–H value defined by $\gamma = \frac{1}{3} [(C–C–H_s) + 2(C–C–H_a)]$.
b) The tilt angle of the CCH_3 group defined by $\theta = \frac{2}{3} [(C–C–H_a) – (C–C–H_s)]$.

Hayashi, M., Adachi, M.: J. Mol. Struct. **78** (1982) 53.   II/15(**3**,690)

$C_4H_{10}O$	*t*–Butyl alcohol 2–Methyl–2–propanol	II/21(**3**,414)
$C_4H_{10}O_2$	3–Methoxy–1–propanol	II/21(**3**,415)
$C_4H_{10}O_2$	1,2–Dimethoxyethane	II/15(**3**,691)
$C_4H_{10}O_3$	Trimethoxymethane	II/15(**3**,692)

C$_4$H$_{10}$O$_4$S$_2$	1,2–Bis(methylsulfonyl)ethane	II/21(**3**,416)
C$_4$H$_{10}$SSi	3,3–Dimethyl–1–thia–3–silacyclobutane	
	3,3–Dimethyl–1,3–thiasilethane	II/15(**3**,693), II/21(**3**,417)
C$_4$H$_{10}$Si	Cyclobutylsilane	II/15(**3**,694)
C$_4$H$_{10}$Si	Silacyclopentane	II/15(**3**,695)
	Tetramethylenesilane	
C$_4$H$_{10}$Zn	Diethylzinc	II/15(**3**,696)
C$_4$H$_{11}$N	N,N–Dimethylethylamine	II/15(**3**,697)
C$_4$H$_{11}$N	Diethylamine	II/23(**3**,347)
C$_4$H$_{11}$N	t–Butylamine	II/21(**3**,418), II/23(**3**,348)
C$_4$H$_{11}$P	Ethyldimethylphosphine	II/21(**3**,419)
C$_4$H$_{11}$P	Trimethyl(methylene)phosphorane	II/15(**3**,698)
C$_4$H$_{11}$P	t–Butylphosphine	II/15(**3**,699)
C$_4$H$_{12}$Al$_2$Cl$_2$	Dimethylaluminum chloride dimer	II/7(**3**,556)
C$_4$H$_{12}$Al$_2$Cl$_4$N$_2$	Bis(μ–dimethylamido)–bis[dichloroaluminum]	II/15(**3**,700)
C$_4$H$_{12}$As$_2$	Tetramethyldiarsine	II/23(**3**,349)
C$_4$H$_{12}$AuP	Methyl(trimethylphosphine)gold(I)	II/21(**3**,420)
C$_4$H$_{12}$B$_2$O	Dimethylborinic anhydride	II/15(**3**,701)
C$_4$H$_{12}$B$_2$S$_2$	Bis(dimethylboryl)disulfane	II/15(**3**,702)
C$_4$H$_{12}$ClN$_2$P	Bis(dimethylamino)chlorophosphine	II/15(**3**,703)
	Tetramethylphosphorodiamidous chloride	
C$_4$H$_{12}$Cl$_2$Ga$_2$	Di–μ–chloro–bis(dimethylgallium)	II/23(**3**,350)
C$_4$H$_{12}$Cl$_2$OSi$_2$	1,3–Dichloro–1,1,3,3–tetramethyldisiloxane	II/15(**3**,704)
C$_4$H$_{12}$Cl$_2$Si$_2$	1,2–Dichloro–1,1,2,2–tetramethyldisilane	II/15(**3**,705)

ED **C$_4$H$_{12}$Ge** **Tetramethylgermane** T$_d$ (heavy atom skeleton)
Ge(CH$_3$)$_4$

r_g	Å [a])		θ [b])	deg [a])
C–H	1.111(3)		Ge–C–H	110.7(2)
Ge–C	1.958(4)		C–Ge–C	109.47 [c])

Local C$_{3v}$ symmetry for the CH$_3$ groups was assumed.
The methyl torsional barrier V_0 was estimated to be
1.3 kJ mol^{-1} on the basis of an effective angle of torsion
τ = 23.0(15)°, from the staggered form, yielded directly
by the analysis.
The nozzle was at room temperature.

[a]) Estimated total errors.
[b]) Undefined, possibly θ_α.
[c]) Assumed.

Csákvári, E., Rozsondai, B., Hargittai, I.: J. Mol. Struct. **245** (1991) 349.

II/15(**3**,706), II/23(**3**,351)

C$_4$H$_{12}$N$_2$	Tetramethylhydrazine	II/15(**3**,707)
C$_4$H$_{12}$N$_2$OS	N,N,N',N'–Tetramethylsulfurous diamide	II/7(**3**,557)
C$_4$H$_{12}$N$_2$O$_2$S	N,N,N',N'–Tetramethylsulfuric diamide	II/7(**3**,558)

C$_4$H$_{12}$N$_2$S	N,N′−Thiobis(dimethylamine)	II/7(**3**,559)
C$_4$H$_{12}$N$_2$Sn	Bis(dimethylamino)tin(II)	II/23(**3**,352)
C$_4$H$_{12}$OSi	Methoxytrimethylsilane	II/15(**3**,708)
C$_4$H$_{12}$O$_3$Si	Methyltrimethoxysilane	II/15(**3**,709)
C$_4$H$_{12}$O$_4$Si	Tetramethoxysilane	II/15(**3**,710)
C$_4$H$_{12}$P$_2$	Tetramethyldiphosphine	II/7(**3**,560)
C$_4$H$_{12}$Pb	Tetramethyllead Tetramethylplumbane	II/7(**3**,561)
C$_4$H$_{12}$Si	Tetramethylsilane	II/7(**3**,562)
C$_4$H$_{12}$Sn	Tetramethyltin Tetramethylstannane	II/7(**3**,563)
C$_4$H$_{13}$NSi	Dimethyl(dimethylamino)silane N−(Dimethylsilyl)dimethylamine	II/15(**3**,711)
C$_4$H$_{13}$NSi	(Diethylamino)silane N,N−Diethylsilylamine	II/21(**3**,421)
C$_4$H$_{13}$PSi	Trimethyl(silylmethylene)phosphorane	II/15(**3**,712)
C$_4$H$_{14}$Al$_2$	Dimethylaluminum hydride dimer	II/7(**3**,564)
C$_4$H$_{14}$B$_2$	Tetramethyldiborane	II/7(**3**,565)
C$_4$H$_{14}$Ga$_2$	Di−μ−hydrido−bis(dimethylgallium)	II/21(**3**,422), II/23(**3**,353)
C$_4$H$_{14}$N$_2$Si	Bis(dimethylamino)silane	II/21(**3**,423)
C$_4$H$_{14}$OSi$_2$	1,1,3,3−Tetramethyldisiloxane Oxybis[dimethylsilane] Bis(dimethylsilyl) ether	II/15(**3**,712a)
C$_4$H$_{14}$SSi$_2$	Bis(dimethylsilyl) sulfide	II/23(**3**,354)
C$_4$H$_{15}$NSi$_2$	Bis(dimethylsilyl) amine	II/15(**3**,713)
C$_4$H$_{15}$NSi$_2$	N−t−Butyl−N−silylsilanamine 2−(t−Butyl)disilazane	II/21(**3**,424)
C$_4$H$_{16}$B$_{10}$	C,C′−Dimethyl carborane C,C′−Dimethyl−1,7−dicarba−*closo*−dodecaborane(12)	II/7(**3**,566)
C$_4$H$_{16}$B$_{10}$	1,12−Dimethyl−1,12−dicarba−*closo*−dodecaborane(12)	II/15(**3**,714)
C$_4$H$_{16}$Ga$_2$N$_2$	Dimeric (dimethylamino)gallium dihydride Bis−μ−dimethylamido−bis[dihydridogallium(III)]	II/21(**3**,425)
C$_4$H$_{22}$B$_{20}$	1,1′−Bi−1,12−dicarba−*closo*−dodecaborane(12) 1,1′−Bi−1,12−dicarbadodecaborane(12)	II/21(**3**,426)
C$_4$H$_{22}$B$_{20}$Hg	Bis(1,12−dicarbadodecacarboran−1−yl)mercury(II)	II/23(**3**,355)
C$_4$H$_{22}$B$_{20}$Hg	Bis(1,12−dicarbadodecarboran(12)−2−yl)−mercury(II)	II/23(**3**,356)

IR **C$_4$N$_2$**   **2−Butynedinitrile**   **D$_{\infty h}$**

Dicyanoacetylene   N≡C−C≡C−C≡N
Acetylenedicarbonitrile

r_0	Å
C≡N	1.1606(50) [a]
C≡C	1.2223(50) [a]
C−C	1.3636(50) [a]

(*continued*)

C$_4$N$_2$ (continued)

Bond lengths transferred from the nitrile side of HC$_5$N predict a B_0 value extremely close to the experimental one (0.05 %). A change of 0.0004 Å in the distances is sufficient to reproduce the experimental B_0. Therefore, the geometric parameters are well within 0.005 Å of the transferred distances. This is further confirmed by *ab initio* predictions.

[a]) Distances are taken from HC$_5$N.

Winther, F., Schönhoff, M., LePrince, R., Guarnieri, A., Bruget, D.N., McNaughton, D.:
   J. Mol. Spectrosc. **152** (1992) 205.
See also: McNaughton, D., Bruget, D.N.: J. Mol. Struct. **273** (1992) 11.

II/21(**3**,427), II/23(**3**,357)

MW | **C$_4$N$_4$** | **Cyanogen dimer** (weakly bound complex) | C$_{2v}$ (T–shaped) (effective symmetry class) (NCCN)$_2$

r_0	Å [a])	θ_0	deg [a])
N(2)···C(1)	3.206(3)	θ_1 [b])	90 [c])
R_{cm}	4.987(3)	θ_2 [b])	13.9(15)

Cyanogen (1) is the one at the bottom of the figure, and cyanogen (2) is the one bonded through N.

[a]) Uncertainties were about three times those of the original data.
[b]) θ_1, θ_2: angles between the *a* axis and the molecular axes of cyanogens (1) and (2), respectively.
[c]) Assumed.

Suni, I.I., Lee, S.H., Klemperer, W.: J. Phys. Chem. **95** (1991) 2859.

II/23(**3**,358)

| C$_4$N$_4$O$_4$Si | Silicon tetraisocyanate | II/7(**3**,469) |
| C$_4$NiO$_4$ | Tetracarbonylnickel(0) | II/15(**3**,715) |

IR | **C$_5$** | **1,2,3,4–Pentatetraenediylidene** | D$_{\infty h}$ C=C=C=C=C

r_0	Å
C=C	1.2833(5)

The assumption has been made that all C=C bonds are equivalent and have equal lengths.

Heath, J.R., Cooksy, A.L., Gruebele, M.H.W., Schmuttenmaer, C.A., Saykally, R.J.:
   Science **244** (1989) 564.
Bernath, P.F., Hinkle, K.H., Keady, J.J.: Science **244** (1989) 562.
Moazzen-Ahmadi, N., MeKellar, A.R.W., Amano, T.: J. Chem. Phys. **91** (1989) 2140.

II/21(**3**,428)

C$_5$Br$_3$GeMnO$_5$	Pentacarbonyltribromogermylmanganese	II/7(**3**,567)
C$_5$Cl$_6$	Perchlorocyclopentadiene	II/7(**3**,568)
C$_5$F$_3$MnO$_5$Si	Pentacarbonyltrifluorosilylmanganese	II/15(**3**,716)
C$_5$F$_3$MoO$_5$P	Pentacarbonyltrifluorophosphinemolybdenum	II/7(**3**,570)
	Pentacarbonyl(phosphorus trifluoride)molybdenum	

C₅F₈	Perfluorocyclopentene		II/7(**3**,569)
C₅F₈	Perfluorospiro[2.2]pentane		II/15(**3**,717)
C₅F₁₂	(Perfluoro)neopentane		II/15(**3**,718)
C₅FeO₅	Iron pentacarbonyl		II/7(**3**,571)
C₅HMnO₅	Manganese pentacarbonyl hydride		II/7(**3**,573), II/15(**3**,719)
	Hydridopentacarbonylmanganese(I)		
C₅HN	Cyanobutadiyne, 2,4–Pentadiynenitrile		II/15(**3**,720)
C₅H₂F₆N₂	3,5–Bis(trifluoromethyl)pyrazole		II/23(**3**,359)
C₅H₂F₆O₂	1,1,1,5,5,5–Hexafluoro–2,4–pentanedione		
	Hexafluoroacetylacetone		II/7(**3**,574), II/23(**3**,360)
C₅H₃ClOS	2–Thiophenecarbonyl chloride		II/21(**3**,429)
C₅H₃ClO₂	2–Furancarbonyl chloride, 2–Furoyl chloride		II/21(**3**,430)
C₅H₃F₂N	2,6–Difluoropyridine		II/15(**3**,721)
C₅H₃GeMnO₅	Pentacarbonylgermylmanganese		II/15(**3**,722)
C₅H₃GeO₅Re	Pentacarbonylgermylrhenium		II/15(**3**,723)
C₅H₃MnO₅Si	Pentacarbonylsilylmanganese		II/15(**3**,724)
C₅H₃O₅ReSi	Pentacarbonylsilylrhenium		II/15(**3**,725)
C₅H₄	1,3–Pentadiyne		II/7(**3**,575), II/15(**3**,725a)
C₅H₄	1,4–Pentadiyne		II/15(**3**,726)

LIF | **C₅H₄Cl** | | **Chlorocyclopentadienyl radical** | | C_{2v} |

State		$\tilde{X}\ ^2B_2$	$\tilde{B}\ ^2B_2$
Energy [eV]		0.00	3.619
r_0 [Å]	C(0)–C(1)	1.425(9)	1.429(1)
	C(1)–C(2)	1.406(9)	1.491(1)
	C(2)–C(–2)	1.437(9)	1.390(1)
	C–H(ring)	1.083 [a]	1.069 [a]
	C–Cl	1.638(3)	1.629(5)

The structure was determined by fitting the rotational constants to two parameters, (i) the length of the C–Cl bond, and (ii) distortion of the ring ΔR_{CC} which was assumed to obey the relation

$$R_{CC}^{(k)}(\text{ring}) = R_{CC}(\text{ring}) + \Delta R_{CC} \cos(4k\pi/5),$$

where $k = 0, \pm 1, \pm 2$ correspond to the bonds opposite the carbon atoms labeled above. The values of R_{CC}(ring) in the ground and excited states were assumed to be 1.420 and 1.446 Å, respectively, as in C₅H₅.

[a]) Fixed at the corresponding C₅H₅ values.

Cullin, D.W., Yu, L., Williamson, J.M., Miller, T.A.: J. Phys. Chem. **96** (1992) 89.

II/23(**3**,361)

C₅H₄ClN	2–Chloropyridine	II/7(**3**,576)
C₅H₄ClNO	4–Chloropyridine *N*–oxide	II/15(**3**,727)
C₅H₄Cl₄	1–Chloro–1–(trichlorovinyl)cyclopropane	II/21(**3**,431)

	C₅H₄F		Fluorocyclopentadienyl radical		C$_{2v}$
LIF					
	State		$\tilde{X}\ ^2B_2$	$\tilde{B}\ ^2B_2$	
	Energy [eV]		0.00	3.813	
	r_0 [Å]	C(0)–C(1)	1.446(2)	1.452(6)	
		C(1)–C(2)	1.353(2)	1.431(6)	
		C(2)–C(–2)	1.503(2)	1.464(6)	
		C–H(ring)	1.083 [a]	1.069 [a]	
		C–F	1.283(1)	1.298(3)	

Structural diagram: F at C(0), H atoms at C(±1) and C(±2) positions of the cyclopentadienyl ring.

The structure was determined by fitting the rotational constants to two parameters, (i) the length of the C–F bond, and (ii) distortion of the ring ΔR_{CC} which was assumed to obey the relation

$$R_{CC}^{(k)}(\text{ring}) = R_{CC}(\text{ring}) + \Delta R_{CC} \cos(4k\pi/5),$$

where $k = 0, \pm 1, \pm 2$ correspond to the bonds opposite the carbon atoms labelled above. The values of R_{CC}(ring) in the ground and excited states were assumed to be 1.420 and 1.446 Å, respectively, as in C₅H₅.

[a] Fixed at the corresponding C₅H₅ values.

Cullin, D.W., Yu, L., Williamson, J.M., Miller, T.A.: J. Phys. Chem. **96** (1992) 89.

II/23(**3**,362)

C₅H₄N₂		1*H*–Pyrrole–3–carbonitrile	II/23(**3**,363)
C₅H₄N₂		Diazocyclopentadiene	II/23(**3**,364)
C₅H₄N₂O₃		4–Nitropyridine *N*–oxide	II/15(**3**,728)
C₅H₄OS		2–Thiophenecarbaldehyde	II/21(**3**,432)
		2–Thiophenecarboxaldehyde	
C₅H₄OS		4*H*–Pyran–4–thione	II/15(**3**,729)
C₅H₄O₂		2–Furancarbaldehyde	II/15(**3**,730)
C₅H₄O₂		4–Cyclopentene–1,3–dione	II/15(**3**,731)
C₅H₄O₂		4*H*–Pyran–4–one	II/15(**3**,732)
C₅H₄O₂S		1,6–Dioxa–6a–thia(6a–S^{IV})pentalene	II/21(**3**,433)
C₅H₄S₂		4*H*–Thiopyran–4–one	II/15(**3**,733)
C₅H₄S₃		1,6,6a–Trithia(6a–S^{IV})pentalene	II/15(**3**,734)

	C₅H₅		Cyclopentadienyl radical		D$_{5h}$
LIF					(CH)₅
	State		$\tilde{X}\ ^2E''_1$	$\tilde{A}\ ^2A''_2$	
	Energy [eV]		0.0	3.661	
	r_0 [Å]	C–H	1.088(5)	1.075(5)	
		C–C	1.420(1)	1.446(1)	

Rotational analysis of C₅H₅ and C₅D₅. Two sets of structural parameters are given in [1] depending on whether the inertial defect in the ground state or in the excited state is assumed to be zero. The mean values are given here and the error limits embrace the two sets of values.

[1] Yu, L., Williamson, J.M., Miller, T.A.: Chem. Phys. Lett. **162** (1989) 431.
[2] Yu, L., Foster, S.C., Williamson, J.M., Heaven, M.C., Miller, T.A.: J. Phys. Chem. **92** (1988) 4263.

II/21(**3**,434)

C_5H_5ArN	Argon – pyridine (1/1)	II/21(**3**,435)
C_5H_5As	Arsabenzene, Arsenin	II/7(**3**,577), II/15(**3**,735)
$C_5H_5BBr_3N$	Pyridine – boron tribromide (1/1)	II/21(**3**,436)
$C_5H_5BCl_3N$	Pyridine – boron trichloride (1/1)	II/21(**3**,437)
$C_5H_5BF_3N$	Pyridine – boron trifluoride (1/1)	II/21(**3**,438)
C_5H_5BeBr	Cyclopentadienylberyllium bromide	II/7(**3**,578)
C_5H_5BeCl	Cyclopentadienylberyllium chloride	II/7(**3**,579)
$C_5H_5Br_3Ti$	Cyclopentadienyltitanium tribromide	II/7(**3**,580)
$C_5H_5Cl_2N$	1,1–Dichloro–2–(cyanomethyl)cyclopropane	II/21(**3**,439)
$C_5H_5F_3O_2$	1,1,1–Trifluoroacetylacetone	II/7(**3**,581)
C_5H_5He	Cyclopentadienyl – helium (1/1)	II/23(**3**,365)
$C_5H_5He_2$	Cyclopentadienyl – helium (1/2)	II/23(**3**,366)
C_5H_5In	Cyclopentadienyl indium	II/7(**3**,582)
C_5H_5KrN	Krypton – pyridine (1/1)	II/21(**3**,440)
C_5H_5N	Pyridine	II/7(**3**,583), II/15(**3**,736), II/21(**3**,441)
C_5H_5N	1–Cyanobicyclo[1.1.0]butane	II/23(**3**,367)
C_5H_5NNiO	(η–Cyclopentadienyl)nitrosylnickel	II/7(**3**,584), II/15(**3**,737)
C_5H_5NO	2–Hydroxypyridine, 2–Pyridinol	II/23(**3**,368)
C_5H_5NO	Pyridine N–oxide	II/7(**3**,585), II/15(**3**,738)
$C_5H_5NO_2S$	Pyridine – sulfur dioxide (1/1)	II/21(**3**,442), II/23(**3**,369)
C_5H_5Ne	Cyclopentadienyl – neon (1/1)	II/23(**3**,370)
C_5H_5P	Phosphabenzene, Phosphorin	II/7(**3**,586)
C_5H_5Sb	Stibabenzene, Antimonin	II/15(**3**,739)
C_5H_5Tl	(η–Cyclopentadienyl)thallium	II/7(**3**,587)
C_5H_6	2–Methyl–1–buten–3–yne	II/7(**3**,588)
C_5H_6	1,2,4–Pentatriene, Vinylallene	II/15(**3**,740)
C_5H_6	Cyclopropylacetylene Ethynylcyclopropane	II/7(**3**,589), II/15(**3**,741)
C_5H_6	Cyclopentadiene	II/7(**3**,590), II/15(**3**,742)
C_5H_6	Bicyclo[2.1.0]pent–2–ene	II/7(**3**,591)
C_5H_6	[1.1.1]Propellane Tricyclo[1.1.1.0^{1,3}]pentane	II/21(**3**,443)
C_5H_6Be	(η–Cyclopentadienyl)hydroberyllium	II/15(**3**,743)
$C_5H_6N_2$	Pentanedinitrile, Glutaronitrile	II/21(**3**,444)
C_5H_6O	3–Cyclopenten–1–one	II/7(**3**,592)
C_5H_6O	2–Methylfuran	II/7(**3**,593)
C_5H_6O	3–Methylfuran	II/7(**3**,594)
C_5H_6O	Bicyclo[1.1.1]pentanone	II/21(**3**,445)
C_5H_6OS	2–Furanmethanethiol	II/15(**3**,744)
C_5H_6S	3–Methylthiophene	II/7(**3**,595)
C_5H_7Cl	1–Chlorospiro[2.2]pentane	II/7(**3**,596)
C_5H_7Cl	1–Chlorobicyclo[1.1.1]pentane	II/7(**3**,597)
C_5H_7ClO	3–Methyl–2–butenoyl chloride	II/21(**3**,446)
C_5H_7ClO	Cyclobutanecarbonyl chloride	II/7(**3**,598), II/21(**3**,447)
C_5H_7FO	Cyclobutanecarbonyl fluoride	II/21(**3**,448)
C_5H_7N	3–Methyl–2–butenenitrile	II/15(**3**,745)

C$_5$H$_7$N	N–Methylpyrrole	II/7(**3**,599)
C$_5$H$_7$N	Cyclobutanecarbonitrile Cyanocyclobutane	II/21(**3**,449)
C$_5$H$_7$NO	Furfurylamine	II/23(**3**,371)
C$_5$H$_7$N$_2$OP	4–Methyl–2–acetyl–1,2,3–phosphadiazole	II/7(**3**,600)
C$_5$H$_7$N$_2$OP	2–Acetyl–5–methyl–2H–1,2,3–diazaphosphole	II/21(**3**,450)
C$_5$H$_8$	3–Methyl–1,2–butadiene 1,1–Dimethylallene	II/7(**3**,601)
C$_5$H$_8$	2–Methyl–1,3–butadiene	II/7(**3**,602), II/15(**3**,746)
C$_5$H$_8$	1,4–Pentadiene	II/15(**3**,747), II/21(**3**,451)
C$_5$H$_8$	1–Pentyne	II/7(**3**,603)
C$_5$H$_8$	3–Methyl–1–butyne	II/7(**3**,604)
C$_5$H$_8$	Vinylcyclopropane	II/7(**3**,605), II/21(**3**,452)
C$_5$H$_8$	3,3–Dimethylcyclopropene	II/15(**3**,748)
C$_5$H$_8$	Methylenecyclobutane	II/15(**3**,749), II/23(**3**,372)
C$_5$H$_8$	Cyclopentene	II/7(**3**,607)
C$_5$H$_8$	Spiro[2.2]pentane	II/7(**3**,608)
C$_5$H$_8$	Bicyclo[1.1.1]pentane	II/7(**3**,609)
C$_5$H$_8$	Bicyclo[2.1.0]pentane	II/7(**3**,610), II/15(**3**,750)
C$_5$H$_8$Br$_4$	Tetrakis(bromomethyl)methane	II/15(**3**,751)
C$_5$H$_8$Cl$_4$	1,3–Dichloro–2,2–bis(chloromethyl)propane Tetrakis(chloromethyl)methane	II/7(**3**,611), II/21(**3**,453)
C$_5$H$_8$N$_2$	N–Cyanopyrrolidine	II/21(**3**,454)
C$_5$H$_8$N$_2$	3,5–Dimethylpyrazole	II/23(**3**,373)
C$_5$H$_8$N$_2$	2,3–Diazabicyclo[2.2.1]hept–2–ene	II/15(**3**,752)
C$_5$H$_8$N$_2$O	1,1′–Carbonylbisaziridine	II/21(**3**,455)
C$_5$H$_8$N$_6$O	3,3–Bis(azidomethyl)oxetane	II/21(**3**,456)
C$_5$H$_8$O	4–Pentyn–1–ol	II/21(**3**,457)
C$_5$H$_8$O	Cyclopropyl methyl ketone Acetylcyclopropane	II/7(**3**,612)
C$_5$H$_8$O	Cyclobutanecarbaldehyde Cyclobutanecarboxaldehyde	II/23(**3**,374)
C$_5$H$_8$O	Cyclopentanone	II/7(**3**,613), II/15(**3**,753), II/21(**3**,458)
C$_5$H$_8$O	3,4–Dihydro–2H–pyran	II/21(**3**,459)
C$_5$H$_8$O	3,6–Dihydro–2H–pyran	II/15(**3**,754)
C$_5$H$_8$O	Cyclopentene oxide 1,2–Epoxycyclopentane	II/7(**3**,614), II/15(**3**,755)
C$_5$H$_8$OS	Tetrahydro–4H–thiopyran–4–one 1–Thiacyclohexan–4–one	II/15(**3**,756)
C$_5$H$_8$O$_2$	Acetylacetone 2,4–Pentanedione	II/7(**3**,615), II/21(**3**,460)
C$_5$H$_8$O$_2$	Tetrahydro–4H–pyran–4–one	II/15(**3**,757)
C$_5$H$_8$O$_3$	Cyclopentene ozonide 6,7,8–Trioxabicyclo[3.2.1]octane	II/15(**3**,758)
C$_5$H$_8$S	5–Thiabicyclo[2.1.1]hexane	II/7(**3**,616), II/15(**3**,759)
C$_5$H$_8$Si	Silylcyclopentadiene	II/7(**3**,617)

C$_5$H$_9$BBe	Cyclopentadienylberyllium − boron hydride (1/1)	II/7(**3**,618)
C$_5$H$_9$Br	1−Bromo−3−methyl−2−butene	II/23(**3**,375)
C$_5$H$_9$Cl	Cyclopentyl chloride	II/15(**3**,760), II/23(**3**,376)
C$_5$H$_9$ClGe	1−Chloro−2−(trimethylgermyl)acetylene 1−Chloro−2−(trimethylgermyl)ethyne	II/7(**3**,620)
C$_5$H$_9$ClSi	1−Chloro−2−(trimethylsilyl)acetylene 1−Chloro−2−(trimethylsilyl)ethyne	II/7(**3**,621)
C$_5$H$_9$Cl$_3$	1,1,1−Tris(chloromethyl)ethane	II/7(**3**,619)
C$_5$H$_9$N	t−Butyl cyanide Pivalonitrile	II/7(**3**,622)
C$_5$H$_9$N	t−Butyl isocyanide	II/21(**3**,461)
C$_5$H$_9$NO	1−Pyrrolidinecarbaldehyde 1−Pyrrolidinecarboxaldehyde	II/21(**3**,462)
C$_5$H$_9$P	(2,2−Dimethylpropylidyne)phosphine	II/15(**3**,761), II/23(**3**,377)
C$_5$H$_{10}$	2−Methyl−1−butene	II/7(**3**,623)
C$_5$H$_{10}$	cis−2−Pentene	II/15(**3**,762)
C$_5$H$_{10}$	trans−2−Pentene	II/15(**3**,763)
C$_5$H$_{10}$	1,1−Dimethylcyclopropane	II/15(**3**,764)
C$_5$H$_{10}$	trans−1,2−Dimethylcyclopropane	II/15(**3**,765)
C$_5$H$_{10}$	Cyclopentane	II/7(**3**,624)
C$_5$H$_{10}$Cl$_2$	1,3−Dichloro−2,2−dimethylpropane	II/7(**3**,625)
C$_5$H$_{10}$Cl$_2$Si	1,1−Dichlorosilacyclohexane	II/15(**3**,766)
C$_5$H$_{10}$FN	t−Butyl cyanide − hydrogen fluoride (1/1)	II/15(**3**,767)
C$_5$H$_{10}$O	4−Penten−1−ol	II/21(**3**,463)
C$_5$H$_{10}$O	3−Methyl−2−butanone Isopropyl methyl ketone	II/21(**3**,464)
C$_5$H$_{10}$O	Pivalaldehyde	II/23(**3**,378)
C$_5$H$_{10}$O	trans−2−Methylcyclopropanemethanol	II/23(**3**,379)
C$_5$H$_{10}$O	Tetrahydropyran	II/15(**3**,768)
C$_5$H$_{10}$OS	2−Methyltetrahydrothiophene 1−oxide	II/21(**3**,465)
C$_5$H$_{10}$OS	Tetrahydro−2H−thiopyran 1−oxide	II/23(**3**,380)
C$_5$H$_{10}$O$_2$	2,2−Dimethyl−1,3−dioxolane	II/21(**3**,466)
C$_5$H$_{10}$O$_3$S	4,6−Dimethyl−1,3,2−dioxathiane 2−oxide 2,4−Pentanediyl sulfite	II/7(**3**,626)
C$_5$H$_{10}$S	Thiane, Tetrahydro−2H−thiopyran	II/15(**3**,769), II/21(**3**,467)
C$_5$H$_{10}$Si	(Trimethylsilyl)acetylene (Trimethylsilyl)ethyne	II/7(**3**,627)
C$_5$H$_{10}$Sn	(Trimethylstannyl)acetylene Ethynyltrimethylstannane	II/15(**3**,770), II/21(**3**,468)
C$_5$H$_{11}$N	N−Methylpyrrolidine	II/21(**3**,469)
C$_5$H$_{11}$N	Piperidine	II/15(**3**,771)
C$_5$H$_{11}$N	Acetylene − trimethylamine (1/1)	II/21(**3**,470)
C$_5$H$_{12}$	Pentane	II/7(**3**,628)
C$_5$H$_{12}$	2,2−Dimethylpropane, Neopentane	II/7(**3**,629), II/15(**3**,772)
C$_5$H$_{12}$BClN$_2$	2−Chloro−1,3−dimethyl−1,3−diaza−2−boracyclohexane 2−Chlorohexahydro−1,3−dimethyl−1,3,2−diazaborine	II/15(**3**,773)

	C$_5$H$_{12}$BF$_3$N$_2$	Bis(dimethylamino)(trifluoromethyl)borane	II/21(**3**,471)
	C$_5$H$_{12}$BN	t–Butyl isocyanide – borane (1/1)	II/15(**3**,774)
	C$_5$H$_{12}$Ge	Germacyclohexane	II/23(**3**,381)
	C$_5$H$_{12}$N$_2$	Acetone dimethylhydrazone	II/15(**3**,775)
	C$_5$H$_{12}$N$_2$O	Tetramethylurea	II/7(**3**,630), II/15(**3**,776)
	C$_5$H$_{12}$N$_2$S	Tetramethylthiourea	II/15(**3**,777)
	C$_5$H$_{12}$O	t–Butyl methyl ether	II/21(**3**,472), II/23(**3**,382)
	C$_5$H$_{12}$OSi	Trimethylsilyl vinyl ether	II/15(**3**,778)
	C$_5$H$_{12}$O$_2$	2,2–Dimethoxypropane	II/15(**3**,779)
	C$_5$H$_{12}$O$_3$	1,1,1–Trimethoxyethane	II/15(**3**,780)
	C$_5$H$_{12}$O$_4$	Tetramethoxymethane	II/7(**3**,631)
	C$_5$H$_{12}$SSi	3,3–Dimethyl–1–thia–3–silacyclopentane	II/23(**3**,383)
	C$_5$H$_{12}$Si	1,1–Dimethylsilacyclobutane	II/23(**3**,384)
	C$_5$H$_{12}$Si	Cyclopentylsilane	II/21(**3**,473)
	C$_5$H$_{12}$Si	Silacyclohexane	II/15(**3**,781)
	C$_5$H$_{14}$N$_2$	N,N,N′,N′–Tetramethylmethanediamine Bis(dimethylamino)methane	II/15(**3**,782)
	C$_5$H$_{14}$P$_2$	Bis(dimethylphosphino)methane	II/15(**3**,783)
	C$_5$H$_{15}$AlO	Trimethylaluminum – dimethylether (1/1)	II/15(**3**,784)
	C$_5$H$_{15}$AlS	Trimethylaluminum – dimethyl sulfide (1/1)	II/15(**3**,785)
	C$_5$H$_{15}$GaO	Trimethylgallium – dimethyl ether (1/1)	II/21(**3**,474)
	C$_5$H$_{15}$NSi	(Dimethylamino)trimethylsilane N,N–Dimethyltrimethylsilylamine	II/21(**3**,475)
	C$_5$H$_{16}$B$_4$Si	2–(Trimethylsilyl)–2,3–dicarba–*nido*–hexaborane(8)	II/21(**3**,476)
	C$_5$H$_{17}$B$_3$Zr	(η–Cyclopentadienyl)zirconium tris(tetrahydroborate)	II/23(**3**,385)
	C$_5$H$_{17}$NSi$_2$	N,N–Bis(dimethylsilyl)methylamine	II/21(**3**,477)
	C$_5$N$_4$	Tetracyanomethane	II/7(**3**,572)
	C$_5$O$_5$Os	Pentacarbonylosmium Osmium pentacarbonyl	II/21(**3**,478)

ED | **C$_5$O$_5$Ru** | **Pentacarbonylruthenium** | **D$_{3h}$** |
Ru(CO)$_5$

r_g	Å [a]
Ru–C(ax)	1.950(9)
Ru–C(eq)	1.969(3)
C≡O	1.143(2)

The measurement was made at room temperature.

[a]) Twice the estimated standard errors.

Huang, J., Hedberg, K., Davis, H.B., Pomeroy, R.K.: Inorg. Chem. **29** (1990) 3923.

II/23(**3**,386)

C$_6$Br$_3$N$_3$O$_6$	1,3,5–Tribromo–2,4,6–trinitrobenzene	II/23(**3**,387)
C$_6$Br$_6$	Hexabromobenzene	II/7(**3**,632)
C$_6$ClF$_5$O$_2$S	Pentafluorobenzenesulfonyl chloride	II/15(**3**,786)

C₆Cl₂F₅N	N,4–Dichloro–2,3,4,5,6–pentafluoro–2,5–cyclo-hexadien–1–imine	II/15(**3**,787)
C₆Cl₃N₃O₆	1,3,5–Trichloro–2,4,6–trinitrobenzene	II/23(**3**,388)
C₆Cl₄O₂	Tetrachloro–p–benzoquinone	II/15(**3**,788)
C₆Cl₄O₂	Tetrachloro–o–benzoquinone	II/15(**3**,789)
C₆Cl₆	Perchloro–1,5–hexadien–3–yne	II/15(**3**,790)
C₆Cl₆	1,2–Dichloro–3,4–bis(dichloromethylene)cyclobutene	II/7(**3**,633)
C₆Cl₆	Hexachlorofulvene	II/7(**3**,634)
C₆Cl₆	Hexachlorobenzene	II/7(**3**,635)
C₆F₃MnO₅	Pentacarbonyl(trifluoromethyl)manganese	II/15(**3**,791)
C₆F₄FeO₄	Tetracarbonyl(tetrafluoroethylene)iron	II/7(**3**,639)
C₆F₄O₂	Tetrafluoro–p–benzoquinone	II/15(**3**,792)
C₆F₆	Hexafluorobenzene	II/7(**3**,636)
C₆F₆	Perfluorobicyclo[2.2.0]hexa–2,5–diene Perfluoro Dewar benzene	II/7(**3**,637)
C₆F₁₂	Dodecafluorocyclohexane	II/7(**3**,638)
C₆F₁₅N	Tris(pentafluoroethyl)amine Perfluorotriethylamine	II/23(**3**,389)
C₆F₁₈Ge₂O	Bis[tris(trifluoromethyl)germyl] oxide	II/21(**3**,479)
C₆HCl₃N₂O₄	1,3,5–Trichloro–2,4–dinitrobenzene	II/23(**3**,390)
C₆HF₅	Pentafluorobenzene	II/15(**3**,792a)
C₆H₂	1,3,5–Hexatriyne Triacetylene	II/23(**3**,391)

C₆H₂Cl₄   **1,2,4,5–Tetrachlorobenzene**   **D₂ₕ**

ED and liquid–
crystal NMR

r_α^0	Å ᵃ)	θ_α^0	deg ᵃ)
C–C (mean)	1.3966(11)	C(1)–C(2)–C(3)	119.80(15)
δ ᵇ)	0.023(6)	C(2)–C(1)–Cl	120.85(10)
C–Cl	1.7166(12)	C(2)–C(3)–C(4) ᶜ)	120.39(29)
C–H	1.082(5)		
C(1)–C(2) ᶜ)	1.412(4)		
C(2)–C(3) ᶜ)	1.389(2)		

The nozzle temperature was 450 K.

ᵃ) Unidentified, possibly estimated standard errors.
ᵇ) The difference [C(1)–C(2)] – [C(2)–C(3)].
ᶜ) Dependent parameter.

Andersen, D.G., Blake, A.J., Blom, R., Cradock, S., Rankin, D.W.H.: Acta Chem. Scand. **45** (1991) 158.   II/7(**3**,644), II/23(**3**,392)

C₆H₂Cl₄	1,2,3,5–Tetrachlorobenzene	II/23(**3**,393)
C₆H₂F₄	1,2,3,4–Tetrafluorobenzene	II/15(**3**,792b)
C₆H₂F₄	1,2,3,5–Tetrafluorobenzene	II/15(**3**,792c)
C₆H₂F₄	1,2,4,5–Tetrafluorobenzene	II/15(**3**,793)

ED **C₆H₂F₄O₂** **Tetrafluorohydroquinone** C_{2v}, C_{2h}

r_g [Å] [a]	C_{2v}	C_{2h}
C–C(mean)	1.394(3)	1.392(3)
C–F(mean)	1.346(5)	1.349(6)
C–O	1.353(9)	1.352(10)
O–H	0.952(10)	0.955(10)
Δ(C–F) [c]	0.007(22)	0.007(28)

θ [b] [deg] [a]	C_{2v}	C_{2h}
C(6)–C(1)–C(2)	117.0(5)	117.1(6)
C(1)–C(2)–C(3)	121.5(2)	121.5(3)
C(1)–C(6)–C(5)	121.5(2)	121.5(3)
C–O–H	98.2(24)	98.6(25)
C(3)–C(2)–F	122.1(17)	121.7(16)
C(5)–C(6)–F	119.6(9)	119.2(16)
tilt [d]	2.1(12)	2.0(13)

A mixture of two planar conformers with C_{2v} and C_{2h} symmetry is likely. The results in the two columns in the tables were obtained in separate refinements assuming the presence of either C_{2v} or C_{2h} model only.
The nozzle was at 156 °C.

[a]) Estimated total errors.
[b]) Undefined, possibly $θ_a$.
[c]) (C(2)–F) – (C(6)–F).
[d]) Angle between the C–C–C bisector and the C–O bond to make the O–H group closer to the fluorine atom.

Vajda, E., Hargittai, I.: J. Phys. Chem. **96** (1992) 5843. II/23(**3**,394)

C₆H₃BrN₂O₄	1–Bromo–2,6–dinitrobenzene	II/21(**3**,480)
C₆H₃Br₃	1,3,5–Tribromobenzene	II/21(**3**,481)
C₆H₃ClN₂O₄	1–Chloro–2,6–dinitrobenzene	II/15(**3**,794)
C₆H₃Cl₃	1,2,3–Trichlorobenzene	II/23(**3**,395)
C₆H₃Cl₃	1,3,5–Trichlorobenzene	II/15(**3**,795), II/23(**3**,396)
C₆H₃F₃	1,2,3–Trifluorobenzene	II/15(**3**,795a)
C₆H₃F₃	1,2,4–Trifluorobenzene	II/15(**3**,795b)
C₆H₃F₃	1,3,5–Trifluorobenzene	II/7(**3**,645), II/15(**3**,796), II/23(**3**,397)
C₆H₃ISn	Triethynyltin iodide	II/15(**3**,797), II/21(**3**,482)
C₆H₃MnO₅	Pentacarbonylmethylmanganese	II/7(**3**,646)
C₆H₃N	5–Cyano–2,4–pentadiyne	II/15(**3**,798)
C₆H₃N₃O₆	1,3,5–Trinitrobenzene	II/15(**3**,799)
C₆H₃O₅Re	Pentacarbonylmethylrhenium	II/15(**3**,800)
C₆H₄AsBr₃	Dibromo(2–bromophenyl)arsine (o–Bromophenyl)dibromoarsine	II/21(**3**,483)
C₆H₄AsBr₃	Dibromo(4–bromophenyl)arsine	II/23(**3**,398)
C₆H₄AsCl₃	Dichloro(2–chlorophenyl)arsine (o–Chlorophenyl)dichloroarsine	II/21(**3**,484)

C$_6$H$_4$BrNO$_2$	1-Bromo-2-nitrobenzene o-Bromonitrobenzene	II/21(**3**,485)
C$_6$H$_4$BrNO$_2$	1-Bromo-3-nitrobenzene m-Bromonitrobenzene	II/21(**3**,486)
C$_6$H$_4$BrNO$_2$	1-Bromo-4-nitrobenzene p-Bromonitrobenzene	II/15(**3**,801)
C$_6$H$_4$BrNO$_2$Se	o-Nitrobenzeneselenenyl bromide	II/15(**3**,802)
C$_6$H$_4$Br$_2$	1,6-Dibromo-1,5-hexadiyne	II/15(**3**,803)
C$_6$H$_4$Br$_2$	o-Dibromobenzene	II/7(**3**,647)
C$_6$H$_4$Br$_2$	p-Dibromobenzene	II/21(**3**,487)
C$_6$H$_4$ClF	1-Chloro-3-fluorobenzene	II/7(**3**,648)
C$_6$H$_4$ClNO$_2$	o-Chloronitrobenzene	II/21(**3**,488)
C$_6$H$_4$ClNO$_2$	m-Chloronitrobenzene	II/15(**3**,804)
C$_6$H$_4$ClNO$_2$	p-Chloronitrobenzene	II/15(**3**,805)
C$_6$H$_4$ClNO$_2$S	o-Nitrobenzenesulfenyl chloride	II/15(**3**,806)
C$_6$H$_4$ClO$_2$P	2-Chloro-1,3,2-benzodioxaphosphole	II/15(**3**,807)
C$_6$H$_4$ClO$_3$P	2-Chloro-1,3,2-benzodioxaphosphole 2-oxide	II/7(**3**,649)
C$_6$H$_4$Cl$_2$	1,2-Dichlorobenzene o-Dichlorobenzene	II/21(**3**,489)

C$_6$H$_4$Cl$_2$ **1,3-Dichlorobenzene** C$_{2v}$

ED, MW and liquid-crystal NMR   m-Dichlorobenzene

r_α^0	Å [a]	θ_α^0	deg [a]
C(1)–C(2)	1.388(4)	C(1)–C(2)–C(3)	118.1(4)
C(3)–C(4)	1.392(3)	C(2)–C(3)–C(4)	122.3(2)
C(4)–C(5)	1.404(3)	C(3)–C(4)–C(5) [b]	118.1(2)
C–Cl	1.7355(14)	C(4)–C(5)–C(6) [b]	121.1(4)
C(2)–H	1.091(6)	C(2)–C(3)–Cl	118.9(3)
C(4)–H	1.085(2)	C(3)–C(4)–H	121.2(2)
C(5)–H	1.091(7)		

Temperature not stated, probably room temperature.

[a]) Uncertainties unidentified, possibly estimated standard errors.
[b]) Dependent parameter.

Anderson, D.G., Cradock, S., Liescheski, P.B., Rankin, D.W.H.:
  J. Mol. Struct. **216** (1990) 181.

MW

r_0	Å	θ_0	deg
C(2)–H	1.108(42)	C(1)–C(2)–C(3)	118.9(44)
C(4)–H	1.091(38)	C(2)–C(3)–C(4)	120.9(41)
C(5)–H	1.083(57)	C(2)–C(3)–Cl	119.0(29)
C(3)–Cl	1.727(35)	C(3)–C(4)–H	118.4(44)
C(2)–C(3)	1.395(47)	C(4)–C(3)–Cl	120.1 [a]
C(3)–C(4)	1.394(51)	C(3)–C(4)–C(5)	119.2(36)
C(4)–C(5)	1.389 [a]	C(5)–C(4)–H	122.4 [a]
		C(4)–C(5)–C(6)	121.0 [a]
		C(1)–C(2)–H	120.5 [a]
		C(4)–C(5)–H	119.5 [a]

(*continued*)

$C_6H_4Cl_2$ (continued)

a) Derived parameter.

Onda, M., Atsuki, M., Yamaguchi, J., Suga, K., Yamaguchi, I.: J. Mol. Struct. **295** (1993) 101.
II/21(**3**,490), II/23(**3**,399)

$C_6H_4Cl_2$	p–Dichlorobenzene	II/15(**3**,808), II/23(**3**,400)
$C_6H_4F_2$	1,2–Difluorobenzene	II/7(**3**,650), II/15(**3**,808a), II/21(**3**,493)
	o–Difluorobenzene	
$C_6H_4F_2$	1,3–Difluorobenzene	II/7(**3**,651), II/15(**3**,808b), II/21(**3**,494)
	m–Difluorobenzene	
$C_6H_4F_2$	p–Difluorobenzene	II/15(**3**,809)
$C_6H_4F_2O$	2,6–Difluorophenol	II/23(**3**,401)
$C_6H_4F_4O_2P_2$	m–Bis(difluorophosphinooxy)benzene	II/21(**3**,495)
$C_6H_4F_4O_2P_2$	p–Bis(difluorophosphinooxy)benzene	II/21(**3**,496)
$C_6H_4FeO_4$	Tetracarbonyl(η^2–ethylene)iron	II/7(**3**,652)
$C_6H_4INO_2$	1–Iodo–2–nitrobenzene	II/23(**3**,402)
$C_6H_4INO_2$	1–Iodo–3–nitrobenzene	II/23(**3**,403)
$C_6H_4INO_2$	1–Iodo–4–nitrobenzene	II/23(**3**,404)
C_6H_4N	Cyanocyclopentadienyl radical	II/23(**3**,405)
$C_6H_4N_2O_4$	o–Dinitrobenzene	II/15(**3**,810)
$C_6H_4N_2O_4$	m–Dinitrobenzene	II/15(**3**,811)
$C_6H_4N_2O_4$	p–Dinitrobenzene	II/15(**3**,812)
$C_6H_4O_2$	p–Benzoquinone	II/7(**3**,653)
$C_6H_4O_3S$	o–Phenylene sulfite	II/15(**3**,813)
C_6H_5ArF	Argon – fluorobenzene (1/1)	II/23(**3**,406)
$C_6H_5AsBr_2$	Dibromophenylarsine	II/23(**3**,407)
$C_6H_5BCl_2$	Dichlorophenylborane	II/15(**3**,814)
$C_6H_5BF_2$	Difluorophenylborane	II/15(**3**,815)
C_6H_5Br	1–Bromo–1,5–hexadiyne	II/15(**3**,816)
C_6H_5Br	Bromobenzene	II/7(**3**,654), II/21(**3**,497)
C_6H_5BrHg	Phenylmercury bromide	II/7(**3**,655)
C_6H_5BrO	4–Bromophenol	II/21(**3**,498)
	p–Bromophenol	
C_6H_5BrSe	Benzeneselenenyl bromide	II/15(**3**,817)
C_6H_5Cl	Chlorobenzene	
	II/7(**3**,656), II/15(**3**,818), II/21(**3**,499), II/23(**3**,408)	
C_6H_5ClO	o–Chlorophenol	II/15(**3**,819)
C_6H_5ClO	4–Chlorophenol	II/21(**3**,500)
	p–Chlorophenol	
$C_6H_5ClO_2S$	Benzenesulfonyl chloride	II/15(**3**,820)
C_6H_4ClS	Benzenesulfenyl chloride	II/15(**3**,821)
$C_6H_5Cl_2OP$	Phenylphosphonic dichloride	II/7(**3**,657)
$C_6H_5Cl_2P$	Phenyldichlorophosphine	II/7(**3**,658)
$C_6H_5Cl_3Si$	Phenyltrichlorosilane	II/7(**3**,659)
$C_6H_5CoO_3$	(η^3–Allyl)tricarbonylcobalt	II/7(**3**,660)
C_6H_5F	Fluorobenzene	II/7(**3**,661), II/15(**3**,822)

C$_6$H$_5$F (C$_6$H$_4$DF)		Fluorobenzene–d_1	II/21(**3**,500A)
C$_6$H$_5$FO		2–Fluorophenol o–Fluorophenol	II/21(**3**,501), II/23(**3**,409)
C$_6$H$_5$FO		4–Fluorophenol p–Fluorophenol	II/21(**3**,502)
C$_6$H$_5$F$_2$N		2,6–Difluoroaniline 2,6–Difluorobenzenamine	II/23(**3**,410)
C$_6$H$_5$F$_2$P		Difluorophenylphosphine	II/15(**3**,823)
C$_6$H$_5$F$_3$Si		Trifluorophenylsilane	II/15(**3**,824)
C$_6$H$_5$F$_4$P		Tetrafluorophenylphosphorane	II/15(**3**,825)
C$_6$H$_5$I		Iodobenzene	II/23(**3**,411)
C$_6$H$_5$N		1–Cyano–1,3–cyclopentadiene	II/21(**3**,503)
C$_6$H$_5$NO		Nitrosobenzene	II/7(**3**,662)
C$_6$H$_5$NO		2–Pyridinecarbaldehyde	II/15(**3**,826)
C$_6$H$_5$NO		3–Pyridinecarbaldehyde Nicotinaldehyde	II/15(**3**,827)
C$_6$H$_5$NO		4–Pyridinecarbaldehyde Isonicotinaldehyde	II/15(**3**,828)
C$_6$H$_5$NO$_2$		Nitrobenzene	II/15(**3**,829), II/23(**3**,412)
C$_6$H$_5$NO$_4$		2–Nitroresorcinol	II/23(**3**,413)

IR **C$_6$H$_6$**   **Acetylene trimer**   **D$_{3h}$ or C$_{3h}$**
(weakly bound complex)   (effective symmetry class)

r_0	Å
R_{cm}	4.354

The distance between the centers of mass of the monomer mole-cules is designated R_{cm}. The structure of the monomer units was assumed to be unchanged in the complex. The assumption was also made that $2C_0 = B_0$. The spectrum observed is that of a planar oblate symmetric rotor having either C$_{3h}$ or D$_{3h}$ symmetry, but the exact angular configuration of the monomer groups can not be determined.

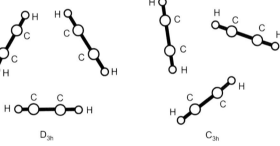

Prichard, D., Muenter, J.S., Howard, B.J.: Chem. Phys. Lett. **135** (1987) 9.   II/21(**3**,504)

C$_6$H$_6$		1,2,4,5–Hexatetraene	II/7(**3**,663)
C$_6$H$_6$		1,2–Hexadien–5–yne Propargylallene	II/15(**3**,830)
C$_6$H$_6$		1,5–Hexadien–3–yne Divinylacetylene	II/15(**3**,831)
C$_6$H$_6$		1,5–Hexadiyne Bipropargyl	II/15(**3**,832)
C$_6$H$_6$		Trimethylenecyclopropane	II/7(**3**,664)
C$_6$H$_6$		3,4–Dimethylenecyclobutene	II/7(**3**,665), II/15(**3**,833)

| C$_6$H$_6$ | Fulvene | II/7(3,666) |

C$_6$H$_6$ **Benzene**   **D$_{6h}$** (CH)$_6$

IR

r_e	Å
C–H	1.0862(15)
C–C	1.3902(2)

Derived from five isotopomers.

Pliva, J., Johns, J.W.C., Goodman, L.: J. Mol. Spectrosc. **148** (1991) 427.

IR

r_m	Å [a]
C–H	1.0859(10)
C–C	1.3894(10)

[a] Uncertainties were not given in original paper.

Pliva, J., Johns, J.W.C., Goodman, L.: J. Mol. Spectrosc. **140** (1990) 214.

UV

State	$\tilde{A}\ ^1B_{2u}$	$\tilde{a}\ ^3B_{1u}$
Symmetry	D$_{6h}$	?
Energy [eV]	4.722	3.660
Reference	[1][3][4]	[2]
r_v [Å] C–C	1.432(3) [a]	1.426(5) [b]
	1.4331(9) [c]	
C–H	1.074(9) [a]	1.07(1) [b]
	1.069(1) [c]	

[a] From ground state values obtained from the rotational Raman spectrum and changes on excitation estimated from analysis of rotational fine structure or Franck-Condon factors or Badgers rule. The uncertainties quoted cover a range of value obtained by these methods, and the fact that they apply to vibrational levels other than the zero-point level.

[b] From Franck-Condon analysis of the phosphorescence spectrum. The carbon skeleton in this state at equilibrium may however be distorted from a regular hexagon through a small pseudo- Jahn-Teller effect.

[c] v: $v = 1$ for levels v_{14}, v_{17} and v_{18} (Wilson notation), rotational analysis of two-photon laser absorption spectra of C$_6$H$_6$ and C$_6$D$_6$ [4]. There are a number of other rotational analyses of high precision available, but so far only for C$_6$H$_6$. In one of these an inertial defect ($<I_c>$ = $\{<I_a> + <I_b>\}$) has been measured [5], but the 'slight non-planarity' of the molecule deduced from it refers at best to root-mean-square amplitudes of out-of-plane vibrations, not to equilibrium configurations. The planarity of the latter is deduced from vibrational analysis [1].

[1] Callomon, J.H., Dunn, T.M., Mills, I.M.: Phil. Trans. Roy. Soc. (London) Ser. A **259** (1966) 499.
[2] Burland, D.M., Robinson, G.W.: J. Chem. Phys. **51** (1969) 4548.
[3] Parmenter, C.S., Tang, K., Ware, W.R.: Chem. Phys. **17** (1976) 359.
[4] Lombardi, J.R., Hänsch, T.W., Wallenstein, R., Friedrich, D.M.: J. Chem. Phys. **65** (1976) 2357.
[5] Riedle, E., Neusser, H.J., Schlag, E.W.: J. Chem. Phys. **75** (1981) 4231.

II/7(3,667), II/15(3,834), II/21(3,505), II/23(3,414)

| C$_6$H$_6$ (C$_6$H$_4$D$_2$) | 1,2–Benzene–d_2, 1,3–Benzene–d_2 | II/21(3,505A) |

C₆H₆	Bicyclo[2.2.0]hexa–2,5–diene Dewar benzene	II/7(**3**,668), II/15(**3**,835)
C₆H₆	Tricyclo[3.1.0.0²,⁶]hex–3–ene Benzvalene	II/7(**3**,669), II/15(**3**,836)
C₆H₆Ar	Argon – benzene (1/1)	II/23(**3**,415)
C₆H₆Ar₂	Argon – benzene (2/1)	II/23(**3**,416)
C₆H₆ClN	o–Chloroaniline 2-Chloroaniline	II/15(**3**,836a)
C₆H₆ClN	m–Chloroaniline 3-Chloroaniline	II/15(**3**,836b)
C₆H₆ClN	p–Chloroaniline 4-Chloroaniline	II/15(**3**,836c)
C₆H₆FN	o–Fluoroaniline 2-Fluoroaniline	II/21(**3**,506), II/23(**3**,417)
C₆H₆FN	p–Fluoroaniline	II/7(**3**,670)
C₆H₆He	Benzene – helium (1/1)	II/23(**3**,418)
C₆H₆He₂	Benzene – helium (1/2)	II/23(**3**,419)
C₆H₆Kr	Benzene – krypton (1/1)	II/23(**3**,420)
C₆H₆N₂	1,1–Dicyanocyclobutane	II/21(**3**,507)
C₆H₆N₂	Benzene – dinitrogen (1/1)	II/23(**3**,421)
C₆H₆N₂O	Nicotinamide	II/23(**3**,422)
C₆H₆N₂O₂	p–Nitroaniline	II/15(**3**,837)
C₆H₆Ne	Benzene – neon (1/1)	II/23(**3**,423)

C₆H₆O   **Phenol** C_s

LIF

State	Ground	Excited
Energy [eV]	0.00	4.507
r_0 [Å] C–C	1.3856(3) [a]	1.4313(3)
C–O	1.3543(21)	1.2405(21)

All C–C bond lengths were assumed to be identical, and all C–C–C and C–C–H bond angles were set equal to 120°. The C–O bond was assumed to bisect the external C–C–C angle and the C–O–H bond angle and O–H bond length were set equal to 108.9° and 0.956 Å, respectively. The C–H bond lengths in the ground and excited states were assigned values of 1.079 and 1.084 Å, respectively, as in benzene.

[a]) Error limits based on the 1σ error limits of the rotational constants.

Martinez III, S.J., Alfano, J.C., Levy, D.H.: J. Mol. Spectrosc. **152** (1992) 80.

ED

r_g	Å [a])	θ_a	deg [a])
C–C (average)	1.399(3)	C(2)–C(1)–C(6)	121.6(2)
C(1)–O	1.381(4)	C(1)–C(2)–C(3)	118.8(2)
C–H (average) [b])	1.086(3)	C(2)–C(3)–C(4)	120.6(5)
O–H [b])	0.958(3)	C(3)–C(4)–C(5)	119.7(8)
		C(2)–C(1)–O	121.2(12)
		C(1)–O–H	106.4(37)

The nozzle temperature was ≈ 353 K.

(*continued*)

C_6H_6O (*continued*)

[a]) Estimated limits of error.
[b]) (C–H) – (O–H) was constrained from the MO calculations at the HF/6–31G* level.

Portalone, G., Schultz, Gy., Domenicano, A., Hargittai, I.: Chem. Phys. Lett. **197** (1992) 482.
II/7(**3**,671), II/15(**3**,838), II/23(**3**,424)

$C_6H_6O_2$	Pyrocatechol	II/21(**3**,508), II/23(**3**,425)
	1,2–Benzenediol, (Catechol)	
$C_6H_6O_2S$	Benzene – sulfur dioxide (1/1)	II/21(**3**,509), II/23(**3**,426)
$C_6H_6O_3$	1,3,5–Benzenetriol, Phloroglucinol	II/23(**3**,427)
$C_6H_6S_2$	*p*–Benzenedithiol	II/21(**3**,510)
C_6H_6Xe	Benzene – xenon (1/1)	II/23(**3**,428)
C_6H_7	Methylcyclopentadienyl radical	II/23(**3**,429)
C_6H_7Cl	Benzene – hydrogen chloride (1/1)	II/15(**3**,838a)
C_6H_7ClSi	Phenylchlorosilane	II/7(**3**,672)
C_6H_7F	Benzene – hydrogen fluoride (1/1)	II/15(**3**,838b)
$C_6H_7He_2$	Helium – methylcyclopentadienyl (2/1)	II/23(**3**,430)
C_6H_7N	4–Cyanocyclopentene	II/15(**3**,839)
C_6H_7N	Aniline	II/7(**3**,673), II/21(**3**,511)
	Benzenamine	
C_6H_7NO	4–Methylpyridine *N*–oxide	II/15(**3**,840)
C_6H_7P	Phenylphosphine	II/15(**3**,841), II/21(**3**,512)
C_6H_8	3–Methylene–1,4–pentadiene	II/21(**3**,513)
C_6H_8	1,3,5–Hexatriene	II/7(**3**,674)
C_6H_8	Ethynylcyclobutane	II/21(**3**,514)
C_6H_8	1,2–Bis(methylene)cyclobutane	II/15(**3**,842), II/21(**3**,515)
C_6H_8	1,3–Cyclohexadiene	II/7(**3**,675)
C_6H_8	1,4–Cyclohexadiene	II/7(**3**,676)
C_6H_8	Bicyclo[2.1.1]hex–2–ene	II/7(**3**,677), II/15(**3**,843)
C_6H_8	Bicyclo[3.1.0]hex–2–ene	II/15(**3**,844)
C_6H_8	1,1′–Bicyclopropylidene	II/15(**3**,845)
C_6H_8	Tricyclo[2.2.0.0^{2,6}]hexane	II/15(**3**,846)
C_6H_8	Tricyclo[3.1.0.0^{2,4}]hexane	II/15(**3**,847)
C_6H_8Be	(*μ*–Cyclopentadienyl)methylberyllium	II/7(**3**,678)
$C_6H_8Br_2$	*anti–cis,cis*–2,2′–Dibromo–1,1′–bicyclopropyl	II/7(**3**,679)
$C_6H_8Br_2$	*anti–trans,trans*–2,2′–Dibromo–1,1′–bicyclopropyl	II/7(**3**,680)
$C_6H_9NO_2$	1–Nitrocyclohexene	II/21(**3**,516)
$C_6H_8N_2$	*p*–Phenylenediamine, 1,4–Benzenediamine	II/21(**3**,517)
C_6H_8O	Bicyclo[3.1.0]hexan–3–one	II/15(**3**,848)
C_6H_8OSi	Phenyl silyl ether	II/7(**3**,681)
C_6H_8Si	Phenylsilane	II/7(**3**,682), II/15(**3**,849)
C_6H_8Zn	(Cyclopentadienyl)methylzinc	II/15(**3**,850)
C_6H_9B	Trivinylboron	II/7(**3**,683)
C_6H_9Cl	1–(*t*–Butyl)–2–chloroacetylene	II/7(**3**,684)
	1–Chloro–3,3–dimethyl–1–butyne	
C_6H_9Cl	*cis*– and *trans*–3–Chlorobicyclo[3.1.0]hexane	II/15(**3**,851)

C$_6$H$_9$Cl	1–Chlorocyclohexene	II/15(**3**,852)
C$_6$H$_9$Cl	3–Chlorocyclohexene	II/15(**3**,853)
C$_6$H$_9$Cl	4–Chlorocyclohexene	II/15(**3**,854)
C$_6$H$_9$ClO	2–Chlorocyclohexanone	II/15(**3**,855)
C$_6$H$_9$Ga	Trivinylgallium	II/15(**3**,856)
C$_6$H$_9$N	Cyanocyclopentane	II/15(**3**,857)
C$_6$H$_9$O$_3$P	Trivinyl phosphite	II/7(**3**,685)
C$_6$H$_{10}$	*t*–Butylacetylene	II/7(**3**,686)
C$_6$H$_{10}$	2,3–Dimethyl–1,3–butadiene	II/7(**3**,687)
C$_6$H$_{10}$	2–Cyclopropyl–1–propene	II/15(**3**,858)
C$_6$H$_{10}$	Bicyclopropane	II/7(**3**,688)
C$_6$H$_{10}$	Cyclohexene	II/7(**3**,689)
C$_6$H$_{10}$	2–Methylbicyclo[2.1.0]pentane	II/15(**3**,859)
C$_6$H$_{10}$	Bicyclo[2.2.0]hexane	II/7(**3**,690)
C$_6$H$_{10}$	Bicyclo[2.1.1]hexane	II/7(**3**,691)
C$_6$H$_{10}$	Bicyclo[3.1.0]hexane	II/15(**3**,860)
C$_6$H$_{10}$BrCl	*cis*–1–Bromo–4–chlorocyclohexane	II/7(**3**,692)
C$_6$H$_{10}$Br$_2$	*cis*–1,3–Dibromo–1,3–dimethylcyclobutane	II/21(**3**,518)
C$_6$H$_{10}$Br$_2$	*trans*–1,3–Dibromo–1,3–dimethylcyclobutane	II/21(**3**,519)
C$_6$H$_{10}$F$_2$	1,1–Difluorocyclohexane	II/7(**3**,693)
C$_6$H$_{10}$N$_2$	2,3–Diazabicyclo[2.2.2]oct–2–ene	II/15(**3**,861)
C$_6$H$_{10}$O	Cyclohexanone	II/15(**3**,862)
C$_6$H$_{10}$O	Cyclohexene oxide 1,2–Epoxycyclohexane	II/7(**3**,694), II/15(**3**,863)
C$_6$H$_{10}$O	7–Oxabicyclo[2.2.1]heptane	II/7(**3**,695), II/15(**3**,864)
C$_6$H$_{10}$S	7–Thiabicyclo[2.2.1]heptane	II/7(**3**,696)
C$_6$H$_{11}$Br	Bromocyclohexane	II/21(**3**,520)
C$_6$H$_{11}$Cl	Chlorocyclohexane Cyclohexyl chloride	II/7(**3**,697), II/15(**3**,864a), II/21(**3**,521)
C$_6$H$_{11}$F	Fluorocyclohexane Cyclohexyl fluoride	II/7(**3**,698)
C$_6$H$_{11}$I	Iodocyclohexane Cyclohexyl iodide	II/21(**3**,522)
C$_6$H$_{12}$	3–Methyl–2–pentene	II/7(**3**,699)
C$_6$H$_{12}$	2,3–Dimethyl–2–butene	II/7(**3**,700)

ED **C$_6$H$_{12}$** **Cyclohexane** **D$_{3d}$** (chair form)

r_a	Å [a])		θ_a	deg [a])
C–C	1.534(2)		C–C–C	111.3(2)
C–H	1.110(4)		H–C–H	105.3(23)
			δ [b])	55.1(7)

The temperature of the measurement was not stated, probably room temperature.

[a]) Three times the estimated standard errors.
[b]) Torsional angle about the C–C bond.

(*continued*)

C₆H₁₂ (*continued*)

Ewbank, J. D., Kirsch, G., Schäfer, L.: J. Mol. Struct. **31** (1976) 39.

MW

r_s	Å		θ_s	deg
C–H(ax)	1.1013(9)		H–C–H	106.65(27)
C–H(eq)	1.0933(15)		C–C–C	111.28(30)
C–C	1.5300(32)		C–C–H(ax)	108.83(32)
			C–C–H(eq)	110.55(29)
			C–C–C–C [a]	55.26(82)

The rotational spectra of cyclohexane–1,1–d_2, cyclohexane–^{13}C–1,1–d_1, cyclohexane–d_1 (equatorial and axial), and cyclohexane–1,1,2,2,3,3–d_6 have been measured.

[a]) Dihedral angle.

Dommen, J., Brupbacher, T., Grasi, G., Bauder, A.: J. Am. Chem. Soc. **112** (1990) 953.

II/7(**3**,701), II/15(**3**,865), II/21(**3**,523)

C₆H₁₂	*trans*– and *cis*–3–Hexene	II/15(**3**,866)
C₆H₁₂ClO₂P	4,4,5,5–Tetramethyl–2–chloro–1,3,2–dioxaphospholane	
		II/7(**3**,702)
C₆H₁₂CuN₂S₄	Bis(dimethyldithiocarbamato)copper(II)	II/21(**3**,524)
C₆H₁₂FNO₃Si	1–Fluoro–2,8,9–trioxa–5–aza–1–silabicyclo[3.3.3] undecane	
	1–Fluorosilatrane	II/23(**3**,431)
C₆H₁₂N₂	1,4–Diazabicyclo[2.2.2]octane	II/7(**3**,703)
	Triethylene diamine	
C₆H₁₂N₂	1,5–Diazabicyclo[3.3.0]octane	II/15(**3**,867)
C₆H₁₂N₂S₄Zn	Bis(dimethyldithiocarbamato)zinc(II)	II/21(**3**,525)
C₆H₁₂N₃P	Tris(1–aziridinyl)phosphine	II/7(**3**,704)
C₆H₁₂O	Oxepane	II/15(**3**,868)
C₆H₁₂O₃	2,4,6–Trimethyl–1,3,5–trioxane	II/7(**3**,705)
	Paraldehyde	
C₆H₁₂O₄	Propionic acid dimer	II/7(**3**,408)

C₆H₁₂S₃   **1,4,7–Trithiacyclononane**   C₁

ED and molecular mechanics (MM2) calculations

r_a	Å [a])		θ [b])	deg [a])
S–C	1.820(1)		C–S–C	103.8(7)
C–C	1.533(4)		S–C(2)–C	115.0(5)
C–H	1.109(4)		C–C(3)–S	115.7(5) [c])
			H–C–H	110.0 [d])
			τ[S–C(2)–C(3)–S]	59.5(18)
			τ[C–C(3)–S(4)–C]	74.6(14)
			τ[C–S(4)–C(5)–C]	–102.9(14)
			τ[S–C(5)–C(6)–S]	74.3(10)
			τ[C–C(6)–S(7)–C]	–104.2(13)
			τ[C–S(7)–C(8)–C]	129.6(14)

(*continued*)

Four molecular models with D_3, C_3, C_2 and C_1 symmetry were fitted to the experimental data. The C_1 model gave both the lowest R factor in the ED analysis and the lowest strain energy in the MM2 calculation.
The nozzle temperature was 493 K.

[a]) Uncertainties were unidentified, possibly estimated standard errors.
[b]) Unidentified, possibly θ_a.
[c]) Constrained to be 1.0(8)° larger than S–C(2)–C.
[d]) Assumed.

Blom, R., Rankin, D.W.H., Robertson, H.E., Schröder, M., Taylor, A.: J. Chem. Soc., Perkin Trans. II 1991, 773.

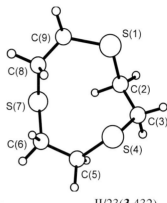

II/23(**3**,432)

$C_6H_{12}Si$	4–Silaspiro[3.3]heptane	II/7(**3**,706), II/15(**3**,869)
$C_6H_{12}Si$	1–Silabicyclo[2.2.1]heptane	II/15(**3**,870)
C_6H_{14}	Hexane	II/7(**3**,707)
C_6H_{14}	1,1,2,2–Tetramethylethane	II/15(**3**,871)
$C_6H_{14}N_2$	1,4–Dimethylpiperazine	II/7(**3**,708)
$C_6H_{14}O$	Dipropyl ether	II/15(**3**,872)
$C_6H_{14}O$	Diisopropyl ether	II/21(**3**,526)
$C_6H_{14}S$	Diisopropyl sulfide	II/21(**3**,527), II/23(**3**,433)
$C_6H_{14}Si$	Cyclohexylsilane	II/23(**3**,434)
$C_6H_{14}Si$	Cyclopropyltrimethylsilane	II/15(**3**,873)
$C_6H_{14}Zn$	Dipropylzinc	II/15(**3**,874)
$C_6H_{15}AlO$	Trimethylaluminum – oxetane (1/1)	II/15(**3**,875)
$C_6H_{15}ClSi$	t–Butyldimethylsilyl chloride	II/21(**3**,528)
$C_6H_{15}N$	Diisopropylamine	II/21(**3**,529)
$C_6H_{15}N$	N–Ethyl–t–butylamine N–Ethyl–1,1–dimethylethanamine	II/23(**3**,435)
$C_6H_{15}N$	Triethylamine	II/23(**3**,436)
$C_6H_{15}O_3P$	Triethyl phosphite	II/7(**3**,709)

$C_6H_{16}Cl_2OSi_2Sn$ 4,4–Dichloro–2,2,6,6–tetramethyl–1–oxa–2,6–stanna–
disila–4–cyclohexane C_s assumed

ED

r_g	Å [a])	θ_α	deg [a])
Si–O	1.693(12)	C–Sn–C	105.0(20)
Si–C(1)	1.897(11)	Cl–Sn–Cl	103.0 [b])
Si–C(3)	1.883(3)	Si–O–Si	135.2(19)
Sn–C	2.159(17)	C(1)–Si–C(2)	112.0 [b])
Sn–Cl	2.326(3)	Si–C–H	113.0 [b])
C–H	1.102(12)	H–C–H	109.0 [b])
		Cl–Sn–C	116.0(11)
		Cl–Sn–C	108.2(8)
		C(3)–Si–O	109.7(14)
		C(3)–Si–C(1)	97.7(7)
		C(3)–Si–C(2)	104.9(5)

(*Table continued*)

C₆H₁₆Cl₂OSi₂Sn (*Table continued*)

θ_α	deg [a]
C(1)–Si–O	106.5(13)
C(2)–Si–O	123.0(17)
Sn–C–Si	107.8(8)
φ(C(3)–Si) [c]	47.4(25)
φ(O–Si) [c]	49.4(62)
φ(C(3)–Sn) [c]	56.3(24)
α(O) [d]	–45.8(57)
α(Sn) [e]	52.6(20)
τ(Cl₂) [f]	7.7(14)
τ((CH₃)₂) [g]	12.0(20)

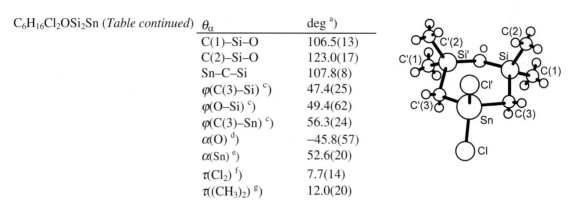

Local C_{3v} symmetry and staggered position of the CH₃ groups were assumed.
The nozzle temperature was 105(10) °C.

[a]) Three times the estimated standard errors.
[b]) Assumed.
[c]) Torsional (dihedral) angle.
[d]) Angle between the planes Si′OSi and Si′SiC(3).
[e]) Angle between the planes C′(3)SnC(3) and C′(3)C(3)Si.
[f]) Tilt angle of the SnCl₂ group in the molecular symmetry plane; 0° if the SnCl₂C₂ fragment has local C_{2v} symmetry.
[g]) Tilt angle of the Si(CH₃)₂ group between the bisector of the C(1)–Si–C(2) angle and the C(3)SiO plane; the methyl groups repel each other.

Belyakov, A.V., Baskakov, A.D., Popik, M.V., Vilkov, L.V., Shiryayev, V.I., Stepina, E.M., Grachev, A.A., Nikitin, V.S.: Metallorg. Khim. **2** (1989) 1311.   II/23(3,437)

C₆H₁₆Si	Triethylsilane	II/15(3,876)
C₆H₁₆Si	*t*–Butyldimethylsilane	II/23(3,438)
C₆H₁₆Si₂	1,1,3,3–Tetramethyl–1,3–disilacyclobutane	II/23(3,439)
C₆H₁₈AlN	Trimethylaluminum – trimethylamine (1/1)	II/7(3,711)
C₆H₁₈AlP	Trimethylaluminum – trimethylphosphine (1/1)	II/15(3,877)
C₆H₁₈Al₂	Trimethylaluminum dimer	II/7(3,710)
C₆H₁₈Al₂S₂	Dimethylaluminum methanethiolate dimer	II/15(3,878)
C₆H₁₈BN	Trimethylamine – trimethylborane (1/1)	II/15(3,879)
C₆H₁₈BN₃	Tris(dimethylamino)borane	II/7(3,712)
C₆H₁₈ClN₃Si	Chlorotris(dimethylamino)silane	II/7(3,713)

ED **C₆H₁₈Cl₂P₂Pd**   **Dichlorobis(trimethylphosphine)palladium** C_1
PdCl₂[P(CH₃)₃]₂

r_g	Å [a]	θ_a	deg [a]
Pd–P	2.264(13)	Cl(1)–Pd–Cl(2)	88.9(40)
Pd–Cl	2.357(29)	P(1)–Pd–P(2)	90.1(46)
l [b])	2.310(8)	P(1)–Pd–Cl(1) [d]	91.3(46)
P–C	1.822(4)	Pd–P–C	116.4(7)
C–H	1.126(5)	C–P–C [d]	101.7(8)
Δ [c])	–0.091(24)	P–C–H	103.5(28)
		ϕ [e]	13.9(20)
		C(1)–P(2)–Pd–P(1)	0 [f]
		C(2)–P(1)–Pd–P(2)	180 [f]

(*continued*)

The Pd bond configuration and the P and C configurations were assumed to have C$_2$ and C$_{3v}$ symmetry, respectively. The nozzle temperature was 202 °C.

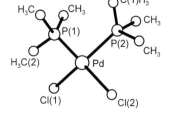

^{a}) Estimated total error.
^{b}) Mean value of Pd–Cl and Pd–P bond lengths.
^{c}) Difference between the Pd–P and Pd–Cl bond lengths.
^{d}) Dependent parameter.
^{e}) Dihedral angle between the PPdP and ClPdCl planes.
^{f}) Assumed.

Schultz, G., Subbotina, N.Yu., Jensen, C.M., Golen, J.A., Hargittai, I.: Inorg. Chim. Acta **191** (1992) 85.
II/23(**3**,440)

C$_6$H$_{18}$F$_2$N$_3$P	Difluorotris(dimethylamino)phosphorane	II/15(**3**,880)
C$_6$H$_{18}$GaN	Trimethylgallium – trimethylamine (1/1)	II/15(**3**,881), II/21(**3**,530)
C$_6$H$_{18}$GaP	Trimethylgallium – trimethylphosphine (1/1)	II/15(**3**,882)
C$_6$H$_{18}$Ge$_2$O	Hexamethyldigermoxane	II/7(**3**,714)
	Bis(trimethylgermyl)ether	
C$_6$H$_{18}$NPSi	Trimethyl(trimethylsilylimino)phosphorane	II/15(**3**,883)
C$_6$H$_{18}$N$_2$SSi$_2$	Bis(trimethylsilyl)sulfur diimide	II/21(**3**,531)
C$_6$H$_{18}$N$_2$Si$_2$	Hexamethylcyclodisilazane	II/21(**3**,532)
C$_6$H$_{18}$N$_3$P	Tris(dimethylamino)phosphine	II/7(**3**,715)
C$_6$H$_{18}$OSi$_2$	Hexamethyldisiloxane	II/15(**3**,884)
C$_6$H$_{18}$OSn$_2$	Hexamethyldistannanoxane	II/7(**3**,717)
	Bis(trimethylstannyl) ether	
C$_6$H$_{18}$O$_2$Si$_2$	Bis(trimethylsilyl) peroxide	II/15(**3**,885)
C$_6$H$_{18}$O$_3$Si$_3$	Hexamethylcyclotrisiloxane	II/7(**3**,716)
C$_6$H$_{18}$O$_6$W	Tungsten hexamethoxide	II/21(**3**,533)
C$_6$H$_{18}$SSi$_2$	Bis(trimethylsilyl) sulfide	II/23(**3**,441)
C$_6$H$_{18}$Si$_2$	Hexamethyldisilane	II/7(**3**,718)
C$_6$H$_{18}$Sn$_2$	Hexamethyldistannane	II/23(**3**,442)

ED **C$_6$H$_{18}$W**   **Hexamethyltungsten**   **D$_{3h}$**
W(CH$_3$)$_6$

r_a	Å ^{a})	θ_a	deg ^{a})
W–C	2.146(3)	α ^{b})	52.0(3)
C–H	1.119(4)	C(1)–W–C(2)	86.0(2)
W···H	2.682(8)	C(1)–W–C(4)	76.1(6)
		W–C–H	106.1(7)

The nozzle was at room temperature.

^{a}) Estimated standard errors.
^{b}) The angle between the W–C(1) bond and the 3-fold symmetry axis.

Haaland, A., Hammel, A., Rypdal, K., Volden, H.V.: J. Am. Chem. Soc. **112** (1990) 4547.
II/23(**3**,443)

C$_6$H$_{19}$NSi$_2$	Hexamethyldisilazane	II/7(**3**,719), II/15(**3**,886)

	$C_6H_{19}O_2PSi_2$	Bis(trimethylsilyl) phosphinate	II/21(**3**,534)
	$C_6H_{21}AlN_2$	Bis(trimethylamine)trihydroaluminum	II/15(**3**,887)
	$C_6H_{21}N_3Si_3$	2,2,4,4,6,6–Hexamethylcyclotrisilazane	II/15(**3**,888)
	$C_6H_{24}Al_3N_3$	cyclo–Tris(μ–dimethylamido)–tris(dihydroaluminum)	II/15(**3**,889)
		cyclo–2,4,6–Hexahydridotrialumina–1,3,5–hexamethyltriazine	
	C_6MoO_6	Hexacarbonylmolybdenum(0)	II/7(**3**,640)
	C_6N_4	Tetracyanoethene	II/7(**3**,641)
		Tetracyanoethylene	
	C_6O_6V	Hexacarbonylvanadium(0)	II/7(**3**,642)
	C_6O_6W	Hexacarbonyltungsten(0)	II/7(**3**,643)

	C_7	**Heptacarbon**	$D_{\infty h}$
IR			C=C=C=C=C=C=C

r_0	Å
C=C	1.2736(4) [a]

Assumes all C=C bonds are of identical length, whereas *ab initio* calculations predict three different values, in a range of 0.016 Å.

[a] Twice the estimated standard error.

Heath, J.R., Sheeks, R.A., Cooksy, A.L., Saykally, R.J.: Science **249** (1990) 895. II/23(**3**,444)

	$C_7CrF_3NO_5$	Pentacarbonyl(trifluoromethyl isocyanide)chromium	
			II/21(**3**,535)
	C_7F_8	Perfluoronorbornadiene	II/21(**3**,536)
		Perfluorobicylo[2.2.1]hepta–2,5–diene	
	$C_7F_{12}N_2$	4,4,5,5,6,6,–Hexafluoro–3,7–bis(trifluoromethyl)–5,6–dihydro– 4*H*–1,2–diazepine	II/15(**3**,890), II/21(**3**,537)
		Perfluoro[3,7–dimethyl–1,2–diazacyclohepta–2,7–diene]	
	$C_7F_{12}N_2$	Perfluoro(1,5–dimethyl–6,7–diazabicyclo[3.2.0]hept–6–ene)	
			II/21(**3**,538)
	C_7H_4ClN	4–Chlorobenzonitrile	II/21(**3**,539)
		p–Chlorobenzonitrile	
	C_7H_4FNS	2–Fluorophenyl isothiocyanate	II/21(**3**,540)
	$C_7H_4FeO_3$	Tricarbonyl(η^4–cyclobutadiene)iron(0)	II/7(**3**,720)
	C_7H_4OS	Benzothiet–2–one	II/23(**3**,445)
	C_7H_5ClO	Benzoyl chloride	II/21(**3**,541)
	C_7H_5ClO	*o*–Chlorobenzaldehyde	II/15(**3**,891)
	C_7H_5ClO	*m*–Chlorobenzaldehyde	II/15(**3**,892)
	$C_7H_5CoO_2$	Dicarbonyl(η^5–2,4–cyclopentadienyl)cobalt	
			II/15(**3**,893), II/23(**3**,446)
	C_7H_5FO	2–Fluorobenzaldehyde	II/21(**3**,542)
	C_7H_5FO	3–Fluorobenzaldehyde	II/21(**3**,543)
	$C_7H_5F_3$	Benzotrifluoride	II/15(**3**,894)
		Benzylidyne trifluoride	
	C_7H_5N	Benzonitrile	II/7(**3**,721), II/21(**3**,544)
		Phenyl cyanide	

C₇H₅N	Phenyl isocyanide	II/7(**3**,722)
C₇H₅NO	Phenyl cyanate	II/23(**3**,447)
C₇H₅NO	Phenyl isocyanate	II/15(**3**,895)
C₇H₅NS	Phenyl isothiocyanate	II/15(**3**,896)
C₇H₆	5–Ethenylidene–1,3–cyclopentadiene	II/23(**3**,448)
C₇H₆Be	Cyclopentadienylberyllium acetylide (μ–Cyclopentadienyl)ethynylberyllium	II/7(**3**,723)
C₇H₆FeO₃	Tricarbonyl(trimethylenemethane)iron	II/7(**3**,724)
C₇H₆FeO₃	Tricarbonyl(η–cyclobutene)iron(0) (η^4–1,3–butadiene)tricarbonyliron	II/7(**3**,725), II/23(**3**,449)
C₇H₆N₂	1*H*–Pyrrolo[2,3–*b*]pyridine	II/23(**3**,450)
C₇H₆O	Tropone, 2,4,6–Cycloheptatrien–1–one	II/7(**3**,726)
C₇H₇Br	Benzyl bromide	II/15(**3**,897)
C₇H₇Cl	Benzyl chloride	II/15(**3**,898)
C₇H₇F	Benzyl fluoride	II/15(**3**,899)
C₇H₇N	2–Methyl–1,3–cyclopentadiene–1–carbonitrile	II/21(**3**,545)
C₇H₇N	3–Methyl–1,3–cyclopentadiene–1–carbonitrile	II/21(**3**,546)
C₇H₇N	4–Methyl–1,3–cyclopentadiene–1–carbonitrile	II/21(**3**,547)
C₇H₇NO₂S	1–Nitro–2–(methylthio)benzene Methyl 2–nitrophenyl sulfide	II/21(**3**,548)
C₇H₈	1,3,5–Cycloheptatriene	II/7(**3**,727)
C₇H₈	Toluene	II/7(**3**,728), II/15(**3**,900)
C₇H₈	Spiro[2.4]hepta–4,6–diene	II/7(**3**,729), II/15(**3**,901)
C₇H₈	Bicyclo[2.2.1]hepta–2,5–diene Norbornadiene	II/7(**3**,730), II/21(**3**,549)
C₇H₈	Quadricyclane Tetracyclo[3.2.0.0²,⁷.0⁴,⁶]heptane	II/7(**3**,731), II/21(**3**,550)
C₇H₈O	Benzyl alcohol	II/15(**3**,902)
C₇H₈O	Anisole Methoxybenzene	II/7(**3**,732), II/21(**3**,551)
C₇H₈O₂S	Methyl phenyl sulfone	II/15(**3**,903)
C₇H₈O₂S	Toluene – sulfur dioxide (1/1)	II/23(**3**,451)
C₇H₈S	Thioanisole (Methylthio)benzene	II/15(**3**,904)
C₇H₈Se	Selenoanisole (Methylseleno)benzene	II/15(**3**,905)
C₇H₉Cl	4–Chlorotricyclo[2.2.1]heptane	II/7(**3**,733)
C₇H₁₀	*cis*–2–Methyl–1,3,5–hexatriene	II/7(**3**,734)
C₇H₁₀	*trans*–2–Methyl–1,3,5–hexatriene	II/7(**3**,735)
C₇H₁₀	*trans*–1,2–Divinylcyclopropane	II/21(**3**,552)
C₇H₁₀	4–Methylenecyclohexene	II/21(**3**,553)
C₇H₁₀	Bicyclo[3.2.0]hept–6–ene	II/7(**3**,736)
C₇H₁₀	Bicyclo[4.1.0]hept–2–ene	II/7(**3**,737)
C₇H₁₀	Norbornene	II/15(**3**,906)
C₇H₁₀	1,3–Cycloheptadiene	II/7(**3**,738)
C₇H₁₀	Tricyclo[2.2.1.0²,⁶]heptane Nortricyclene	II/7(**3**,739)

C_7H_{10}	Tricyclo[4.1.0.0^{1,3}]heptane	II/15(**3**,907)
$C_7H_{10}Cl_2$	1,4–Dichlorobicyclo[2.2.1]heptane 1,4–Dichloronorbornane	II/7(**3**,740)
$C_7H_{10}Cl_2$	7,7–Dichlorobicyclo[4.1.0]heptane	II/7(**3**,741)
$C_7H_{10}O_3$	2,8,9–Trioxaadamantane	II/7(**3**,742)
$C_7H_{11}Al$	Dimethyl(cyclopentadienyl)aluminum	II/7(**3**,743)
$C_7H_{11}N$	Cyclohexyl cyanide Cyclohexanecarbonitrile	II/15(**3**,908)
C_7H_{12}	Cycloheptene	II/21(**3**,554)
C_7H_{12}	Norbornane	II/7(**3**,744), II/15(**3**,909)
C_7H_{12}	Bicyclo[3.1.1]heptane	II/7(**3**,745)
C_7H_{12}	Bicyclo[3.2.0]heptane	II/15(**3**,910)
C_7H_{12}	Bicyclo[4.1.0]heptane	II/7(**3**,746)
$C_7H_{12}O$	1–Methoxycyclohexene	II/7(**3**,747)
$C_7H_{12}O$	Cyclohexanecarbaldehyde	II/15(**3**,911)
$C_7H_{12}O$	Cycloheptanone	II/15(**3**,912)
$C_7H_{12}O$	Cycloheptene oxide 1,2–Epoxycycloheptane	II/15(**3**,913)
$C_7H_{13}ClSi$	1–Chloro–1–silabicyclo[2.2.2]octane	II/15(**3**,914)
$C_7H_{13}N$	1–Azabicyclo[2.2.2]octane	II/15(**3**,915)
C_7H_{14}	1–Methyl–1–isopropylcyclopropane	II/23(**3**,452)
C_7H_{14}	trans–1,2–Dimethylcyclopentane	II/23(**3**,453)
C_7H_{14}	Methylcyclohexane	II/7(**3**,748), II/15(**3**,916)
C_7H_{14}	Cycloheptane	II/15(**3**,917)
$C_7H_{14}O$	2,4–Dimethyl–3–pentanone Diisopropyl ketone	II/23(**3**,454)
$C_7H_{14}Si$	1–Methyl–1–silabicyclo[2.2.1]heptane	II/15(**3**,918)
$C_7H_{14}Si$	3–Silabicyclo[3.2.1]octane	II/15(**3**,919)
$C_7H_{15}NO_3Si$	1–Methyl–2,8,9–trioxa–5–aza–1–silabicyclo[3.3.3]undecane Methyl silatrane	II/15(**3**,920)
C_7H_{16}	2,4–Dimethylpentane	II/23(**3**,455)
C_7H_{16}	Heptane	II/7(**3**,749)
$C_7H_{16}Ge$	1,1–Dimethylgermacyclohexane	II/23(**3**,456)
$C_7H_{16}O_2Si$	1,1–Dimethoxysilacyclohexane	II/15(**3**,921)
$C_7H_{18}Ge_2N_2$	Bis(trimethylgermyl)carbodiimide	II/23(**3**,457)
$C_7H_{18}N_2Si_2$	Bis(trimethylsilyl)carbodiimide	II/23(**3**,458)
$C_7H_{18}P_2$	Hexamethylcarbodiphosphorane Methanetetraylbis[trimethylphosphorane]	II/15(**3**,922)
$C_7H_{19}LiSi_2$	[Bis(trimethylsilyl)methyl]lithium	II/15(**3**,923), II/21(**3**,555)
$C_7H_{20}Si_2$	Bis(trimethylsilyl)methane	II/15(**3**,924)
$C_7H_{21}NSi_2$	N,N–Bis(trimethylsilyl)methylamine	II/21(**3**,556)
$C_8F_{12}Mo_2O_8$	Tetrakis[μ–trifluoroacetato]–dimolybdenum(II) Tetrakis[μ–trifluoroacetato–O:O']dimolybdenum(Mo–Mo)	II/15(**3**,925)
$C_8F_{18}S_6$	Hexakis(trifluoromethylthio)ethane	II/15(**3**,926)
$C_8H_4N_2$	Phthalonitrile o–Dicyanobenzene	II/21(**3**,557)
$C_8H_4N_2$	p–Dicyanobenzene	II/15(**3**,927), II/21(**3**,558)

$C_8H_4N_2$		p–Diisocyanobenzene	II/15(**3**,928), II/21(**3**,559)
C_8H_4Sn		Tetraethynyltin	II/15(**3**,929), II/21(**3**,560)
		Tetraethynylstannane	
C_8H_6		Phenylacetylene	II/15(**3**,930)
$C_8H_6Cl_4$		1,2,6,6–Tetrachloro–4,5–dimethylspiro[2.3]hexa–1,4–diene	
			II/15(**3**,931)
$C_8H_6N_4$		2,2′–Bipyrimidine	II/15(**3**,932)
$C_8H_6O_2$		Phenylglyoxal	II/23(**3**,459)
$C_8H_6O_2$		Terephthalaldehyde	II/21(**3**,561)
$C_8H_7Cl_2P$		Dichloro(styryl)phosphine	II/21(**3**,562)
		Styryldichlorophosphine	
C_8H_7F		2–Fluorostyrene	II/21(**3**,563)
C_8H_7F		3–Fluorostyrene	II/21(**3**,564)
C_8H_7N		Phenylacetonitrile, Benzyl cyanide	II/23(**3**,460)
C_8H_7N		Indole	II/23(**3**,461)
C_8H_8		Acetylene tetramer	II/21(**3**,565)
C_8H_8		Styrene, Vinylbenzene	II/21(**3**,566)
C_8H_8		3,6–Bis(methylene)–1,4–cyclohexadiene	II/15(**3**,933)
		p–Quionodimethane	
C_8H_8		Cyclooctatetraene	II/7(**3**,750)
C_8H_8		Bicyclo[2.2.2]octa–2,5,7–triene	II/15(**3**,934)
		Barrellene	
C_8H_8		Tricyclo[3.3.0.0^{2,8}]octa–3,6–diene	II/7(**3**,751)
		Semibullvalene	

C_8H_8   **Cubane** O_h

MW, ED

r_0	Å [a]
C–C	1.5708(10)
C–H	1.097(5)

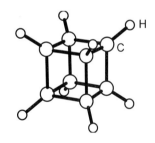

The rotational spectrum of cubane–d was observed.

[a]) The uncertainties in the two parameters are related as
δr (C–H) = –5.68 δr (C–C) and were not estimated in the original paper.

Hirota, E., Endo, Y., Fujitake, M., Della, E.W., Pigon, P.E., Chickos, J.S.: J. Mol. Struct. **190** (1988) 235.

ED, MW

r_g	Å [b]		r_α^0	Å [b]		r_e	Å [b]
C–C	1.573(2)		C–C	1.571(2)		C–C	1.562(4)
C–H	1.114(6)		C–H	1.098(6)		C–H	1.096(13)

The nozzle temperature was 77 °C.

(*continued*)

C_8H_8 (*continued*)

[b]) Twice the estimated standard errors including systematic errors.

Hedberg, L., Hedberg, K., Eaton, P.E., Nodari, N., Robiette, A.G.: J. Am. Chem. Soc. **113** (1991) 1514.

II/15(**3**,935), II/21(**3**,567), II/23(**3**,462)

$C_8H_8O_2$	*p*–Methoxybenzaldehyde	II/21(**3**,568)
	p–Anisaldehyde	
$C_8H_8O_2Si$	Di–2–furylsilane	II/21(**3**,569)
C_8H_8S	Phenyl vinyl sulfide	II/15(**3**,936)
C_8H_9ClNOP	3–Phenyl–2–chloro–1,3,2–oxazaphospholane	II/15(**3**,937)
C_8H_9N	*N*–Benzylidenemethylamine	II/15(**3**,938)
C_8H_9N	*N*–Phenylaziridine	II/7(**3**,752)
C_8H_{10}	6,6–Dimethylfulvene	II/7(**3**,753)
C_8H_{10}	Ethylbenzene	II/15(**3**,939)
C_8H_{10}	*o*–Xylene	II/7(**3**,754)
C_8H_{10}	*p*–Xylene	II/15(**3**,940)
C_8H_{10}	Bicyclo[2.2.2]octa–2,5–diene	II/7(**3**,755)
C_8H_{10}	Tricyclo[3.3.0.0^{2,6}]oct–3–ene	II/7(**3**,756)
$C_8H_{10}O$	4–Ethylphenol	II/23(**3**,463)
$C_8H_{10}Si$	1–Methyl–1–silabicyclo[2.2.2]octa–2,5,7–triene	II/15(**3**,941)
$C_8H_{11}N$	*N,N*–Dimethylaniline	II/7(**3**,757)
$C_8H_{11}P$	Dimethylphenylphosphine	II/15(**3**,942)
C_8H_{12}	Ethynylcyclohexane	II/15(**3**,943)
C_8H_{12}	1,3–Cyclooctadiene	II/7(**3**,758)
C_8H_{12}	(Z,Z)–1,5–Cyclooctadiene	II/15(**3**,946)
C_8H_{12}	Cyclooctyne	II/7(**3**,759), II/21(**3**,570)
C_8H_{12}	1,1′–Bicyclobutylidene	II/15(**3**,944)
C_8H_{12}	Bicyclo[2.2.2]oct–2–ene	II/7(**3**,760)
C_8H_{12}	Bicyclo[3.3.0]oct–1(5)–ene	II/21(**3**,571)
C_8H_{12}	Tricyclo[5.1.0.0^{2,4}]octane	II/15(**3**,945)
C_8H_{12}	Tricyclo[3.3.0.0^{2,6}]octane	II/7(**3**,761)
C_8H_{12}	Tricyclo[4.2.0.0^{2,5}]octane	II/7(**3**,762)
$C_8H_{12}Cr_2O_8$	Dichromium tetraacetate	II/21(**3**,572)
	Tetra–μ–acetato–dichromium(II)	
$C_8H_{12}Mo_2O_8$	Dimolybdenum tetraacetate	II/15(**3**,947)
$C_8H_{12}Si$	Tetravinylsilane	II/15(**3**,948)
$C_8H_{12}Sn$	Tetravinylstannane	II/15(**3**,949)
C_8H_{14}	3,4–Dimethyl–2,4–hexadiene	II/7(**3**,763)
C_8H_{14}	1,1′–Bicyclobutyl	II/7(**3**,764)
C_8H_{14}	*trans*–Cyclooctene	II/7(**3**,765), II/15(**3**,950)
C_8H_{14}	1,1′–Dimethyl–1,1′–bicyclopropane	II/21(**3**,573)
C_8H_{14}	Bicyclo[2.2.2]octane	II/7(**3**,766)
C_8H_{14}	Bicyclo[3.2.1]octane	II/15(**3**,951)
C_8H_{14}	Bicyclo[4.2.0]octane	II/15(**3**,952)

C$_8$H$_{14}$Ge		(Trimethylgermyl)cyclopentadiene	II/7(**3**,767)
		(Cyclopentadienyl)trimethylgermane	
C$_8$H$_{14}$O$_2$		1,2:7,8–Diepoxyoctane	II/7(**3**,768)
C$_8$H$_{14}$Si		(Cyclopentadienyl)trimethylsilane	II/7(**3**,769)
C$_8$H$_{14}$Sn		(Trimethylstannyl)cyclopentadiene	II/7(**3**,770)
		(Cyclopentadienyl)trimethylstannane	
C$_8$H$_{14}$Zn		Di–3–butenylzinc	II/15(**3**,953)
C$_8$H$_{16}$		1–*t*–Butyl–1–methylcyclopropane	II/23(**3**,464)
C$_8$H$_{16}$		Cyclooctane	II/7(**3**,771), II/21(**3**,574)
C$_8$H$_{16}$		1,1–Dimethylcyclohexane	II/7(**3**,772)
C$_8$H$_{16}$Si		3–Methyl–3–silabicyclo[3.2.1]octane	II/15(**3**,955)
C$_8$H$_{18}$		Hexamethylethane	II/15(**3**,956)
C$_8$H$_{18}$Be		Di–*t*–butylberyllium	II/7(**3**,773)
C$_8$H$_{18}$ClP		Di–*t*–butylchlorophosphine	II/15(**3**,957)
C$_8$H$_{18}$Cl$_2$Si		Di–*t*–butyldichlorosilane	II/23(**3**,465)
C$_8$H$_{18}$Cl$_2$Sn		Di–*t*–butyldichlorostannane	II/21(**3**,575)
C$_8$H$_{18}$FP		Di–*t*–butylfluorophosphine	II/15(**3**,958)
C$_8$H$_{18}$F$_2$Si		Di–*t*–butyldifluorosilane	II/21(**3**,576)
C$_8$H$_{18}$Ge$_2$O		Bis(trimethylgermyl)ketene	II/7(**3**,774)
C$_8$H$_{18}$NO		Di–*t*–butyl nitroxide radical	II/7(**3**,775)
C$_8$H$_{18}$O		Di–*t*–butyl ether	II/21(**3**,577)
C$_8$H$_{18}$OSi$_2$		Bis(trimethylsilyl)ketene	II/15(**3**,959)
C$_8$H$_{18}$O$_2$		Di–*t*–butyl peroxide	II/15(**3**,960)
C$_8$H$_{18}$O$_2$Zn		Bis(3–methoxypropyl)zinc	II/15(**3**,961)
C$_8$H$_{18}$S		Di–*t*–butyl sulfide	II/21(**3**,578)
C$_8$H$_{18}$S$_2$Zn		Bis(3–mercaptopropyl)zinc	II/21(**3**,579)
C$_8$H$_{18}$Sn$_2$		Bis(trimethylstannyl)acetylene	II/15(**3**,962)
C$_8$H$_{19}$ClSi		Di–*t*–butylchlorosilane	II/23(**3**,466)
C$_8$H$_{19}$N		Di–*t*–butylamine	II/21(**3**,580)
C$_8$H$_{22}$CdN$_2$		Dimethyl(*N*,*N*,*N*′,*N*′–tetramethyl–1,2–ethanediamine)–cadmium(II)	II/23(**3**,467)
C$_8$H$_{24}$Al$_4$F$_4$		Dimethylaluminum fluoride tetramer	II/7(**3**,776), II/15(**3**,963)
C$_8$H$_{24}$N$_4$Sn		Tetrakis(dimethylamino)stannane	II/7(**3**,777)

ED **C$_8$H$_{24}$N$_4$Sn$_2$** **Bis(μ–dimethylamido)–bis[dimethylamidotin(II)]** **C$_{2h}$**

r_g	Å [a)]	θ [b)]	deg [a)]
C–H	1.20(1) [c)]	Sn–N(r)–Sn	96.1(5)
C–N (average)	1.483(3) [c)]	N(r)–Sn–N(r)	83.9 [d)]
Sn–N(s)	1.996(4) [c)]	X–Sn–N(s) [e)]	156.1(5)
Sn–N(r)	2.293(5)	X–N(r)–C	135(1)
		Sn–N(s)–C	118.1(3) [c)]
		N–C–H	109 [c)][f)]
		C–N(s)–C	123.8 [c)][d)]
		Sn–N(r)–C	118 [d)]
		C–N(r)–C	90 [d)]
		N(r)–Sn–N(s)	133 [d)]

(*continued*)

C₈H₂₄N₄Sn₂ (*continued*)

Vaporized at 385 K.

[a]) Estimated standard errors including systematic errors.
[b]) Undefined, possibly θ_a.
[c]) Average for monomer and dimer (essentially monomer, 74(3) % at 385 K); for the angle Sn–N–C averaging excludes angle Sn–N(r)–C of dimer ring N atom.
[d]) Dependent parameter.
[e]) X is the center of the planar SnNSnN ring; N(r) and N(s) are ring and side-chain N atoms, respectively.
[f]) Assumed.

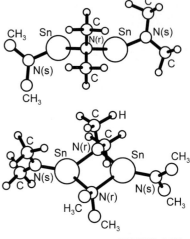

Beagley, B., Scott, N.G., Schmidling, D.: J. Mol. Struct. **221** (1990) 15.

II/23(**3**,468)

ED | **C₈H₂₄N₄Ti** | **Tetrakis(dimethylamino)titanium(IV)** | **S₄**
| | | Ti[N(CH₃)₂]₄

r_a	Å [a])
Ti–N	1.917(3)
N–C	1.461(2)
C–H	1.120(4)

θ_a	deg [a])
Ti–N–C	124.3(3)
N–C–H	109.3(7)
N–Ti–N* [b])	114.2(17)
N–Ti–N** [b])	107.2(9)
ϕ(N*–Ti–N–C) [c])[b])	51(1)
ϕ(Ti–N–C–H) [d])	20(2)

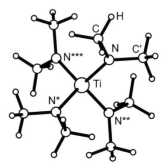

Local C_{3v} symmetry of NCH₃ fragments and C_2 symmetry of TiN(CH₃)₂ fragments were assumed.
The nozzle temperature was 152···155 °C.

[a]) Three times the estimated standard errors including systematic errors.
[b]) See figure for the definition of N* and N**.
[c]) Torsional angle, defined as zero when the two NC₂ groups are coplanar and the molecular symmetry is D_{2d}.
[d]) Torsional angle.

Haaland, A., Rypdal, K., Volden, H.V., Andersen, R.A.: J. Chem. Soc., Dalton Trans. 1992, 891.

II/23(**3**,469)

C₈H₂₄N₄V	Tetrakis(dimethylamino)vanadium(IV)	II/23(**3**,470)
C₈H₂₄N₄Zr	Tetrakis(dimethylamido)zirconium(IV)	II/21(**3**,581)
C₈H₂₄O₄Si₄	Octamethylcyclotetrasiloxane	II/7(**3**,778)
C₈H₂₄Si₃	Octamethyltrisilane	II/15(**3**,964)
C₈H₂₄Si₄	Octamethylcyclotetrasilane	II/23(**3**,471)

IR | **C₉** | **Nonacarbon** | **D∞h**
| | | C=C=C=C=C=C=C=C=C

r_0	Å
C=C	1.27868(275) [a])

(*continued*)

Assumes all C=C bonds are of identical length, whereas *ab initio* calculations predict three different values, in a range of 0.02 Å.

[a]) Three times the estimated standard error.

Heath, J.R., Saykally, R.J.: J. Chem. Phys. **93** (1990) 8392.   II/23(**3**,472)

$C_9F_{17}N$	Perfluoroquinolizine		II/21(**3**,582)
$C_9F_{21}N$	Tris(heptafluoropropyl)amine		II/21(**3**,583)
	Perfluorotripropylamine		
$C_9H_2F_8$	3,3,3–Trifluoro–2–pentafluorophenyl–1–propene		II/15(**3**,965)
$C_9H_3N_3$	1,3,5–Benzenetricarbonitrile		II/23(**3**,473)
	1,3,5–Tricyanobenzene		
$C_9H_6CrO_3$	Tricarbonyl(benzene)chromium(0)		II/15(**3**,966)
C_9H_7	Indenyl radical		II/7(**3**,779)
C_9H_8	Indene		II/7(**3**,780)
$C_9H_8O_2$	1–Phenyl–1,2–propanedione		II/23(**3**,474)
$C_9H_9N_2P$	5–Methyl–2–phenyl–2H–1,3,2–diazaphosphole		
		II/7(**3**,781),	II/21(**3**,584)
C_9H_{10}	Phenylcyclopropane		II/7(**3**,782)
$C_9H_{10}N_2$	4–(Dimethylamino)benzonitrile		II/23(**3**,475)
$C_9H_{10}ArN_2$	4–(Dimethylamino)benzonitrile – argon (1/1)		II/23(**3**,476)
$C_9H_{10}O$	*p*–Ethylbenzaldehyde		II/21(**3**,585)
C_9H_{12}	Cumene		II/7(**3**,783)
	Isopropylbenzene		
C_9H_{12}	3,3–Dimethyl–6–methylene–1,4–cyclohexadiene		II/15(**3**,967)
C_9H_{12}	1,3,5–Trimethylbenzene		II/15(**3**,968)
	Mesitylene		
C_9H_{12}	1,2,6–Cyclononatriene		II/15(**3**,969)

ED   **C_9H_{12}**   **Triasterane**   **D_{3h}**
       Tetracyclo[3.3.1.0^{2,8}.0^{4,6}]nonane

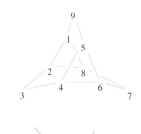

r_a	Å [a])		θ_α	deg [a])
C(1)–C(2)	1.508(5)		C(2)–C(3)–C(4)	112.2(3)
C(2)–C(3)	1.520(4)		M–C(1)–H [b])	124.8(44)
C(1)–H	1.087(10)		H–C–H	107.4(32)
C(3)–H	1.127(12)		M–C(1)–C(9) [b])	123.9(2) [c])
			C(1)–C(2)–C(8)	118.9(3) [c])
			C(9)–C(1)–H	111.3(44) [c])

The nozzle temperature was 65 °C.

[a]) Twice the estimated standard errors.
[b]) M–C bisects the cyclopropyl ring.
[c]) Dependent angle.

Ahlquist, B., Almenningen, A., Benterud, B., Trætteberg, M.,
   Bakken, P., Lüttke, W.: Chem. Ber. **125** (1992) 1217.   II/23(**3**,477)

$C_9H_{12}N_2O$	4–(Dimethylamino)benzonitrile – water (1/1)	II/23(**3**,478)
C_9H_{14}	1,2–Cyclononadiene	II/15(**3**,970)

C₉H₁₄	Cyclononyne	II/15(**3**,971)
C₉H₁₄O	Bicyclo[3.3.1]nonan-9-one	II/21(**3**,586)
C₉H₁₆	1-Methyl-*trans*-cyclooctene	II/15(**3**,972)
C₉H₁₆	Bicyclo[3.3.1]nonane	II/15(**3**,973)
C₉H₁₆	*trans*-Bicyclo[4.3.0]nonane	II/15(**3**,974)
C₉H₁₆NO₂	2,2,6,6-Tetramethyl-4-oxo-1-piperidinyloxyl	II/7(**3**,784)
C₉H₁₆O	2,2,6-Trimethylcyclohexanone	II/15(**3**,975)
C₉H₁₇NO₂	2,2,6,6-Tetramethyl-1-hydroxy-4-piperidinone	II/7(**3**,785)
C₉H₁₈	Cyclononane	II/23(**3**,479)
C₉H₂₀	2,2,4,4-Tetramethylpentane	II/15(**3**,976)
C₉H₂₁N	Triisopropylamine	II/23(**3**,480)
C₉H₂₇Al₃O₃	Methoxydimethylaluminum trimer	II/7(**3**,786)
C₉H₂₇NSi₃	Tris(trimethylsilyl)amine	II/23(**3**,481)
C₉H₂₇NSn₃	Tris(trimethylstannyl)amine	II/15(**3**,977)
C₉H₂₇PSi₃	Tris(trimethylsilyl)phosphine	II/23(**3**,482)
C₁₀F₁₈	*cis*-Perfluorodecalin	II/21(**3**,587)
C₁₀F₁₈	*trans*-Perfluorodecalin	II/21(**3**,588)
C₁₀H₂CuF₁₂O₄	Bis(1,1,1,5,5,5-hexafluoro-2,4-pentanedionato)copper(II)	II/15(**3**,978)
C₁₀H₈	Naphthalene	II/7(**3**,789), II/15(**3**,979)
C₁₀H₈	Azulene	II/7(**3**,790)
C₁₀H₈N₂	2,2′-Bipyridine 2,2′-Bipyridyl	II/7(**3**,791), II/21(**3**,589)
C₁₀H₈N₂	4,4′-Bipyridine 4,4′-Bipyridyl	II/7(**3**,792)
C₁₀H₈N₂S	Di-2-pyridyl sulfide	II/15(**3**,980)
C₁₀H₁₀	3-Methyl-3-phenyl-1-cyclopropene	II/21(**3**,590)
C₁₀H₁₀	Tricyclo[3.3.2.0⁴,⁶]deca-2,7,9-triene Bullvalene	II/7(**3**,793)
C₁₀H₁₀	Basketene Bishomocubene Pentacyclo[4.4.0.0²,⁵.0³,⁸.0⁴,⁷]dec-9-ene	II/15(**3**,981)
C₁₀H₁₀Be	Bis(η-cyclopentadienyl)beryllium Beryllocene	II/7(**3**,794), II/15(**3**,982)
C₁₀H₁₀Cl₂Ti	Dichlorobis(η-cyclopentadienyl)titanium(IV)	II/7(**3**,795), II/15(**3**,983)
C₁₀H₁₀Cl₂Zr	Dichlorobis(η-cyclopentadienyl)zirconium(IV)	II/7(**3**,796), II/15(**3**,984)
C₁₀H₁₀Co	Bis(η-cyclopentadienyl)cobalt(II) Cobaltocene	II/15(**3**,985)
C₁₀H₁₀Cr	Bis(η-cyclopentadienyl)chromium(II) Chromocene	II/7(**3**,797), II/15(**3**,986)
C₁₀H₁₀Fe	Bis(η-cyclopentadienyl)iron(II) Ferrocene	II/7(**3**,798)
C₁₀H₁₀Mg	Bis(η-cyclopentadienyl)magnesium(II)	II/7(**3**,799)
C₁₀H₁₀Mn	Bis(η-cyclopentadienyl)manganese(II) Manganocene	II/7(**3**,800), II/15(**3**,987)

$C_{10}H_{10}N_2O_2$		2(1H)–Pyridinone dimer	II/23(**3**,483)
		2–Pyridone dimer	
		2–Hydroxypyridine dimer	
$C_{10}H_{10}Ni$		Bis(η–cyclopentadienyl)nickel(II)	II/7(**3**,801)
		Nickelocene	
$C_{10}H_{10}Pb$		Bis(η–cyclopentadienyl)lead(II)	II/7(**3**,802)
$C_{10}H_{10}Ru$		Bis(η–cyclopentadienyl)ruthenium	II/7(**3**,803)
		Ruthenocene	
$C_{10}H_{10}Sn$		Bis(η–cyclopentadienyl)tin(II)	II/7(**3**,804)
$C_{10}H_{10}V$		Bis(η–cyclopentadienyl)vanadium(II)	II/15(**3**,988)
$C_{10}H_{12}$		Phenylcyclobutane	II/7(**3**,805)
$C_{10}H_{12}O$		p–Isopropylbenzaldehyde	II/21(**3**,591)
$C_{10}H_{12}O_2$		Tetramethyl–p–benzoquinone	II/15(**3**,989)
		Duroquinone	
$C_{10}H_{14}BTi$		Bis(η–cyclopentadienyl)titanium(1+) tetrahydroborate	
			II/15(**3**,990)
$C_{10}H_{14}BeO_4$		Bis(acetylacetonato)beryllium	II/15(**3**,991)
$C_{10}H_{14}CuO_4$		Bis(acetylacetonato)copper(II)	II/15(**3**,992)
$C_{10}H_{14}NiO_4$		Bis(acetylacetonato)nickel(II)	II/15(**3**,993)
$C_{10}H_{14}O_4Zn$		Bis(acetylacetonato)zinc(II)	II/15(**3**,994)

ED **$C_{10}H_{14}O_5V$**   **Oxobis(2,4–pentanedionato)vanadium(IV)** C_{2v} assumed
Oxobis(acetylacetonato)vanadium(IV)

$VO[CH_3COCHCOCH_3]_2$

r_g	Å [a]	θ_α	deg [a]
V–O(5)	1.586(17)	O(1)–V–O(2)	87.2(15)
V–O(1)	2.005(6)	O(1)–V–O(3)	85.0(16)
C(1)–O(1)	1.293(8)	O(5)–V–C(2)	111.5(24)
C(1)–C(2)	1.425(10)	V–O(1)–C(1)	129.6(14)
C(1)–C(4)	1.547(14)	O(1)–C(1)–C(4)	112.7(15)
C–H	1.100(18)	C(1)–C(4)–H	105.4(56)
		θ_1 [b]	0.0 [c]
		θ_2 [d]	0.0 [c]

Each of the CCOCHCOC fragments was assumed to have local C_{2v} symmetry and to be planar except for the methyl H atoms. The methyl groups were assumed to have local C_{3v} symmetry with no tilt.

The nozzle temperature was 222···239 °C.

[a] Twice the estimated standard errors including systematic errors.
[b] O(1)–C(1)–C(4)–H torsional angle; one of the H atoms on the methyl groups bound to C(1) is eclipsed with respect to the C(1)–O(1) vector.
[c] Determined by R-factor optimization.
[d] Angle between the O(1)–V–O(2) plane and the ligand plane.

Forsyth, G.A., Rice, D.A., Hagen, K.: Polyhedron **9** (1990) 1603.   II/23(**3**,484)

$C_{10}H_{15}Br$		1–Bromoadamantane	II/7(**3**,806)
$C_{10}H_{15}Cl$		1–Chloroadamantane	II/7(**3**,807)

$C_{10}H_{15}ClGe$	Chloro(η–pentamethylcyclopentadienyl)germanium(II)	
		II/15($\mathbf{3}$,994a)
$C_{10}H_{15}F$	1–Fluoroadamantane	II/7($\mathbf{3}$,808)
$C_{10}H_{15}FeP_5$	(Cyclopentaphosphorus)–	II/21($\mathbf{3}$,592)
	(η–pentamethylcyclopentadienyl)iron(II)	
$C_{10}H_{15}I$	1–Iodoadamantane	II/7($\mathbf{3}$,809)
$C_{10}H_{15}In$	(η^5–Pentamethylcyclopentadienyl)indium	II/21($\mathbf{3}$,593)
$C_{10}H_{15}O_3Re$	Trioxo(η^5–pentamethylcyclopentadienyl)rhenium(VII)	
		II/23($\mathbf{3}$,485)
$C_{10}H_{15}Tl$	(η^5–Pentamethylcyclopentadienyl)thallium	II/21($\mathbf{3}$,594)
$C_{10}H_{16}$	3–Carene	II/7($\mathbf{3}$,810)
	3,7,7–Trimethylbicyclo[4.1.0]hept–3–ene	
$C_{10}H_{16}$	1,6–Cyclodecadiene	II/7($\mathbf{3}$,811)
$C_{10}H_{16}$	α–Pinene	II/7($\mathbf{3}$,812)
	2,7,7–Trimethylbicyclo[3.1.1]hept–2–ene	
$C_{10}H_{16}$	β–Pinene	II/7($\mathbf{3}$,813)
	6,6–Dimethyl–2–methylenebicyclo[3.1.1]heptane	
$C_{10}H_{16}$	Adamantane	II/7($\mathbf{3}$,814)
	Tricyclo[3.3.1.1^{3,7}]decane	
$C_{10}H_{16}Cl_2$	3α,4β–Dichlorocarene	II/7($\mathbf{3}$,815)
$C_{10}H_{16}Mg$	η–Cyclopentadienyl(neopentyl)magnesium	II/21($\mathbf{3}$,595)
$C_{10}H_{16}O$	3–Carene oxide	II/7($\mathbf{3}$,816)
	3,4–Epoxy–3–carene	
$C_{10}H_{16}O$	α–Pinene oxide	II/7($\mathbf{3}$,817)
$C_{10}H_{16}O$	trans–2–Decalone	II/7($\mathbf{3}$,818)
$C_{10}H_{16}S$	3,3,6,6–Tetramethylthiacyclohept–4–yne	II/7($\mathbf{3}$,819)
$C_{10}H_{18}$	2,3,4,5–Tetramethyl–2,4–hexadiene	II/15($\mathbf{3}$,995)
$C_{10}H_{18}$	Decalin	II/7($\mathbf{3}$,820), II/15($\mathbf{3}$,996)
	cis– and trans–Bicyclo[4.4.0]decane, Decahydronaphthalene	
$C_{10}H_{18}Al_2$	Dimethyl(1–propynyl)aluminum dimer	II/15($\mathbf{3}$,997)
$C_{10}H_{18}B_2$	Hexamethyltetracarbahexaborane(6)	II/7($\mathbf{3}$,821)
$C_{10}H_{18}Ga$	Dimethyl(μ–1–propynyl)gallium dimer	II/15($\mathbf{3}$,998)
	Bis(μ–1–propynyl)–bis[dimethylgallium(III)]	
$C_{10}H_{18}Ge$	1–Methyl–1–germaadamantane	II/15($\mathbf{3}$,999)
$C_{10}H_{18}In$	Dimethyl(μ–1–propynyl)indium dimer	II/15($\mathbf{3}$,1000)
	Bis(μ–1–propynyl)–bis[dimethylindium(III)]	
$C_{10}H_{18}Si$	1–Methyl–1–silaadamantane	II/15($\mathbf{3}$,1001)
$C_{10}H_{18}Zn$	Di–4–pentenylzinc	II/15($\mathbf{3}$,1002)
$C_{10}H_{19}Cl$	4–t–Butyl–1–chlorocyclohexane	II/7($\mathbf{3}$,822)
$C_{10}H_{20}$	Cyclodecane	II/7($\mathbf{3}$,823)
$C_{10}H_{20}N_2$	N,N'–Di–t–butyl–1,2–ethanediimine	II/15($\mathbf{3}$,1003)
$C_{10}H_{21}Re$	Hexahydrido(η^5–pentamethylcyclopentadienyl)rhenium(VII)	
		II/23($\mathbf{3}$,486)
$C_{10}H_{22}Mg$	Bis(neopentyl)magnesium	II/15($\mathbf{3}$,1004)
$C_{10}H_{22}Mn$	Bis(neopentyl)manganese	II/21($\mathbf{3}$,596)
$C_{10}H_{22}O_2Zn$	Bis(4–methoxybutyl)zinc	II/15($\mathbf{3}$,1005)

$C_{10}H_{24}N_2Zn$	Bis[3–(dimethylamino)propyl]zinc	II/21(**3**,597)
$C_{10}H_{24}N_4$	Tetrakis(dimethylamino)ethylene	II/23(**3**,487)
$C_{10}H_{27}Cl_3Si_4$	Tris(trimethylsilyl)(trichlorosilyl)methane	II/21(**3**,598)
$C_{10}H_{28}Si_3$	Tris(trimethylsilyl)methane	II/15(**3**,1006)
$C_{10}H_{29}PSi_3$	[Tris(trimethylsilyl)methyl]phosphine	II/15(**3**,1007)
$C_{10}H_{30}N_5Ta$	Pentakis(dimethylamido)tantalum	II/23(**3**,488)
$C_{10}H_{30}O_5Si_5$	Decamethylcyclopentasiloxane	II/7(**3**,824)
$C_{10}H_{30}O_7Si_6$	Decamethylbicyclo[5.5.1]hexasiloxane	II/15(**3**,1008)
$C_{10}Mn_2O_{10}$	Decacarbonyldimanganese(0)	II/7(**3**,787)
$C_{10}O_{10}Re_2$	Decacarbonyldirhenium(0)	II/7(**3**,788)
$C_{11}H_{15}N$	1–Cyanoadamantane	II/7(**3**,825)
$C_{11}H_{16}Cl_2$	8,8–Dichloro–1,4,4–trimethyltricyclo[5.1.0.0^{3,5}]octane	
		II/7(**3**,826)
$C_{11}H_{18}O$	4a–Methyl–*trans*–2–decalone	II/15(**3**,1009)
$C_{11}H_{20}$	Bicyclo[3.3.3]undecane	II/15(**3**,1010)
	Manxane	
$C_{11}H_{22}Si_2$	5,5–Bis(trimethylsilyl)–1,3–cyclopentadiene	II/7(**3**,827)
$C_{11}H_{22}Sn_2$	5,5–Bis(trimethylstannyl)–1,3–cyclopentadiene	II/7(**3**,828)
$C_{12}F_{10}$	Decafluorobiphenyl	II/7(**3**,829), II/21(**3**,599)
$C_{12}F_{12}Sn$	Tetrakis(3,3,3–trifluoropropynyl)tin	II/15(**3**,1011)
$C_{12}H_6Br_4$	3,3′,5,5′–Tetrabromobiphenyl	II/21(**3**,600)
$C_{12}H_7Br_3$	3,4′,5–Tribromobiphenyl	II/21(**3**,601)
$C_{12}H_8$	Biphenylene	II/7(**3**,830)
$C_{12}H_8Br_2$	3,3′–Dibromobiphenyl	II/21(**3**,602)
$C_{12}H_8Cl_2$	2,2′–Dichlorobiphenyl	II/7(**3**,831)
$C_{12}H_8Cl_2$	4,4′–Dichlorobiphenyl	II/21(**3**,603)
$C_{12}H_8F_2$	4,4′–Difluorobiphenyl	II/21(**3**,604)
$C_{12}H_8S_2$	Thianthrene	II/15(**3**,1012)
$C_{12}H_9Cl$	4–Chlorobiphenyl	II/21(**3**,605)
$C_{12}H_9F$	4–Fluorobiphenyl	II/21(**3**,606)

ED **$C_{12}H_9NO_2$**   **2–Nitrobiphenyl** C_1

r_g	Å [a]		θ [b]	deg [a]
C(1)–C(2)	1.528(10)		C–N=O	117.0(5)
C–C (ring)	1.393(2)		φ_1 [d]	–32.8(106)
C–N	1.466 [c]		φ_2 [e]	64.5(18)
N=O	1.206(4)			
C–H	1.107 [c]			

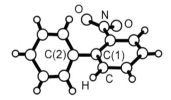

The benzene rings were assumed to have D_{6h} symmetry. The N atom was assumed to lie in the ring plane. The C–H and C–N bonds were assumed to coincide with the CCC bisector.
The nozzle temperature was 100(10) °C.

[a]) Estimated standard errors.
[b]) Unidentified, possibly θ_a.
[c]) Assumed.

(*continued*)

C$_{12}$H$_9$NO$_2$ (continued)

^{d}) Angle of rotation of the NO$_2$ group around the C–N bond, defined as zero for a planar conformation and as positive for clockwise rotation looked from N to C.
^{e}) Angle of rotation of the C$_6$H$_5$ group around the C–C bond, defined as zero for a planar conformation and as positive for clockwise rotation looked from C(2) to C(1).

Levit, P.B., Belyakov, A.V., Tselinskii, I.V., Golubinskii, A.V., Vilkov, L.V., Shlyapochnikov, V.A.: Zh. Fiz. Khim. **65** (1991) 1946; Russ. J. Phys. Chem. (Engl. Transl.) **65** (1991) 1031.

II/23(**3**,489)

C$_{12}$H$_{10}$	Biphenyl		II/7(**3**,832), II/21(**3**,607)
C$_{12}$H$_{10}$Cl$_2$Si	Diphenyldichlorosilane		II/7(**3**,833)
C$_{12}$H$_{10}$Hg	Diphenylmercury		II/7(**3**,834)
C$_{12}$H$_{10}$N$_2$	trans–Azobenzene		II/15(**3**,1013)
C$_{12}$H$_{10}$O	Diphenyl ether		II/15(**3**,1014)
C$_{12}$H$_{10}$OS	Diphenyl sulfoxide		II/15(**3**,1015)
C$_{12}$H$_{10}$O$_2$S	Diphenyl sulfone		II/15(**3**,1016)
C$_{12}$H$_{10}$O$_3$Si	Tri–2–furylsilane		II/21(**3**,608)
C$_{12}$H$_{10}$S	Diphenyl sulfide		II/15(**3**,1017)
C$_{12}$H$_{10}$S$_3$	4,4'–Thiobis(benzenethiol) Bis(4–mercaptophenyl) sulfide		II/21(**3**,609)
C$_{12}$H$_{11}$P	Diphenylphosphine		II/15(**3**,1018)
C$_{12}$H$_{12}$Cr	Bis(benzene)chromium(0)		II/7(**3**,835)
C$_{12}$H$_{14}$Ge	1,1'–Dimethylgermanocene Bis(1–methylcyclopentadienyl)germanium(II)		II/15(**3**,1019)
C$_{12}$H$_{14}$Mn	1,1'–Dimethylmanganocene Bis(1–methyldicyclopentadienyl)manganese(II)		II/15(**3**,1020)
C$_{12}$H$_{14}$Sn	1,1'–Dimethylstannocene Bis(1–methyldicyclopentadienyl)tin(II)		II/15(**3**,1021)
C$_{12}$H$_{16}$	Tetraspiro[2.0.2.0.2.0.2.0]dodecane [4]–Rotane		II/15(**3**,1022)
C$_{12}$H$_{18}$	Hexamethylbenzene		II/15(**3**,1023)
C$_{12}$H$_{18}$	Hexamethylbicyclo[2.2.0]hexa–2,5–diene		II/7(**3**,836)
C$_{12}$H$_{18}$	Hexamethylprismane		II/15(**3**,1024)
C$_{12}$H$_{20}$O	1,1–Dimethyl–trans–2–decalone		II/15(**3**,1025)

C$_{12}$H$_{22}$ **1,1'–Bicyclohexyl**   C$_{2h}$, C$_2$

ED vibrational spectroscopy
and molecular mechanics
calculations MM3

r_a	Å ^{a})	θ_a	deg ^{a})
C(ring)–C(ring)	1.535(2)	C(2)–C(1)–C(6)	110.7(10)
C(1)–C(1')	1.550(14)	C–C–H	109.1(6)
C–H	1.102(3)	α ^{b})	133.6(22)
		φ ^{c})	105.1(32)
		C(2)–C(1)–C(1') ^{d})	113.1
		H–C–H ^{d})	109.8
		θ $^{d})^{e}$)	56.9

(continued)

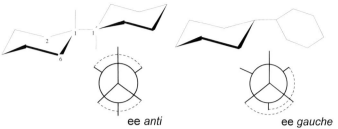

ee *anti* ee *gauche*

The ED results were interpreted in terms of a mixture of C_{2h} (ee *anti*) and C_2 (ee *gauche*) conformers with a ratio of 52.6(90) % and 47.4(90) %, respectively.

The following assumtions were made: The chair fragments have D_{3d} symmetry, the –CH$_2$– units have C_{2v} symmetry, and *anti* and *gauche* conformers have the same structure except for the torsional angle H–C(1)–C(1′)–H.

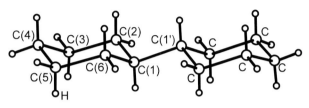

The nozzle temperature was 112(2) °C.

[a]) Three times the estimated standard errors including the experimental scale error.
[b]) Angle between the C(1)–C(1′) bond and the bisector of the C(2)–C(1)–C(6) bond angle.
[c]) H–C(1)–C(1′)–H torsion angle of the C_2 conformer; $\varphi = 0°$ in the *anti* conformer.
[d]) Dependent parameter.
[e]) Torsion angle in the cyclohexane ring related to the endocyclic C–C–C bond angle as $\cos\theta = (C-C-C)/[1 + \cos(C-C-C)]$.

Dorofeeva, O.V., Mastryukov, V.S., Almenningen, A., Horn, A., Klaeboe, P., Yang, L., Allinger, N.L.: J. Mol. Struct. **263** (1991) 281. II/23(**3**,490)

$C_{12}H_{22}Si_2$	1,3–Bis(trimethylsilyl)benzene	II/23(**3**,491)
$C_{12}H_{22}Si_2$	1,4–Bis(trimethylsilyl)benzene	II/15(**3**,1026)
$C_{12}H_{24}$	Cyclododecane	II/21(**3**,610)
$C_{12}H_{27}BrSn$	Bromotri–*t*–butylstannane	II/21(**3**,611)
	Tri–*t*–butylbromostannane	
$C_{12}H_{27}ClSn$	Tri–*t*–butylchlorostannane	II/23(**3**,492)
$C_{12}H_{27}OP$	Tri–*t*–butylphosphine oxide	II/21(**3**,612)
$C_{12}H_{27}P$	Tri–*t*–butylphosphine	II/15(**3**,1027)
$C_{12}H_{28}NP$	P,P,P–Tri–*t*–butylphosphine imide	II/21(**3**,613)
$C_{12}H_{28}Si$	Tetraisopropylsilane	II/23(**3**,493)
$C_{12}H_{28}Si$	Tri–*t*–butylsilane	II/15(**3**,1028)
$C_{12}H_{30}Al_2O_2$	Dimethylaluminum *t*–butoxide dimer	II/15(**3**,1029)
$C_{12}H_{36}Al_3P_3$	*cyclo*–Tris–*μ*–dimethylphosphido–tris(dimethylaluminum)	II/21(**3**,614)
	Trimeric dimethyl(dimethylphosphino)aluminum	
$C_{12}H_{36}BeN_2Si_4$	Bis[bis(trimethylsilyl)amido]beryllium	II/7(**3**,837)
$C_{12}H_{36}CdN_2Si_4$	Bis[bis(trimethylsilyl)amido]cadmium(II)	II/21(**3**,615)
$C_{12}H_{36}CoN_2Si_4$	Bis[bis(trimethylsilyl)amido]cobalt(II)	II/21(**3**,616)
$C_{12}H_{36}FeN_2Si_4$	Bis[bis(trimethylsilyl)amido]iron(II)	II/21(**3**,617)
$C_{12}H_{36}GeN_2Si_4$	Bis[bis(trimethylsilyl)amido]germanium	II/15(**3**,1030)
$C_{12}H_{36}HgN_2Si_4$	Bis[bis(trimethylsilyl)amido]mercury(II)	II/21(**3**,618)
$C_{12}H_{36}Li_2N_2Si_4$	Bis(trimethylsilyl)amidolithium dimer	II/15(**3**,1031)
$C_{12}H_{36}MgN_2Si_4$	Bis[bis(trimethylsilyl)amido]magnesium	II/15(**3**,1032)
$C_{12}H_{36}MnN_2Si_4$	Bis[bis(trimethylsilyl)amido]manganese(II)	II/21(**3**,619)

$C_{12}H_{36}N_2PbSi_4$	Bis[bis(trimethylsilyl)amido]lead	II/15(**3**,1033)
$C_{12}H_{36}N_2Si_4Sn$	Bis[bis(trimethylsilyl)amido]tin	II/15(**3**,1034)
$C_{12}H_{36}N_2Si_4Zn$	Bis[bis(trimethylsilyl)amido]zinc	II/15(**3**,1035)
$C_{12}H_{36}N_6W$	Hexakis(dimethylamido)tungsten(VI)	II/21(**3**,620)
$C_{12}H_{36}O_6Si_6$	Dodecamethylcyclohexasiloxane	II/7(**3**,838)
$C_{12}H_{36}Si_5$	Tetrakis(trimethylsilyl)silane	II/7(**3**,839)
$C_{13}H_8OS$	9H–Thioxanthen–9–one Thioxanthone	II/21(**3**,621)
$C_{13}H_8O_2$	9H–Xanthen–9–one, Xanthone	II/21(**3**,622)
$C_{13}H_{10}O$	Xanthene	II/23(**3**,494)
$C_{13}H_{10}S$	Thioxanthene	II/23(**3**,495)
$C_{13}H_{11}N$	N–Benzylideneaniline	II/15(**3**,1036)
$C_{13}H_{24}Ti$	Trimethyl(η^5–pentamethylcyclopentadienyl)titanium(IV)	
		II/23(**3**,496)
$C_{13}H_{28}$	Tri–t–butylmethane	II/7(**3**,840)
$C_{13}H_{36}Si_4$	Tetrakis(trimethylsilyl)methane	II/21(**3**,623)
$C_{14}H_8N_2O_8$	6,6′–Dinitrobiphenyl–2,2′–dicarboxylic acid	II/15(**3**,1037)
$C_{14}H_8O_2$	9,10–Anthraquinone	II/15(**3**,1038)
$C_{14}H_{10}$	Diphenylacetylene Tolan	II/21(**3**,624)
$C_{14}H_{10}$	Anthracene	II/7(**3**,841), II/15(**3**,1039)
$C_{14}H_{10}O_2$	Benzil	II/21(**3**,625)
$C_{14}H_{12}$	cis–Stilbene	II/15(**3**,1040)
$C_{14}H_{12}$	trans–Stilbene	II/15(**3**,1041)
$C_{14}H_{16}$	Heptacyclo[6.6.0.0^{2,6}.0^{3,13}.0^{4,11}.0^{5,9}.0^{10,14}]tetradecane	
		II/21(**3**,626)
$C_{14}H_{20}$	1,8–Cyclotetradecadiyne	II/7(**3**,842)
$C_{14}H_{24}$	Perhydroanthracene	II/7(**3**,843)
$C_{14}H_{28}$	1,4–Di–t–butylcyclohexane	II/7(**3**,844)
$C_{14}H_{38}GeSi_4$	Bis[bis(trimethylsilyl)methyl]germanium(II)	
		II/15(**3**,1042), II/21(**3**,627)
$C_{14}H_{38}Si_4Sn$	Bis[bis(trimethylsilyl)methyl]tin(II)	II/15(**3**,1043), II/21(**3**,628)
$C_{14}H_{38}MnSi_4$	Bis[bis(trimethylsilyl)methyl]manganese	II/23(**3**,497)
$C_{15}H_3AlF_{18}O_6$	Tris(1,1,1,5,5,5–hexafluoro–2,4–pentanedionato)aluminum(III)	
		II/15(**3**,1044)
$C_{15}H_3CrF_{18}O_6$	Tris(1,1,1,5,5,5–hexafluoro–2,4–pentanedionato)chromium(III)	
		II/15(**3**,1045)
$C_{15}H_{10}O_3$	1,3–Diphenyl–1,2,3–propanetrione	II/23(**3**,498)
$C_{15}H_{15}F$	4′–Fluoro–2,4,6–trimethylbiphenyl	II/15(**3**,1046)
$C_{15}H_{30}Si_3$	1,3,5–Tris(trimethylsilyl)benzene	II/23(**3**,499)
$C_{16}H_{16}U$	Bis(η^8–cyclooctatetraene)uranium(IV)	II/23(**3**,500)
$C_{16}H_{20}$	4,4,4′,4′–Tetramethyl–1,1′–bi(2,5–cyclohexadienylidene)	
		II/15(**3**,1047)
$C_{16}H_{26}Si_2Zn$	(η–Trimethylsilylcyclopentadienyl)– [σ–(1–trimethylsilyl–2,4–cyclopentadienyl)]zinc	II/21(**3**,629)
$C_{16}H_{34}$	Hexadecane, Cetane	II/15(**3**,1048)

$C_{16}H_{36}CrO_4$	Chromium tetra–t–butoxide	II/21(**3**,630)
$C_{16}H_{36}O_4V$	Vanadium(IV) t–butoxide	II/23(**3**,501)
$C_{18}H_{16}Si$	Triphenylsilane	II/21(**3**,631)
$C_{18}H_{30}$	4,4,8,8,12,12–Hexamethyltrispiro[2.1.2.1.2.1]dodecane	
		II/21(**3**,632)
$C_{18}H_{54}CeN_3Si_6$	Tris[bis(trimethylsilyl)amido]cerium(III)	II/21(**3**,633)
$C_{18}H_{54}N_3PrSi_6$	Tris[bis(trimethylsilyl)amido]praseodymium(III)	II/21(**3**,634)
$C_{18}H_{54}N_3ScSi_6$	Tris[bis(trimethylsilyl)amido]scandium(III)	II/21(**3**,635)
$C_{19}H_{15}$	Triphenylmethyl radical	II/7(**3**,845)
$C_{19}H_{16}$	Triphenylmethane	II/7(**3**,846)
$C_{20}H_{12}$	Perylene	II/7(**3**,847)
$C_{20}H_{30}$	1,1′,2,2′,3,3′,4,4′,5,5′–Decamethylbi–2,4–cyclopentadien–1–yl Decamethylbicyclopentadienyl	II/21(**3**,636)
$C_{20}H_{30}Ba$	Bis(η^5–pentamethylcyclopentadienyl)barium(II)	
		II/21(**3**,637), II/23(**3**,502)
$C_{20}H_{30}Ca$	Bis(η^5–pentamethylcyclopentadienyl)calcium(II)	II/21(**3**,638)
$C_{20}H_{30}Fe$	Decamethylferrocene	II/15(**3**,1049)
$C_{20}H_{30}Ge$	Decamethylgermanocene	II/15(**3**,1050)
$C_{20}H_{30}Mg$	Bis(η^5–pentamethylcyclopentadienyl)magnesium(II)	II/21(**3**,639)
$C_{20}H_{30}Mn$	Decamethylmanganocene	II/15(**3**,1051)
$C_{20}H_{30}Si$	Bis(η^5–pentamethylcyclopentadienyl)silicon(II)	II/23(**3**,503)
$C_{20}H_{30}Sr$	Bis(η^5–pentamethylcyclopentadienyl)strontium(II)	
		II/21(**3**,640), II/23(**3**,504)
$C_{20}H_{30}Yb$	Bis(η^5–pentamethylcyclopentadienyl)ytterbium(II)	II/21(**3**,641)
$C_{20}H_{30}Zn$	(η^5–Pentamethylcyclopentadienyl)– [η^5–(pentamethyl–2,4–cyclopentadienyl)]zinc	II/21(**3**,642)
$C_{20}H_{44}Cr$	Tetrakisneopentylchromium	II/23(**3**,505)
$C_{22}H_{38}CuO_4$	Bis(dipivaloylmethanato)copper(II)	II/23(**3**,506)
$C_{24}H_{12}$	Coronene	II/7(**3**,848)
$C_{24}H_{20}Ge$	Tetraphenylgermane	II/23(**3**,507)
$C_{24}H_{20}Si$	Tetraphenylsilane	II/23(**3**,508)
$C_{24}H_{20}Sn$	Tetraphenylstannane Tetraphenyltin	II/15(**3**,1052), II/23(**3**,509)
$C_{33}H_{57}DyO_6$	Tris(dipivaloylmethanato)dysprosium(III)	II/21(**3**,643)
$C_{33}H_{57}ErO_6$	Tris(dipivaloylmethanato)erbium(III)	II/21(**3**,644)
$C_{33}H_{57}EuO_6$	Tris(dipivaloylmethanato)europium(III)	II/21(**3**,645)
$C_{33}H_{57}GdO_6$	Tris(dipivaloylmethanato)gadolinium(III)	II/21(**3**,646)
$C_{33}H_{57}HoO_6$	Tris(dipivaloylmethanato)holmium(III)	II/21(**3**,647)
$C_{33}H_{57}O_6Pr$	Tris(dipivaloylmethanato)praseodymium(III)	II/21(**3**,648)
$C_{33}H_{57}O_6Sm$	Tris(dipivaloylmethanato)samarium(III)	II/21(**3**,649)
$C_{33}H_{57}O_6Tb$	Tris(dipivaloylmethanato)terbium(III)	II/21(**3**,650)
$C_{33}H_{57}O_6Y$	Tris(dipivaloylmethanato)yttrium(III)	II/21(**3**,651)

C$_{60}$ [5,6]Fullerene–C$_{60}$ I_h

ED Buckminsterfullerene

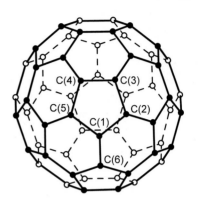

r_g	Å [a]
C(1)–C(2) [b]	1.458(6)
C(1)–C(6) [c]	1.401(10)
r [d]	1.439(2)
Δr [e]	0.057(6)
d [f]	7.113(10)

The nozzle was at 730 °C.

[a]) Twice the estimated standard errors.
[b]) Bond length within the five-membered ring (that is for the bond fusing five- and six-membered rings).
[c]) Connecting five-membered rings (the bond fusing six-membered rings).
[d]) Weighted average of the two bond lengths.
[e]) Difference between the two bond lengths.
[f]) Diameter of the icosahedral sphere.

Hedberg, K., Hedberg, L., Bethune, D.S., Brown, C.A., Dorn, H.C., Johnson, R.D., de Vries, M.: Science **254** (1991) 410.

Errata

Inner Cover		Remarks to r_{av} and r_0: replace v with υ
Preface		Line 13: assicstance *should read* assistance
		Line 19: iniative *should read* initiative
14		Paragraph 4, line 5: continuosly *should read* continuously
24		Remarks to r_{av} and r_0: replace v with υ
26	ArH_3N	Reference: Kiemperer *should read* Klemperer
28	BH_2	Table: II *should read* Π
38	Cl_2Mg	chloride *should read* dichloride
40	Cl_6Ga_2	Title, right side: Cl_6Ga_2 *should read* Ga_2Cl_6
45	FO_3S	Footnote c: $\tilde{X}\,^2A_2 - {}^2E(2)$ *should read* ${}^2E(2) - \tilde{X}\,^2A_2$
		Footnote d: Electronic ab initio *should read* Ab initio
53	Ga_2O	nonoxide *should read* monoxide
55	HOS	MW Reference: Sito *should read* Saito
61	H_2P	Title: λ^5–Phosphane *should read* λ^2-Phosphane
62	H_2S_2	Dihydrogendisulfide *should read* Dihydrogen disulfide
68	H_4O_2	Footnote b: The Eulerian angles should be printed italic.
69	H_5ISi_2	II/15(2,198) *should read* II/15(2,298)
76	O_2S	ED Table, Footnote b: S = 0 *should read* S=O
78	O_6Sb_4	hexaoxides *should read* hexaoxide
82	CCl_2OS	Carbonylthiohypochlorite *should read* Carbonyl chloride thiohypochlorite
90	$CHNS$	Reference: Winnewisser, G.: *should read* Winnewisser, G.,
93	CH_2	Title: remove C_{2v}
94	CH_2	Footnote a: replace v with υ
96	$CH_2F_4P_2S_2$	...phosphonothionyl)... *should read* ...phosphonothioyl)...
98	CH_2O	Footnotes c and e: replace v with υ
101	CH_2S	Footnote c: replace v with υ
121	$C_2Cl_2O_2$	Insert *trans* below figures, delete upper figure
122	$C_2F_6S_4$	...methyl) tetrasulfane *should read* ...methyl)tetrasulfane
141	C_2H_6OS	Text: S–O bond *should read* S=O bond
155	C_3H_6 ($C_3H_4D_2$)	Title: D_{3h} (C_{2v}) *should read* C_{2v}
159	$C_3H_9AlCl_3N$	Aluminumtrichloride *should read* Aluminum trichloride
171	C_4H_8	Cyclobutane, Title: $(CH_2)_4$ *should read* C_4H_8
180	C_5H_5	Title: $(CH)_5$ *should read* C_5H_5
190	C_6H_6	Benzene, Title: $(CH)_6$ *should read* C_6H_6
		Footnote c: replace v with υ
201	C_8H_8	*p*–Quionodimethane *should read* *p*–Quinodimethane
211	$C_{12}H_{22}$	Paragraph 2, line 1: assumtions *should read* assumptions